Space and Time in Mediterranean Prehistory

Space and Time in Mediterranean Prehistory addresses these two concepts as interrelated, rather than as separate categories, and as a means for understanding past social relations at different scales. The need for this volume was realized through four main observations: the ever-growing interest in space and spatiality across the social sciences; the comparative theoretical and methodological neglect of time and temporality; the lack in the existing literature of an explicit and balanced focus on both space and time; and the large amount of new information coming from the prehistoric Mediterranean. It focuses on the active and interactive role of space and time in the production of any social environment, drawing equally on contemporary theory and on case studies from Mediterranean prehistory.

Space and Time in Mediterranean Prehistory seeks to break down the space-time continuum, often assumed rather than inferred, into space-time units and to uncover the varying and variable interrelations of space and time in prehistoric societies across the Mediterranean. The volume is a response to the dissatisfaction with traditional views of space and time in prehistory and revisits these concepts to develop a timely integrative conceptual and analytical framework for the study of space and time in archaeology.

Stella Souvatzi received her Ph.D. in Archaeology from the University of Cambridge. She is Adjunct Professor at the School of Humanities, Hellenic Open University and currently holds a British Institute at Ankara – Koç University Research Center for Anatolian Civilizations Senior Residential Fellowship at Istanbul. She has taught in Britain, Greece, and Cyprus and has conducted extensive archaeological fieldwork all over Greece. Her research interests include the Neolithic archaeology of Greece, eastern Mediterranean, and southeastern Europe, and the theory of archaeology and anthropology. Her publications include *A Social Archaeology of Households in Neolithic Greece: An Anthropological Approach* (Cambridge University Press, 2008).

Athena Hadji teaches Prehistoric Archaeology for the Program *Plato's Academy*, launched by the National University of Athens and also an e-learning program on Prehistoric Aegean Art. On occasion, she curates art exhibitions. She holds a Ph.D. in Mediterranean Archaeology from the University of California Berkeley. She has taught Greek Mythology, Archaeology, Museum Studies, History of Art, and Anthropology of Space and Art in Berkeley, Greece and Cyprus. She has contributed to journals, edited volumes, and international conferences with subjects ranging from Early Cycladic art to cultural heritage. Her current research projects are Early Cycladic sculpture and contemporary Athenian graffiti.

Routledge Studies in Archaeology

1 **An Archaeology of Materials**
Substantial Transformations in Early Prehistoric Europe
Chantal Conneller

2 **Roman Urban Street Networks**
Streets and the Organization of Space in Four Cities
Alan Kaiser

3 **Tracing Prehistoric Social Networks through Technology**
A Diachronic Perspective on the Aegean
Edited by Ann Brysbaert

4 **Hadrian's Wall and the End of Empire**
The Roman Frontier in the 4th and 5th Centuries
Rob Collins

5 **U.S. Cultural Diplomacy and Archaeology**
Soft Power, Hard Heritage
Christina Luke and Morag M. Kersel

6 **The Prehistory of Iberia**
Debating Early Social Stratification and the State
Edited by Maria Cruz Berrocal, Leonardo García Sanjuán, and Antonio Gilman

7 **Materiality and Consumption in the Bronze Age Mediterranean**
Louise Steel

8 **Archaeology in Environment and Technology**
Intersections and Transformations
Edited by David Frankel, Jennifer M. Webb, and Susan Lawrence

9 **An Archaeology of Land Ownership**
Edited by Maria Relaki and Despina Catapoti

10 **From Prehistoric Villages to Cities**
Settlement Aggregation and Community Transformation
Edited by Jennifer Birch

11 **Space and Time in Mediterranean Prehistory**
Edited by Stella Souvatzi and Athena Hadji

Space and Time in Mediterranean Prehistory

Edited by Stella Souvatzi and Athena Hadji

Routledge
Taylor & Francis Group
LONDON AND NEW YORK

First published 2014
by Routledge

2 Park Square, Milton Park, Abingdon, Oxfordshire OX14 4RN
711 Third Avenue, New York, NY 10017

Routledge is an imprint of the Taylor & Francis Group, an informa business

First issued in paperback 2018

Library of Congress Cataloging-in-Publication Data

Space and time in mediterranean prehistory / [edited by] Stella Souvatzi, Athena Hadji.
pages cm — (Routledge studies in archaeology)
1. Prehistoric peoples—Mediterranean Region. 2. Social archaeology—Mediterranean Region. 3. Space and time. 4. Mediterranean Region—Antiquities. I. Souvatzi, Stella G., 1966–
GN848.S73 2013
937—dc23 2013010152

ISBN: 978-0-415-83732-3 (hbk)
ISBN: 978-1-138-37784-4 (pbk)

Typeset in Sabon
by Apex CoVantage, LLC

For my mother, Anastasia (SS)
To the ones I love (AH)

Contents

Figures

Tables

Preface

This volume addresses the concepts of space and time as interrelated, rather than as two separate categories, and as a means for understanding past social relations at different scales. It draws equally on contemporary social theory and on case studies from Mediterranean prehistory, spanning from the earlier Neolithic through to the Late Bronze Age, and extending from Spain to Turkey and Israel. The need for this volume was realized through four main observations: (1) the ever-growing interest in space and spatiality across the social sciences; (2) the comparative neglect of time and temporality; (3) the lack in the existing literature of an explicit and balanced focus on *both* space and time; and (4) the large amount of new information coming from the prehistoric Mediterranean.

The volume is not intended as an overall synthesis of the prehistoric Mediterranean. Rather, it focuses on space and time, with Mediterranean prehistory serving as the empirical archaeological background for application and detailed analysis. Its ultimate aim and greater ambition is to contribute toward the development of an integrative conceptual and analytical framework for the study of space and time in archaeology, drawing on interdisciplinary perspectives.

In spite of a growing number of monographs and edited volumes on issues of space, and to a lesser extent on time and on Mediterranean archaeology in the last decade or so, no volume focuses explicitly on both space and time as interrelated, much less on the interplay between different scales of space and time, and very few volumes include both European and Near Eastern evidence as well as information from the Mediterranean in its full West-East expanse. Another strength of the volume as a whole is the multitude of space and time scales and issues that the contributing chapters address and the range of key themes of interdisciplinary interest that they cover. Furthermore, contributors include not only archaeologists, but also architects and anthropologists, and come, in addition, from diverse national and academic backgrounds. The diversity of regions and case studies subsumed within the volume further lends itself to regionally specific interests such as those in the prehistoric Mediterranean, Europe, or Southwest Asia.

All of the chapters are newly commissioned and are invited in their majority. There are also a few selected and substantially revised versions of papers from a session entitled (Scales of) Space and Time in Mediterranean Prehistory, organized by the editors at the Sixteenth Annual Meeting of the European Association of Archaeologists (2010) in the Hague. In order to ensure coherence and consistency across the volume and to facilitate discussion at the end of it, we have provided the contributors with specific guidelines with regard to chapter content and asked them to address in any way they think appropriate the central conceptual question of the volume—that is, the idea of space and time dynamics as interrelated phenomena—while allowing them of course to focus on one or the other theme most strongly. This provided a stimulus to think about core aspects that they might not have included in their original presentations or proposals.

Space and Time in Mediterranean Prehistory is quite unique in that it brings together key topics and regions (space, time, Mediterranean archaeology, prehistoric Europe, and the Near East) that have often been considered separately until now. Thus, although most of its case studies are from the prehistoric Mediterranean, the volume makes clear the implications of its materials and cross-disciplinary insights for archaeologists and social scientists in diverse other areas and regions, aspiring to provide an important resource for reading, further research, and above all, dialogue and exchange.

We wish to thank the contributors to this volume for the enthusiasm, time, and effort they invested in preparing their chapters. We are also grateful to the two anonymous reviewers for their constructive comments. Finally, we are happy to thank Routledge, and especially our editor, Laura Stearns, for placing her faith in our book and for her patience and understanding through difficult stages.

1 Space and Time in Mediterranean Prehistory and Beyond

Athena Hadji and Stella Souvatzi

INTRODUCTION

Space and time, besides their undeniable and unavoidable physical properties, are fundamental in the formation and transformation of any social and cultural environment and integral parts of any aspect of life. Their many dimensions and levels of analysis give them the potential for a dynamic theoretical and analytical interface of a host of themes, issues, and domains topical in a wide range of disciplines, including archaeology, anthropology, architecture, geography, history, philosophy, and sociology. Indeed, after the postmodernist critique in the 1980s, and particularly since the 1990s, space and time reflect growing interdisciplinary interests and are emerging as subjects of renewed focus, self-reflexive criticisms, and theoretical stimulus across the social sciences and the humanities (see Souvatzi 2012b for a review and references).

In archaeology, space and time have always been central themes of inquiry and of methodologically innovative research, which has, in addition, incorporated advances and ideas from other social sciences along the way. However, the full theoretical and analytical potential of space-time has not been utilized thus far, whereas their interrelationship and the interplay between different scales of space and time have rarely been considered, if at all. First, space and time have more often than not been approached independently of each other. Their dialectical relationship has rarely been applied in practice, that is, in empirical analysis and in case studies, in which the two concepts continue to be examined separately. Thus, many archaeological volumes on space are focused either on various theoretical themes, in which time remains just a strand among many, or on very specific methodological themes such as the use of new technologies for spatial analysis, or on specific spatial scales, most often large ones—especially in prehistoric archaeology. On the other hand, volumes on time explore the diversity of archaeological approaches to time, address time as a theoretical concept, or examine the perception of time and history in past societies, drawing on empirical evidence from a variety of regions and periods, but space and its dialectical relationship with time are not foregrounded, if they are considered at all.

In addition, space has habitually been taken for granted as a backdrop for social interaction, whereas time has been treated mostly as a device for the construction of chronological sequences against which things simply happen. A third problem is that there is compartmentalization and often, and most unfortunately, polarization between different archaeological theoretical approaches.

The Mediterranean offers an ideal and critical area for research on these issues for many reasons. First, its strategic location between three continents, in conjunction with its fragmented yet connecting geography, with its sea, mainlands, and islands, has always made it a locus for the exchange of cultural traits and knowledge, offering therefore a great potential for the study of networks of interaction on many different scales of space and time. Second, the Mediterranean provides a very clear picture of cultural plurality, diversity, and elaboration, reflecting different engagements of people with the physical and social environment at different times. This is crucial for understanding sociocultural and spatiotemporal variability and beyond that, some of the factors that produce it. Third, the Mediterranean, particularly the eastern part of it, has attracted the attention of numerous researchers—professional and amateur—at least since the Renaissance, while the intensification of research from the late nineteenth century onward and the various theoretical and political backgrounds of the researchers—local and foreign—involved, as well as earlier and contemporary politics, have equally contributed to the ways archaeological knowledge has been produced. Today, the breadth and interdisciplinary nature of the issues addressed by Mediterranean archaeology, the rapid increase of prehistoric evidence in recent years, and the increasing concern to relate research to wider archaeological themes, as well as the 'multi-cultural world of Mediterranean archaeologies' (Knapp and Blake 2005: 15) all underline the relevance of the Mediterranean within modern society.

However, Mediterranean prehistoric research is characterized by considerable fragmentation, regional specialism or 'hyper-specialization' (Cherry 2004b: 235–236), and disparate case studies, which not only discourages comparisons of social and cultural phenomena in different regions (cf. Athanassopoulos and Wandsnider 2004b: 1–6; Cherry 2004a: 27–33; Knapp and Blake 2005: 2–3; Knapp and van Dommelen 2010: 3–4) but impedes theorization of space and time. Collections of papers on or syntheses of Mediterranean prehistory (or on Mediterranean archaeology) variously give overviews of general or specific themes and several strands of evidence in various regions. There are rarely explicit unifying themes across the Mediterranean or comparisons between geographical areas or a discussion forum (for an exception, see van Dommelen and Knapp 2010). At the same time, the traditional frontier between 'West' and 'East', favoring always the eastern Mediterranean, seems to be maintained. Last, many Mediterranean prehistorians, especially of the west and central Mediterranean, seem to have little knowledge of the relevant eastern and Near Eastern material, and conversely, much Near Eastern work is read only by Near Eastern specialists, who have little knowledge of or interest in wider Mediterranean or European contexts.

This volume attempts to take a new look at space and time as interrelated, foregrounding rather than backgrounding the spatial and temporal dimensions of society and exploring different scales of interpretation. It focuses on the active and interactive role of space and time as a means for understanding past social relations, experiences, and actions, drawing equally on contemporary social theory and case studies from Mediterranean prehistory.

The volume is quite unique in that it brings together key themes and regions that have typically been considered separately until now. First, the growing interest in space and time highlights that these are exceptionally active fields, but ones that could greatly benefit from the kind of integrative and interdisciplinary approach advocated here. Second, the large amount of new information coming from prehistoric sites across the Mediterranean necessitates a more effective and more focused synthesis. Third, the crossing of the traditional frontier between 'West' and 'East' is achieved by including both typically Mediterranean and also Near Eastern sites.

Contributors are united by the need to engage in multiple scales of analysis. They come not only from archaeology but also from anthropology and architecture. We believe that a rapprochement of this sort is the start of a constructive dialogue among the social sciences, since the strict academic disciplines divide does not seem to be of particular relevance anymore (e.g., Hadji 2004; Lobao, Hooks and Tickamyer 2007; Shankland 2012). The chapters rearticulate the concepts of space and time and their interrelation through key themes, including interaction, spatiality and temporality, the material-spatial character of the past in the present, identity, memory, experience, mobility and movement, social reproduction and change, and body and embodiment. Contributions have been carefully selected in order to secure a balanced representation of (a) theory and data; (b) scope, spanning from the earlier, aceramic Neolithic through to the Late Bronze Age; (c) geographical regions, extending from Spain to Turkey and Israel; and (d) scale, ranging from landscape to domestic buildings and from longer-term patterns to short-term daily activities. Finally, we also provide a concluding chapter with final discussion and remarks.

The ultimate aim, and greater ambition, of the volume is to contribute toward the development of an integrative conceptual and analytical framework for the study of space and time in archaeology. Although it will probably take more than a single volume before we construct such a framework, we can certainly begin to generate discussions that go beyond separate archaeological case studies or experiences and toward broader issues; to enable comparisons of trends that overlap and interconnect; and to consider social and cultural interactions on different scales. We can also hope to be able to set a preliminary agenda for future research.

The remainder of this chapter critically examines issues related to the archaeological theorization of and approach to space and time, bringing in insights from the other social disciplines. First, it discusses useful concepts and ideas in areas of social thought from which archaeologists might not have drawn inspiration as yet. This is followed by a general review of the relevant state of research in archaeology. More specific themes and theoretical and methodological issues are addressed with regard to Mediterranean prehistory. There is also a discussion of the Mediterranean as a concept and as an analytical unit. The chapter concludes with an attempt toward an alternative, integrative approach to space and time in archaeology.

SPACE IN THE SOCIAL SCIENCES

Space is increasingly recognized as fundamental in social analysis and theoretical discourse (Gupta and Ferguson 1992; Hastrup and Olwig 1997; Lobao, Hooks and Tickamyer 2007; Low and Lawrence-Zúñiga 2003a; Rodman 2003; Stevenson 2003). Still, Eriksen (1995), in a book subtitled *An Introduction to Social and Cultural Anthropology*, devotes no chapter to space or time as such. A spatial orientation can be seen only in the last chapter, concerning the local and the global. Until fairly recently, anthropology did not focus on a theoretical analysis of the concept of space systematically and thoroughly, limiting its scope to a consideration of space as a container of human actions and a receiver of human impact. In contemporary anthropological theory, space is no longer void, literally or metaphorically. It is constantly charged, mediated, negotiated, and claimed by different social groups with diverse and often conflicting interests. By the abolition of the neutrality of space, its active role in the formation of social relations is acknowledged and its social character is emphasized in anthropological theory, especially under the influence of French social theorists such as Michel Foucault, and the concepts of heterotopia (Foucault 1986) and Panopticon-space as power of suppression (1977); Henri Lefebvre, space as social production (1991) and rhythmanalysis (2004 [1992]); Gaston Bachelard, intimate spaces, the domestic as poetry (1964); Pierre Bourdieu, *habitus* (1977 [1972]); but also Walter Benjamin, who was associated with the Frankfurt School, although his influence on the study of space in anthropology transcends the boundaries of the school's theoretical premises (Benjamin 2002; Hadji 2009, 2010a, 2010b). Space in anthropology is now viewed as a dynamic entity, a force that shapes social actions to the extent that it is also shaped by them. The Euclidian properties of space are not negated; they are simply not considered definitive to the transformation of space into place (Menin 2003).

In the theory of space, three crucial terms are employed, which until recently were left undefined and unscrutinized and were often confused and conflated: space, place, and landscape. Accordingly, if space is a generic term employed to describe a geometrically defined, quantitatively measured, possibly vast, and conceptually neutral entity, then place is understood as a specific part of geometrically defined space. A basic principle of place is that it is not defined by sheer size; it is rather determined with regard to the qualities assigned to it by human action: place is a space mediated by human intervention, whether literally, through construction, utilization of natural resources, occasional use, and occupation, or metaphorically, through intention or the projection of intentions on space. Landscape is place as spectacle, usually perceived as a χάρμα βροτοίσιν (Hom. *Il.* 14.312), a pleasant view, a place as viewed and as perceived by its spectator. Landscape, otherwise "space as meaning, space as invention" (Perec 1997: 23), is constituted as a signification of place by individual or collective projections upon it (see also Ingold 1993, with a slightly different stance).

A discussion on the anthropology of space cannot fail to mention Edward Hall and the theory of proxemics. Proxemics (Hall 1959, 1963, 1966, 1976), the theory that focused on the cultural delimitation of *personal space,* based to a large extent on Hall's observations of different contemporary cultures as he ventured around the globe, had and still has wide-reaching applications in diverse fields such as marketing, communication studies, sociology, and social psychology. Since a significant number of anthropologists were involved in the study of space per se, not merely as a given entity for the unfolding of human traits, Hall emphasized the idea of embodied space, that space is first and foremost perceived through the existence of the body and its moving in space and interacting with other bodies in space. Ingold (1993) also attributes primal importance to the positioning, movement, and interaction of bodies in space.

In sum, anthropology has shifted its focus on spatial analysis, incorporating the political, ideological, and sociological factors into the use, signification, and understanding of the so-called physical space. Current trends in the study of space include, but are not limited to, space as phenomenology, gender, class, politics, clash, conflict, and above all claims to space appropriation. The last is a pressing issue in urban settings as they are reshaped in the light of rapid shifts in the mega-cities of the twenty-first century as well as the new global setting due to a far-reaching fiscal but also sociocultural crisis affecting spatial norms.

TIME IN THE SOCIAL SCIENCES

Although there is a considerable body of work on the anthropology of space, there is no coherent body of theory or unanimity as to what social time actually is (Adam 1990; Fabian 1983; Gell 1992; Gingrich, Ochs and Swedlund 2002; Gosden 1994; Munn 1992; Rabinow 2007). For the most part, time emerged as an object of inquiry as part of broader issues under study, such as kinship, production and exchange, memory and tradition, conflict, gender, technology, and even religion and art. A unified theme in most theoretical positions is the idea of time as lived duration.

Two fundamental elements of time are *rhythm* and *duration* (Adam 1990; Appadurai 1981; Bailey 1983; Barden 1973; Lüscher 1974). Barden (1973) considers duration as having two aspects, namely the duration of a given event and the duration of the act of watching as perceived by the observer of the event. Different observers can have different temporal perceptions of the *same* event under observation; thus, duration is but one of "three fundamental features of time experience", the other two being succession and simultaneity. In particular, the equation of duration with unity is a notion that accounts for the mistaken equation of the passage of time with change, which is one of the most prominent misconceptions in the study of archaeological time, as discussed below.

Central to the study of time and time perception is the concept of temporalization. Johannes Fabian distinguishes various levels of temporalization, "a provision of encoding time" (1983: 74, following multiple order theory, after Pepper 1935: 18; Hadji 2004). There have been different conceptualizations of time, each resulting in a compartmentalization of time, whether it is accomplished in a listing of various 'forms' or 'orders' of time, or with the invention of antithetic categories of time.

Prominent among antithetic time categories, and most relevant to the present discussion, is the idea of linear versus cyclical time (Hadji 2004: 63–64, 72; Helms 1978: 6; Sato 1991, 2001: 15687; Silverman 2001: 15684). This distinction is an entirely artificial conceptual barrier, a mere construct, especially given that linear time is identified with Western thought, and cyclical with 'traditional' or 'primitive' societies (Gurvich's peasant class with critique from Gell 1992: chapter 7), ultimately following the pattern of Us and the Other (Adam 1990: Table 1; Fabian 1983; Gell 1992; Leach 1961; Munn 1992). Also, the dichotomy between subjective and objective time is false, since personal time operates in accordance with social time, whether this is clock time, ritual time, calendar time, or otherwise constructed social time (Movahedi 1985: 397; Silverman 2001: 15684; Zerubavel 1982: 13). It is a common realization that time does not always or solely move in a linear fashion, and the increments in which it moves are by no means equal. There are points when time is stretched and seems forever lasting, and points when time is compact, short, and simply not enough, and this experience is beyond any notion of objective counting of time.

Toward a united space-time front, the theories of chronogeography and time-geography have also shaped contemporary anthropological thinking on space and time. Archaeology also did not stand unaffected. Although geography has traditionally been an ahistorical discipline (Gamble 1986), both chronogeography and time-geography have focused on shifting the emphasis in geographic studies from space issues to time-space issues (Glennie and Thrift 2001: 15693; Lobao, Hooks and Tickamyer 2007: 8–13, with references). Time-geography, regarding the use of time and space, gives emphasis to tacit, practical knowledge, rather than to explicitly formulated knowledge, thus reminiscent of Bourdieu's (1977 [1972]) *habitus* axiom.

As Munn (1992: 94) points out, "A further complication appears in the fact that time and space are integral to each other. Although Western theory frequently treats space as time's antithetical 'other', time's 'other' turns out somewhat embarrassingly to be its Other Self. In a lived world, spatial and temporal dimensions cannot be disentangled, and the two commingle in various ways" (also, Elias 1982a; 1982b: 1000–1002, §11; Glennie and Thrift 2001: 15692; Lewis and Weigert 1981: 436). Ingold (1993), along the same lines, is a strong proponent of the 'temporality of the landscape', arguing that landscape is in essence task-scape since the decisive formative factor is not the perception of landscape, mostly the visual, but the movement of bodies through landscape while they engage in activities of living, in *tasks*.

Finally, timing space does not equate to applying a fixed time entity onto space. A space can and does acquire different meanings through different uses by the same or different subjects (Stevenson 2003: 78–79). The evolutionary paradigm (Fabian 1983: 39) can be characterized as time-centered only to the extent that it equates time with change, at least in archaeological thought and practice, as discussed below. However, the passage of time alone is no guarantee that change will take place, nor does it explain change when it does occur (Baert 1992: 127). A key element for a theory that avoids the equation of time with change is an outlook that acknowledges the active role of agents and agency, without at the same time overemphasizing the ability of agents to reshape things as they wish (Hadji, 2004: 31–34, 146–169; Souvatzi 2007: 37–38, 51–52; Souvatzi 2008: 40–45, 237, 247–248).

SPACE IN ARCHAEOLOGY

In archaeology, the study of material remains in space, both mobile and fixed, forms a central theme of inquiry. Architecture and artifacts, their interrelation, and their absolute and relative positioning in space are constantly studied, analyzed, and correlated with data from previous research to form an archaeological narrative that places objects found in space in a series of events that happened in time. Moreover, archaeological data are represented in two dimensions (maps, drawings, photographs) meant to render their three-dimensional properties accessible to the reader.

Archaeologists have long pursued theoretically and methodologically innovative research on space, since the inception of archaeology as a discipline, and have incorporated advances and ideas, both technical (from photography in the nineteenth century to geographic information systems [GIS], 3D modeling, and the study of micromorphology today, to name but a few) and theoretical (including terminology, borrowing somewhat freely from the other social sciences, especially but not solely anthropology, such as the concepts of phenomenology and agency). Archaeology adopted basic space units from geography, such as site, settlement, and region (e.g., Ashmore and Knapp 1999; Barrett 1999; Blake 2007; Glennie and Thrift 2001; Ingold 1993; Lowenthal, 1975; Souvatzi 2012b).

However, space in archaeology was viewed until recently as the backdrop to human actions, activities, and interactions. Thus, space was, and still is, involved only indirectly in the study of issues such as memory, identity, power, gender, community, representation, meaning, connectivity, body, and embodiment. With few exceptions, it was not until the 1990s that a systematic interest in the theory of space in archaeology started to emerge. We will briefly follow this trajectory below.

Through the excavation and interpretation of space, the goal of prehistoric archaeology from its inception was the discovery of 'deep time', following a series of scientific revolutions that were in full bloom in the mid-nineteenth century. Space was a means to assess time. Traditionally, the spatializing of time is first attributed to Newton and his formulation of a theory of time. Newton conceptualized time as "uniform, infinitely divisible, and continuous" (Sorokin and Merton 1937: 616). Archaeology following the lead of geology equated stratigraphical (spatial) sequences with time lapses in the quest for the discovery of deep time, within the cultural-historical paradigm that sought space-bound cultures.

In the 1960s, the emergence of processual (or 'New') archaeology, introduced by Lewis Binford (1962), revolutionized the treatment of space in archaeology, stating and using distinct links with anthropology, adopting a clearly functionalist stance, and under heavy influence from post–World War II positivism. Emphasis was now given to space from an environmental viewpoint as well as the environmental impact and influence on human occupation and vice versa; sophistication of space measurement units and methods; and the development of 'middle-range' theory to account for a direct causality between social habitation theory and observed spatial patterns (Souvatzi 2012b with references).

From the 1980s on and especially since the 1990s archaeology has started to study space per se, as an autonomous inquiry and with a keen interest in it as socially and culturally produced, with direct influences from anthropological theory on space. This change of paradigm is due to the 'postprocessual' methodological and theoretical shift in archaeology within the broader frame of postmodernist theory. In the spirit of a critique of processual archaeology, postprocessualist approaches called for a meaning-bound archaeological space and a multiplicity of interpretations including, but not restricted to, social and cultural/ideational ones (for a concise history of these approaches see Souvatzi 2012b). Self-reflection became prominent among postprocessualists, and a central theme in their spatial agendas was the constitution of space through human acts, projections, intentions, and interactions. Among the most influential space (but also time) theories that have been adopted by and adapted to archaeological interpretation are

Giddens's (1984) theory of structuration with the concept of human agency and Bourdieu's (1977 [1972]) concept of *habitus* as a main social drive starting from and being most evident in space.

Today, archaeology still draws heavily from the other social sciences. Whereas certain theories prove resilient to time (such as agency), other theories of space remain virtually unexplored archaeologically—for example, the theory of Proxemics, which, as mentioned above, exercised a lasting influence outside the realm of anthropology. Archaeology did not pick up on it (also discussed in Vis 2009, esp. 41–45; for notable exceptions, see Moore 1996; Tringham 1973), despite the popularity of the study of the body in prehistoric archaeology (Borić and Robb 2008; Meskell and Joyce 2003).

A more detailed analysis of contemporary archaeological approaches to space follows in the Mediterranean archaeology section below, which to a large extent applies to all prehistoric archaeology.

TIME IN ARCHAEOLOGY

The concept of time has been utilized in a very limited way in archaeological thought, despite the fact that archaeological reasoning extends beyond dating into ordering cultural relations and social articulations in time. Traditionally, archaeology has used time as a device against which to plot other issues (i.e., social change, cultural comparisons, technological development), most often on a macro scale. This treatment of time had two problems. First, time has been deemed to be objective, unquestioned, unchangeable, and beyond any scrutiny or doubt. Second, time has been treated in a simplistic way, its potential for the advancement of archaeological methodology and theory having gone unused due to a constructed homogenization of data which are not of necessity homogeneous (e.g., Bailey 1983; McGlade 1999; Newton-Smith 1980; Sato 2001: 15686; Shanks and Tilley 1987a; Thomas 1996). This has begun to change in recent years, when a renewed interest in time per se emerged among archaeologists (e.g., Bailey 1981; 1983; 1987; Bintliff 1991; 2008; Hadji 2004; Karlsson 2001; Knapp 1992a; Lucas 2005).

The origin of these two problems of time in archaeology can be detected in Victorian notions of time and in the realm of positivism. Victorian conceptualizations of time, such as the idea of time as linear progression and time as moving 'slowly-slowly' (Zimmerman 2007), have been formed, accepted, and disseminated uncritically in prehistoric archaeology, especially with the predominance of technological advancements as chronological, but also as cultural markers. What is evident is an anxiety to create a linear narrative of prehistoric archaeology, to 'historicize' prehistory. A narrative of this nature does not allow for discontinuities setting more emphasis on aspects of time that pertain to *integration,* rather than *conflict* (Lüscher 1974). Any nonhomogeneous data are doomed to be considered as anomalous in an otherwise (artificially) homogeneous record (Gosden 1994; Hadji 2004; McGlade

1999; Shanks and Tilley 1987a, 1987b; Souvatzi 2008), comprised of the construction of artifact sequences and architectural phases and the compilation of chronological charts (Richards 1995; for a detailed account of issues of dating and chronology building in archaeology including methodology and techniques, see Manning 1999, chapter 3; on chronology theory, see Smith 1992 with critique; also, Arnold 2008; for further references, see Hadji 2004). Moreover, the influence of positivism is evident in the insistence upon establishing law-like generalizations as well as the idea that "concepts are put to use only in so far as they are observable and measurable" (Baert, 1992: 26; also Heilbroner 1980: 48, n.11; Holdaway and Wandsnider 2008; Murray 1999; along the same lines, Shanks and Tilley 1987b: 32, who characterize positivism as 'systematic empiricism').

Consequently, archaeology has traditionally restricted the use of time to the construction of cultural/chronological phases known as periods, for example the time of ages and stages or else time as periodization (Fabian 1983; Sato 2001: 15686). The origin and development of the concept of ages and stages of human progress has been discussed extensively elsewhere (Hadji 2004). Whereas in anthropology, periodization does not distinguish qualitatively between the various ages it creates and thus defines, in archaeology certain periods are markedly understudied as compared with others.

Within the various attempts to solve the time problem in archaeology, time perspectivism was perhaps the most systematic (for its history, see Bailey in Holdaway and Wandsnider 2008: 13–30; for critique, see Harding 2005). The idea of time perspectivism, namely that "observations made at different temporal scales differentially make different processes apparent" (Holdaway and Wandsnider 2008: 3), originating in the Annales School, was introduced in archaeology by Geoff Bailey in a series of articles in the 1980s (1981; 1983; 1987). The Holdaway and Wandsnider volume of 2008 revisited the concept of time perspectivism as potentially useful to an archaeological treatment of time.

In the past few decades, archaeologists have discussed time in association with social memory, tradition, and the idea of the past in the present (Bradley 2002; Hodder 2006; Joyce and Gillespie 2000; Knapp 2009; Tringham 2000); structuration theory and the concept of social *agency* (Dobres 2000 Dobres and Hoffman 1994; 1999; Dobres and Robb 2000; Gardner 2004; Johnson 1989; Joyce and Lopiparo 2005; Saitta, 1994).

Also, time has been discussed in relation to style (on types see Adams and Adams 1991; on style Carr and Neitzel 1995; Conkey and Hastorf 1990; Hadji 2004; Lechtman 1977; 1988; Plog 1983). Last but certainly not least, the issue of temporality, in association with memory and the omnipresent past was taken up by Ingold (1993; also, Barrett 1999; Bradley 1987; 2002; Gosden 1994; Harding 2005; Hodder 2006; Thomas 1996), under the influence of the phenomenology of Husserl (1991), Heidegger (1962 [1953]), and Merleau Ponty (2012 [1945]).

In sum, archaeology has mainly focused on space, implying time in its spatial inquiries. Time cannot be equated with chronology. No matter how useful and at times fascinating an orderly narration of human past activity can be, the time dimension eludes us if we maintain what prehistoric archaeology set out to do in the nineteenth century: studying and interpreting timeless spaces. A timeless space is a space empty of human action. Instead, we need to accept the fragmentary nature of the archaeological record. If culture is in the epicenter of the archaeological inquiry, a good starting point would be the acknowledgment that cultures are dynamic, volatile, somewhat elusive, and constantly reshaping and renegotiated; they only exist in the present tense (Bradley 1987; Rowlands 1993).

An archaeology of time is one that recognizes that prehistoric people had different perceptions of time than we do. Ethnographic examples across space and time, without being definitive, indicate that the people whom we have named, for instance, Chalcolithic or Early Bronze Age might very well have thought of their own time in a totally different way than we do on their account (Hadji 2004: 176–178).

THE MEDITERRANEAN: A SPACE, A PLACE, OR A CONCEPT?

The idea of the Mediterranean as a conceptual category is not new: it was already old when the Romans declared it 'our sea' (*mare nostrum*). What has changed over time are the geographical units the Mediterranean encompasses. The Roman *mare nostrum* was the same body of water that prompted Thucydides to refer to the Minoan reign over it as a thalassocracy (supremacy over the sea) already in the introduction to his *History of the Peloponnesian War.*

In recent scholarship, from Braudel's (1972 [1966]) *La Méditerranée et le Monde Méditerranéen* to Horden and Purcell's (2000) *The Corrupting Sea,* the Mediterranean has featured prominently in historical analysis, including dimensions of space and time, often tacitly. Prehistoric archaeology has used the term rather more freely, usually to denote the eastern part of the sea, often neglecting to speak about the western part, with few exceptions (Knapp 1992b).

Braudel's grand synthesis, in the spirit of the Annales School, with its emphasis on the *longue durée* viewed the Mediterranean as a united geographical front against the attacks of time, a timeless sea that unites rather than divides the varied geographical units it comprises (islands, mainland(s), mountain ranges, plains, west and east, south and north borders) (Horden and Purcell 2000; 2006; Purcell 2003). In addition, as rightly noted by Horden and Purcell (2006: 728), until the nineteenth century, the term *Mediterranean* referred to the body of water, not the regions included in it or separated by it or connected through it.

The Mediterranean Sea did hold a special position already in the ancient world because it was a way—the main gateway—for the people who inhabited its lands to connect with each other, as well as the facilitator for the transport of goods, technologies, people, and ideas at least from the Neolithic period onward. Due to this mobility, a *koine* was established during the so-called Mycenaean era throughout the Mediterranean, as perceived today, encompassing its west-east as well as north-south axes in their totality: its connectivity, neither insularity nor the vast expanses of land contained within, is then what is characteristically Mediterranean (Horden and Purcell 2000; 2006; Knapp and van Dommelen 2010; Morris 2003).

Herzfeld (1984; 2005), an anthropologist and "the most outspoken anti-Mediterranianist of all" (Horden and Purcell 2006: 737), rejects the Mediterranean as a concept and critically attacks the concept of 'Mediterraneanism'. According to him, Mediterranianism is a twofold entity, comprising exclusivism and exceptionalism. Mediterranianism arbitrarily and axiomatically claims that certain regions have priority over others, both culturally and as objects of study, thus elevating the Mediterranean into a unique study object in its entirety, when in essence only a small part thereof is implied (Horden and Purcell 2006: 726). However, as Driessen (1999) points out, regardless of where and when *Mediterranean* acquired its contemporary meaning, the idea of uniquely Mediterranean traits is a notion shared by people inhabiting the region throughout its geographical boundaries (also, van Dommelen 1997 on hybridization and the post-colonial critique). This is particularly evident in the current crisis situation in the European Union, where Mediterranean countries (Portugal, Spain, Italy, Greece) are set apart and as a subtle but distinct divide between the poor, corrupt, and erroneous Mediterranean south and the rigid north starts to emerge.

Our stance toward the Mediterranean, then, follows Purcell (2003: 16) in what unfolds as the ancient view toward the Mediterranean:

> They [the ancients] did, however, postulate 'The Corrupting Sea'. As my title suggests, their sea was not a Mediterranean of palm trees or winter rain but a medium resisting homogeneity, order, and social control. It was the source of change and contestation, an essentially unfixed category—and this despite the facts that it could not, for all philosophers' wishful thinking, be ignored or rejected and that the changes, contrasts, and contacts which it bore actually made it possible for cultures to transcend the local and make sense of worlds huge beyond easy individual cognition.

Certainly, Horden and Purcell (2000) examine the Mediterranean from the historian's viewpoint. Archaeologists, however, reach the same conclusion, namely that the Mediterranean is not about homogeneity, but about connectivity (Barker and Mattingly 1999–2000; Blake and Knapp 2005;

Knapp and van Dommelen 2010). Mediterranean prehistory is not an uncontested, unproblematic, idealized field of inquiry; on the contrary, it is a field of practitioners who study regions geographically or climatologically falling under the umbrella-category Mediterranean, with different scopes and from differing standpoints (i.e., King, Proudfoot and Smith 1997). The aim of the current volume is to transcend boundaries, not so much geographical or regional as disciplinary, and it is in this framework that the Mediterranean is examined (Blake and Knapp 2005; Tartaron 2008).

SPACE AND TIME IN MEDITERRANEAN PREHISTORY: CONTEXTS, THEMES, AND STATE OF RESEARCH

In accordance with the wider archaeological research and the larger epistemological shifts starting out with the postmodernist and postcolonial critiques, the literature on Mediterranean prehistory has been undergoing a change in recent years. Since the 1990s and particularly the 2000s it explores many new themes of wider interest, theoretically diverse interpretative schemes, and interdisciplinary perspectives, and increasingly it reflects an awareness of the fluidity and permeability of analytical domains. This change has of course taken various forms and has followed different intellectual histories, depending on the varying theoretical traditions and research agendas across the Mediterranean. Currently, there is a growing tendency for social analysis, including the spatial and temporal dimensions of action, informed by concepts of social agency, theories of practice, and other elements of contemporary social theories. The variety of prominent, and largely interrelated, themes and issues includes identity, memory, mobility, connectivity, meaning, experience, diversity, architecture and representation, materiality, culture contact and colonialism, innovation and tradition, and transition and transformation. All of these notions relate more or less directly to issues raised by contemporary space and time theorizing. However, relatively few studies focus directly or clearly on space and time.

The following discussion draws on this recent development of thought and themes and elaborates on issues and points critical to a more comprehensive view of space and time.

Space, -Scape, and Architecture

While up until the 1980s landscape and settlement research and surveys comprised isolated studies with a narrow regional focus (see Alcock and Cherry 2004a; Athanassopoulos and Wandsnider 2004b; Cherry 2004a; and Sollars 2005, for the history and critiques of earlier landscape and settlement archaeology), recent studies increasingly move toward a concern for context, scale, and diversity and interaction, and for addressing historical or anthropological questions. At the same time, the focus on representation under the influence of postmodernism has resulted in a plethora of spatial metaphors and -scape concepts widely (e.g., Appadurai 1990; Ingold 1993), most usual in Mediterranean archaeology being 'island-scape', 'sea-scape', 'coast-scape', and 'memory-scape'.

Central and recurring themes of landscape research such as the long-term configuration of the human-natural landscape and demographic and environmental questions are now addressed with an awareness of the spatiotemporal scales and dynamics of human ecology. They include the formational complexity of the Mediterranean landscape (e.g., Wandsnider 2004); the history of Mediterranean pastoral and peasant societies and their effect on the landscape (e.g., Barker 2005; Barton et al. 2004); comparisons and contrasts between different settlement histories and settlement patterns in order to gain a broader picture of local, regional, and wider processes (e.g., Hill 2004; Sollars 2005); GIS technology, 3D modeling, geomorphological studies, and network analysis for identifying spatial relationships and links between changing landscape and changing human occupation (e.g., Gillings, Mattingly and van Dalen 2002; Fitzjohn 2007), as well as attempts to systematize or synthesize the many regional survey data of the last decades into a broader macro-regional picture (e.g., Alcock and Cherry 2004b; Athanassopoulos and Wandsnider 2004a).

A distinct trend takes an explicit view of the landscape as a social and historical construct, which so far has been little embraced in Mediterranean studies, advocating greater account of human perception and agency, meaning, subjectivity and the reflexivity of landscape monuments, and a more concentrated engagement with the materialities of the landscape—drawing on social theory from a range of sources, especially geography, architecture, Marxist thought, phenomenology, and agency theory (e.g., Ashmore and Knapp 1999). Specific areas of interest include the contrast between the diversity and fragmentation of landscape and the connectivity of different landscape components, including settlements (e.g., Barker 2005; Gkiasta 2010; Sollars 2005); the location and circulation of material resources and artifacts (Given 2004; Karimali 2005); monument construction, landmarks. and other symbolic markings and ordering of space (see below); the social dimensions of different forms of land use, tenure and tenancy (e.g., Souvatzi 2013); and the range of meanings and sensory experience of landscapes (e.g., Turnbull 2002).

The notions of 'island-scapes' and 'sea-scapes' are increasingly employed to promote a view of islands primarily as social constructs rather than as bounded entities and of insularity or 'isolation as a social choice' (Knapp and Blake 2005: 10; also, Knapp 2007), and generally a more inclusive understanding of sea, island, ports, coast, and land as a shared and interconnected unit (e.g., Antoniadou and Pace 2007; Broodbank 2000; Cherry 2004b; Knapp 2008; Marangou and Della Casa 2008). Reflecting a growing wider interest in island archaeology and islands as a focus of research (see Rainbird 2007: 26–45 for an overview; also, Fitzpatrick 2004), this trend rejects traditional concepts of Mediterranean islands as isolated 'laboratories' (Evans 1973, after Darwin's 'island biogeography') and attempts to take a new look at the islands, with questions ranging from travel technologies and maritime itineraries (e.g., Farr 2006; Manning and Hulin 2005: 276–278; Papageorgiou 2008) to inland-mainland or interisland relations (e.g., Peltenburg 2007; Ramis 2010). There have also been calls for a greater theorization of the Mediterranean sea as well as of the coastlines and a 'coast-scape approach' (Berg 2010; Pullen and Tartaron 2007; also, Abulafia 2011 for a historian's explicitly sea-based approach to the Mediterranean).

Another area of growing focus concerns the role of architecture as a powerful symbolic expression of difference and identity, social order, and cultural elaboration, and as a material mnemonic device, and thus a form of history, most often in relation to monuments and the islands. It is argued that difference, symbolism, boundaries, and representation are distinguishable features of the Mediterranean, and that there is a strong correlation between island insularity and monumental elaboration, probably resulting from heightened social competition for limited land (see discussion in Kolb 2005: 173) or from a need to express distinct island identities and ways of development (Knapp 2007; 2009: 47). Social analysis of construction, elaboration, and meaning of colossal temples, tombs and funerary complexes, palaces and monumental buildings on Cyprus, Crete, the Maltese islands, Sardinia, Corsica, and the Balearic islands attempts to understand in particular how monumental construction was involved in establishing social relations, collective unity, identity and power, and ritual landscapes and 'memory-scapes', as well as ideological and political authority, thus playing an active role in political discourse and social change (e.g., Blake 2005: 110–120; Knapp 2009; Kolb 2005; Robb 2001; Turnbull 2002).

Beyond monuments and larger -scape concepts, the analytical potential and social dynamics of domestic architecture and space, houses, households, and everyday practices—a most important and particularly rich focus of research in both archaeology and anthropology for the past three decades (see Souvatzi 2008: 7–20, and 2012b: 177–179 for review and references)—has received less attention in typically Mediterranean prehistoric research. We return to this point later in this chapter.

Mobility, Connectivity, Interaction, and Identity

Following the reconceptualization of space as fluid and fragmented and of the sea as a connecting factor, starting out with the postcolonial critique of the Mediterranean as a coherent cultural and spatial entity, mobility, 'connectivity' (Horden and Purcell 2000: 123–172) or 'interdependence' (Harris 2006: 19) or interaction, and identity have emerged as the most prominent foci of research in Mediterranean prehistory. They are usually approached as interlinked and through a bundle of new concepts, most of which are derived from anthropology—migration, colonialism (as opposed to 'colonization'—see Knapp and van Dommelen 2010: 3), and colonial encounters, diaspora, entanglement, and hybridity or 'hybridization' (Knapp 2012: 33; Maran and Stockhammer 2012). Emphasizing in particular the movement of people, objects, and ideas and the dynamic role of culture contact and networks of interaction in shaping social identities and co-presence, but also in causing tensions and transformations, these new conceptual frameworks attempt to capture the material, social, and ideological connections throughout the prehistoric Mediterranean and at a variety of scales; the ways in which divergent forms of cultural and social variation, similarity and difference are interrelated; and the contexts in which encounters of people, things, and ideas resulted in new or 'hybrid' material forms and social and symbolic meanings.

For example, the extensive interregional exchange of raw materials and finished objects, one of the main and long-standing areas of study, particularly in the eastern Mediterranean, is now considered with new questions in mind, which also incorporate the exchange of ideas and cultural concepts. These include the social use of commodities, different value systems, and a focus on consumption (e.g., Manning and Hulin 2005; Robb and Farr 2005); the social, symbolic, and political dimensions of craft specialization and of trading exotic goods and materials (e.g., Karimali 2005; Vianello 2011); and object 'biographies' and the role of circulated objects in establishing social relations and cultural understandings (e.g., Chapman et al. 2011; Sherratt 2012).

Approaches to identity have also moved from an almost exclusive earlier focus on gender and ethnicity to the inclusion of age, class, sexuality, and significantly, the impact of migration and colonialism—all with an emphasis on the material media used for identity statements, especially during periods of transition, culture encounter, and change (e.g., van Dommelen and Knapp 2010). In addition to the analysis of monumental architecture as a means of conceptualizing (island) identities discussed above, other main areas of interest include the role of shared social and ritual practices in expressing local and transregional identities and in establishing a sense of group-oriented identity (e.g., Blake 2005; Knapp 2009; Robb 2001); local identities and regional processes, and commonalities and differences between island identities and mainland ones (e.g., Janes 2010; Ramis 2010); and indigenous identities, cultural diffusion, and the formation of new identities resulting from prolonged migrations, object diasporas, colonialism, and hybridization practices, that is, through the social and material entanglements between indigenous inhabitants and newcomers (e.g., Knapp 2012; Panagiotopoulos 2012; Russel 2010).

Continuity, Change, History, and the Issue of Complexity

The impact of tradition and innovation in the formation of social systems over time, social and economic transitions, and questions revolving around issues of social complexity have always held a prominent position in prehistoric

Mediterranean research, usually with regard to later prehistory and the large scale of political, economic, and social interaction, the emergence of 'complex' societies, and state formation, commonly considered as the culmination of ordered progress. In addition to maintaining the traditional focus on long-term continuity, dominant approaches have tended to rely heavily on neo-evolutionary societal typologies; to equate complexity with political centralization; and to classify societies dichotomously, either as simple or complex, static or dynamic, and egalitarian or unequal (see Chapman, this volume). In wider social archaeological research, however, neo-evolutionary reconstructions and models of complexity have been criticized within most theoretical agendas (see Souvatzi 2008: 40–43 for discussion and references), and current debate attempts to move still further toward a reconceptualization of complexity as a historically contingent phenomenon, as inherent in all social forms, and as pertinent to the understanding of all types of society (Chapman 2003; Crumley 1995; 2003; Kohring and Wynne-Jones 2007; Souvatzi 2007).

Whereas neo-evolutionary perspectives and reliance on the grand models of social evolution have continued to shape archaeological reconstructions of the Mediterranean past, Mediterranean archaeology is continually confronted by numerous examples of societies and trajectories that do not fit into such typologies. Equally significant are the critique of idealized continuity and of orientalist notions of timeless and 'authentic' prehistoric Mediterranean communities, the collapse of diffusionism and the *ex oriente lux* idea, and the reconceptualization of cultural spaces as based on flows of people, materials, and locales. They have contributed, among other things, to the breaking down of the static models of core and periphery and the applicability of world systems theory (see Chapman this volume for an extensive discussion; also, Parkinson and Galaty 2010), and they have encouraged archaeologists to explore social transformations in diverse periods and places, changing cycles of continuity and change, and more localized networks of interaction.

Indeed, an increasing number of studies question the idea of change as a steady and uniform path toward 'progress', embrace a view of Mediterranean societies as more dynamic and variable, and argue for a more complex history of Mediterranean prehistory. Chapman (2005) argues that a focus on tension, conflict, and disjunction is key to a deeper understanding of Mediterranean prehistoric societies and can alter considerably how we have so far perceived complexity and the emergence of 'institutionalized inequality' in different regions of the Mediterranean. Studies of Mediterranean agro-pastoral landscapes and settlements indicate numerous local histories, parallel as well as divergent and continuous as well as discontinuous, suggesting, among other things, that agriculture emerges as a complex set of varying histories rather than a simple east-to-west process of colonization and acculturation, and that different theoretical approaches are suited to interpret data within different temporal scales of analysis (Barker 2005; Barton et al. 2004; Hill 2004; Zeder 2008).

Similarly, the habitual overemphasis on metals and the metals trade as the driving force behind economic development and on the Minoan and Mycenaean palatial economies in Greece as the culmination of complexity and prehistoric societal achievement has been criticized as simplistic and particularistic and for disregarding divergent regional trajectories and the coexistence of multiple political formations (Barker 2005: 57–58; Manning and Hulin 2005: 282–286; Schoep et al. 2011). The characterization of the Aegean Bronze Age economies—palatial and pre-palatial—as redistributive systems has also been found inaccurate and misleading, and the specific ways in which different Aegean social institutions were constituted, related to each other and changed over time, as well as the roles of elites and the reasons for their decline and collapse are proposed as more fruitful subjects of research (Nakassis, Parkinson and Galaty 2011; Pullen 2011). Along similar lines, in his analysis of cultural contacts between Cyprus and the Aegean at the end of prehistory, Knapp (2012) argues that connectivity and hybridized material and social practices rather than conquering Mycenaeans, as is commonly assumed, influenced the emergence of new political regimes during the Early Iron Age on Cyprus. Finally, it has been shown that egalitarianism is not synonymous with simplicity, but instead involves highly complex codes of social behavior and considerable variation, and that egalitarian or corporate-based social relations, practices, and ideologies can exist in any society and at different levels (e.g., Pullen 2011: 192–193; Souvatzi 2007; 2008: 41–42, 238–243).

Collectively, these studies reveal a picture of considerable complexity, flux, and ambiguity, serving more than anything to highlight the spatiotemporal variability underpinning social systems and to underline that it is not possible to enclose all kinds of continuities, discontinuities, and changes at different scales and times into generalized models.

TOWARD A MORE INTEGRATIVE FRAMEWORK FOR SPACE AND TIME IN MEDITERRANEAN PREHISTORY AND BEYOND

Far from a single notion or a unified research agenda, space and time exist in prehistoric archaeology, through a collection of theories, concepts, frames of research, and methods, and through what we are attempting to elicit from a variety of categories, terms, and notions. Whereas all of the recent studies have produced useful insights, many can be criticized for their emphasis either on space or on time and their focus on very specific scales of space or time. In Mediterranean prehistory in particular, the important contributions of much recent research have developed in ways that have yet to be fully integrated into the analysis of space and time. Conversely, the issue of space and time has yet to be fully included in prehistoric Mediterranean archaeology's theoretical and interpretative practice. Two consequences of this are that many studies work at cross-purposes and that little has been done by way of integration. In addition, the extent to which the chosen concepts, theories, and scales meet the specific research purposes and stated intentions is sometimes questionable (see also Chapman, this volume).

A most crucial requirement is to recognize the importance of a multiscalar approach to both space *and* time that will explore linkages between a whole range of spatial and temporal relationships. Whether from the point of view of traditional, top-down models or in terms of alternative narratives, Mediterranean prehistoric archaeology has focused heavily on the large spatial scale and the long term. One implication of this is a lack of a sense of short-term and small-scale social action and the bewildering and contradictory complexity of everyday lived reality.

Reluctance to move toward different scales of interpretation is also evident from the striking absence from major syntheses of Mediterranean prehistory of distinct chapters devoted to houses, households, communities, or domestic architecture (as opposed to the focus on a monumental one). Despite increasing general references, relatively few archaeologists and even fewer volumes, and mostly within an eastern Mediterranean or Neolithic context, have actually taken daily life, individual social units, and domestic or intra-settlement space seriously and have sought to make theoretically informed use of their diversity or attempted to effectively integrate them within the wider picture (e.g., see Souvatzi 2008 on households and their interaction with the wider society; Parker and Foster 2012 on current perspectives on household archaeology; Papaconstantinou 2006 on domestic space; Robb 2007 on early village life; and Kuijt 2000 on farming communities; also, papers by Knapp 2003 on imagined communities; Blake 2007: 243–248 on diasporic communities; and Belarte Franco 2009 on domestic architecture and social change). In this volume, several contributions (Stavrides, Harkness, Watkins, Düring, Marketou, Márquez-Romero and Jiménez-Jáimez, and Athanasiou) focus on houses, households, settlements, or communities as dynamic analytical units and as ever-emergent social institutions that generate and are generated by meaningful interactions that are structured by a set of places and within particular spans of time, thus eschewing common views of them as static, homogeneous, and bounded entities.

Another main concern should be with issues of social agency and their articulation into a more comprehensive approach to space (and spatiality) and time (and temporality). With the exception of the recent attempts to incorporate into analysis historical, temporal, and transitional processes, much of Mediterranean prehistoric research still requires accounts that are contextually and historically informed or that effectively acknowledge that space is a locus where the short-, the medium- and the long-term, as well as different rhythms of temporality, conjoin or conflict. A view of architecture—domestic or monumental—as a process of constantly structuring and restructuring space and the temporal aspects of the built environment have also gone largely unexplored (Harkness, Meegan, and Márquez-Romero and Jiménez-Jáimez, this volume). Cultural diversity and differences, social transitions, meaning, identity, and memory—all major themes in contemporary Mediterranean prehistory—and their inscriptions upon space are also intrinsically temporal and historically contingent (Skeates, Miller Bonney, Marketou, Murrieta-Flores, and Yasur-Landau and Cline, this volume). When we reduce them to exogenous factors or when we view them through sets of dichotomies (as opposed to contradictions), we trivialize the dialectics between space and time.

Furthermore, the continuing presence in many works of functionalism and evolutionism, which objectify society and externalize variability and change, makes it very difficult to conceptualize space as a locus of action and contradiction. At the same time, the many current -scape concepts, under the influence of postmodernism, remain rather exploratory. In these approaches, 'space' enters into the picture in many different ways, but this has not served to increase their analytical sharpness and has not resulted in an analytical theory of space. Equally, social approaches, with their sights turned less toward representation and the symbolic interpretation of given architectural forms, might have much more to contribute to a study of space and architecture as something in flux. We should focus more on the social production of space (Stavrides, this volume) and the dialectical relationship between people and space. It is important to realize that built space is not merely a source of stability and of a representation of an unchanging social order or meaning. It is also a vehicle for change, is itself enmeshed in historical processes, and is constructed by as much as it constructs and reconstructs any social environment.

The current conceptualization of time in terms of social memory and identity might also have a lot to gain from an awareness that the meaning ascribed to things and practices may vary or change at different spatial and temporal levels, as well as from a consideration of the links between short-term practices with long-term concepts and memories (e.g., Souvatzi 2012a). Landscapes, settlements, monuments, buildings, and objects can all be enmeshed in strategies of memory- and identity-making—for example, through the repetition of practices and the transmission of social knowledge, as well as the reproduction and transmission of architecture itself and of the objects associated with it—and the meanings and social relations associated with them can be reconsidered and reinterpreted by each generation (Meegan, Watkins, Düring, Miller Bonney, Murrieta-Flores, Márquez-Romero and Jiménez-Jáimez, and Yasur-Landau and Cline, this volume). A further archaeological inquiry on time and memory can relate to the social dynamics of residential mobility and site discontinuity, abandonment, or destruction (natural or deliberate) (Skeates, and Marketou, this volume).

Finally, embodied space, including the material, representational, spatial, and experiential aspects of body space and spatial orientation and movement, focal points of research in archaeology and the social sciences (e.g., Joyce 2005; Munn 2003 [1996]), can serve as another example of what needs further attention in Mediterranean archaeology (Harkness, Meegan, and Skeates, this volume), as does urban space and planning (Athanasiou, this volume).

In conclusion, we need to replace the single focus on one analytical scale by a consideration of the subtlety and variety of social processes on various scales, as well as of their interplay in the creation, maintenance, and change of social reality. Such consideration promises to yield unique insights not only into space and time but also into the complexity and variety of social life as a whole. It requires more systematic theoretical self-reflection and a willingness to explore new paths of ideas and interpretation.

ACKNOWLEDGMENTS

We would like to thank Bernard Knapp for reading and commenting on an earlier version of this chapter faithfully and enthusiastically and for the many discussions and archaeological trips on Cyprus. We are also grateful to Toula Marketou for her warm hospitality on her excavations, her home on the Street of the Knights and the Library of the Archaeological Service on Rhodes, where a draft of the introduction was written.

REFERENCES

Abulafia, D. 2005. Mediterraneans. In: W. V. Harris (ed.), *Rethinking the Mediterranean*, 64–93. Oxford: Oxford University Press.

Abulafia, D. 2011. *The Great Sea: A Human History of the Mediterranean*. Oxford: Oxford University Press.

Adam, B. 1990. *Time and Social Theory*. Cambridge: Polity Press.

Adams, W. Y. and E. W. Adams. 1991. *Pragmatics of Archaeological Typology*. Cambridge: Cambridge University Press.

Alcock, S. and J. F. Cherry. 2004a. Introduction. In: S. Alcock and J. F. Cherry (eds.), *Side by Side Survey: Comparative Regional Studies in the Mediterranean World*, 1–9. Oxford: Oxbow Books.

Alcock, S. and J. F. Cherry, eds. 2004b. *Side by Side Survey: Comparative Regional Studies in the Mediterranean World*. Oxford: Oxbow Books.

Antoniadou, S. and A. Pace, eds. 2007. *Mediterranean Crossroads*. Athens: Pierides Foundation.

Appadurai, A. 1981. The past as a scarce resource. *Man (New Series)* 16: 201–219.

Appadurai, A. 1990. Disjunction and difference in the global cultural economy. *Public Culture* 2: 1–24.

Arnold, P. J. 2008. No time like the present. In: S. Holdaway and L. Wandsnider (eds.), *Time in Archaeology: Time Perspectivism Revisited*, 160–169. Salt Lake City: The University of Utah Press.

Ashmore, W. 2007. Social archaeologies of landscape. In: L. Meskell and R. W. Preucel (eds.), *A Companion to Social Archaeology*, 255–271. Oxford/ Malden: Blackwell Publishing.

Ashmore, W. and A. B. Knapp, eds. 1999. *Archaeologies of Landscape: Contemporary Approaches*. Oxford: Blackwell.

Athanassopoulos, E. F. and L. Wandsnider, eds. 2004a. *Mediterranean Archaeological Landscapes: Current Issues*. Philadelphia: University of Pennsylvania Museum of Archaeology and Anthropology.

Athanassopoulos, E. F. and L. Wandsnider. 2004b. Mediterranean landscape archaeology past and present. In: E. F. Athanassopoulos and L. Wandsnider (eds.),

Mediterranean Archaeological Landscapes: Current Issues, 1–13. Philadelphia: University of Pennsylvania Museum of Archaeology and Anthropology.

Bachelard, G. 1964. *The Poetics of Space.* New York: Orion Press.

Baert, P. 1992. *Time, Self and Social Being.* Avebury Series in Philosophy. Aldershot: Avebury.

Bailey, G. 1981. Concepts, time-scales and explanations in economic prehistory. In: A. Sheridan and G. N. Bailey (eds.), *Economic Archaeology,* 97–117. British Archaeological Reports International Series 96. Oxford: Tempus Reparatum.

Bailey, G. 1983. Concepts of time in Quaternary prehistory. *Annual Review of Anthropology* 12: 165–192.

Bailey, G. 1987. Breaking the time barrier. *Archaeological Review from Cambridge* 6(1): 6–20.

Bailey, G. 2006. Time's arrow: the measurement and theory of archaeological time. *Antiquity* 80: 717–720.

Barden, G. 1973. Reflections of time. *The Human Context* 5: 331–344.

Barker, G. 2005. Agriculture, pastoralism and Mediterranean landscapes in prehistory. In: E. Blake and B. A. Knapp (eds.), *The Archaeology of Mediterranean Prehistory,* 46–75. Oxford: Blackwell.

Barker, G. and D. Mattingly, eds. 1999–2000. *The Archaeology of Mediterranean Landscapes,* 5 vols. Oxford: Oxbow Books.

Barrett, J. C. 1999. Chronologies of landscape. In: P. J. Ucko and R. Layton (eds.), *The Archaeology and Anthropology of Landscape: Shaping Your Landscape,* 21–30. London and New York: Routledge.

Barton, M. C., J. Bernabeu, J. E. Aura, L. Molina and S. Schmich. 2004. Historical contingency, non-linearity, and the Neolithization of the western Mediterranean. In: E. F. Athanassopoulos and L. Wandsnider (eds.), *Mediterranean Archaeological Landscapes: Current Issues,* 99–123. Philadelphia: University of Pennsylvania Museum of Archaeology and Anthropology.

Bender, B. 2002. Time and landscape. *Current Anthropology* 43(Supplement): 103–112.

Bender, B. 2003.The translation of time. In: R. M. Van Dyke and S. E. Alcock (eds.), *Archaeologies of Memory,* 221–227. London: Blackwell Publishing.

Benjamin, W. 2002. *The Arcades Project.* Translated by H. Eiland and K. McLaughlin. Cambridge, MA: Harvard University Press.

Berg, I. 2010. Re-capturing the sea: the past and future of 'island archaeology' in Greece. *Shima: The International Journal of Research into Island Cultures* 4(1): 16–26.

Binford, L. 1962. Archaeology as Anthropology. *American Antiquity* 28, 217–225.

Bintliff, J., ed. 1991. *The Annales School and Archaeology.* Leicester: Leicester University Press.

Bintliff, J. 2008. Time, structure, and agency: the Annales, emergent complexity, and archaeology. In: J. Bintliff (ed.), *A Companion to Archaeology,* 174–194. Malden, MA: Blackwell.

Blake, E. 2005. The material expression of cult, ritual and feasting. In: E. Blake and A. B. Knapp (eds.), *The Archaeology of Mediterranean Prehistory,* 102–129. Oxford: Blackwell.

Blake, E. 2007. Space, spatiality, and archaeology. In: L. Meskell and R. W. Preucel (eds.), *A Companion to Social Archaeology,* 230–254. Oxford and Malden, MA: Blackwell.

Blake, E. and B. A. Knapp, eds. 2005. *The Archaeology of Mediterranean Prehistory.* Oxford: Blackwell.

Bloch, M. 1977. The past in the present and the past. *Man* 12: 278–292.

Borić, D. and J. Robb, eds. 2008. *Past Bodies: Body-Centered Research in Archaeology.* Oxford: Oxbow Books.

Bourdieu, P. 1977 [1972]. *Outline of a Theory of Practice*. Translated by R. Nice. Cambridge: Cambridge University Press.
Bradley, R. 1987. Time regained: the creation of continuity. *Journal of the British Archaeological Association* 140: 1–17.
Bradley, R. 2002. *The Past in Prehistoric Societies*. London: Routledge.
Braudel, F. 1972 [1966]. *The Mediterranean and the Mediterranean World in the Age of Phillip II*. Translated by S. Reynolds. London: Collins.
Broodbank, C. 2000. *An Island Archaeology of the Early Cyclades*. Cambridge: Cambridge University Press.
Carr, C. and J. E. Neitzel. 1995. Integrating approaches to material style in theory and philosophy. In: C. Carr and J. E. Neitzel (eds.), *Style, Society and Person: Archaeological and Ethnological Perspectives*, 3–20. New York: Plenum Press.
Chapman, J., B. Gaydarska, E. Skafida and S. Souvatzi. 2011. Personhood and the life cycle of *Spondylus* rings: an example from Late Neolithic, Greece. In: F. Ifantidis and M. Nikolaidou (eds.), *Spondylus in Prehistory: New Data and Approaches—Contributions to the Archaeology of Shell Technologies*, 139–160. Oxford: Archaeopress, British Archaeological Reports International Series 2216.
Chapman, R. 2003. *Archaeologies of Complexity*. London: Routledge.
Chapman, R. 2005. Changing social relations in the Mediterranean Copper and Bronze Ages. In: E. Blake and A. B. Knapp (eds.), *The Archaeology of Mediterranean Prehistory*, 77–101. Oxford: Blackwell.
Cherry, J. F. 2004a. Cyprus, the Mediterranean, and survey: current issues and future trends. *British School at Athens Studies* 11: 23–55.
Cherry, J. F. 2004b. Mediterranean island prehistory: what's different and what's new? In: S. M. Fitzpatrick (ed.), *Voyages of Discovery: The Archaeology of Islands*, 233–248. New York and London: Praeger.
Clark, G. 1994. *Space, Time and Man: A Prehistorian's View*. Cambridge: Cambridge University Press.
Conkey, M. and C. Hastorf, eds. 1990. *The Uses of Style in Archaeology*. Cambridge and New York: Cambridge University Press.
Connerton, P. 1989. *How Societies Remember*. Cambridge: Cambridge University Press.
Crumley, C. L. 1995. Heterarchy and the analysis of complex societies. In: R. M. Ehrenreich, C. L. Crumley and J. E. Levy (eds.), *Heterarchy and the Analysis of Complex Societies*, 1–5. Archaeological Papers of the American Anthropological Association, no. 6. Arlington, VA: American Anthropological Association.
Crumley, C. L. 2003. Alternative forms of societal order. In: V. L. Scarborough, F. Valdez, Jr., and N. Dunning (eds.), *Heterarchy, Political Economy, and the Ancient Maya: The Three Rivers Region of the East-Central Yucatan Peninsula*, 136–145. Tucson: University of Arizona Press.
Damm, C. B. 2000. Time, gender and production: a critical evaluation of archaeological time concepts. In: M. Donald and L. Hurcombe (eds.), *Gender and Material Culture in Archaeological Perspective*, 110–122. London: Palgrave Macmillan.
Dietler, M. and I. Herbich. 1993. Living on Luo time: reckoning sequence, duration, history and biography in a rural African society. *World Archaeology* 25(2): 248–260.
Dobres, M. A. 2000. *Technology and Social Agency: Outlining a Practice Framework for Archaeology*. Oxford: Blackwell.
Dobres, M. A. and C. R. Hoffman. 1994. Social agency and the dynamics of prehistoric technology. *Journal of Archaeological Method and Theory* 1(3): 211–258.
Dobres, M. A. and C. R. Hoffman, eds. 1999. *The Social Dynamics of Technology: Practice, Politics and World Views*. Washington and London: Smithsonian Institution Press.

Dobres, M. A. and J. E. Robb, eds. 2000. *Agency in Archaeology*. London and New York: Routledge.
Driessen, H. 1999. Pre- and post-Braudelian conceptions of the Mediterranean area: the puzzle of boundaries. *Narodna Umjetnost: Croatian Journal of Ethnology and Folklore Research* 36(1): 53–63.
Durkheim, E. 1965 [1912]. *The Elementary Forms of the Religious Life*. Translated by J. W. Swain. New York: The Free Press.
Edensor, T. 2010. Introduction: thinking about rhythm and space. In: T. Edensor (ed.), *Geographies of Rhythm: Nature, Place, Mobilities and Bodies*, 1–20. Aldershot: Ashgate.
Elias, N. 1982a. Über die Zeit. *Merkur* 36: 841–856.
Elias, N. 1982b. Über die Zeit (II). *Merkur* 36: 998–1016.
Eriksen, T. H.1995. *Small Places, Large Issues: An Introduction to Social and Cultural Anthropology*. London: Pluto Press.
Evans, J. D. 1973. Islands as laboratories for the study of cultural process. In: C. Renfrew (ed.), *The Explanation of Culture Change: Models in Prehistory*, 517–520. London: Duckworth.
Fabian, J. 1983. *Time and The Other: How Anthropology Makes Its Object*. New York: Columbia University Press.
Farr, H. 2006. Seafaring as social action. *Journal of Maritime Archaeology* 1: 85–99.
Feuchtwang, S. and M. Rowlands. 2010. Re-evaluating the long term: civilisation and temporalities. In: D. Garrow and T. Yarrow (eds.), *Archaeology and Anthropology*, 117–136. Oxford: Oxbow Books.
Filippucci, P. 2010. Archaeology and the anthropology of memory: takes on the recent past. In: D. Garrow and T. Yarrow (eds.), *Archaeology and Anthropology*, 69–83. Oxford: Oxbow Books.
Fitzjohn, M. 2007. Viewing places: GIS applications for examining the perception of space in the mountains of Sicily. *World Archaeology* 39(1): 36–50.
Fitzpatrick, S. M., ed. 2004. *Voyages of Discovery: The Archaeology of Islands*. New York and London: Praeger.
Fletcher, R. 1992. Time perspectivism, Annales, and the potential of archaeology. In: A. B. Knapp (ed.), *Archaeology, Annales, and Ethnohistory*, 35–49. Cambridge: Cambridge University Press.
Forenbaher, S. 2008. Archaeological record of the Adriatic offshore islands as an indicator of long-distance interaction in prehistory. *European Journal of Archaeology* 11: 223–244.
Foucault, M. 1977 [1975]. *From Discipline and Punish: The Birth of the Prison*. Translated by Alan Sheridan. New York: Vintage Books.
Foucault, M. 1986. Of other spaces. *Diacritics* 16: 22–27.
Foxhall, L. 2000. The running sands of time: archaeology and the short-term. *World Archaeology* 31(3): 484–498.
Frankenberg, R. 1994. Anthropology and the politics of time. *Time Society* 3: 117–128.
Gamble, C. 1986. Archaeology, geography and time. *Progress in Human Geography* 11: 227–246.
Gardner, A., ed. 2004. *Agency Uncovered: Archaeological Perspectives on Social Agency, Power, and Being Human*. Walnut Creek: Left Coast Press.
Gell, A. 1992. *The Anthropology of Time: Cultural Constructions of Temporal Maps and Images*. Oxford: Berg.
Gellner, E. 1958. Time and theory in social anthropology. *Mind* 67: 182–202.
Giddens, A. 1984. *The Constitution of Society: Outline of the Theory of Structuration*. Oxford: Polity Press.
Gillings, M., D. Mattingly and J. van Dalen, eds. 2002. *Geographical Information Systems and Landscape Archaeology*. Oxford: Oxbow.

Gingrich, A. 1994. Time, ritual, and social experience. In: K. Hastrup and P. Hervik (eds.), *Social Experience and Anthropological Knowledge*. London and New York: Routledge.
Gingrich, A., E. Ochs and A. Swedlund, eds. 2002. *Repertoires of Time-keeping in Anthropology. Current Anthropology* 43, special issue.
Given, M. 2004. From density counts to ideational landscapes: intensive survey, phenomenology, and the Sydney Cyprus Survey Project. In: E. F. Athanassopoulos and L. Wandsnider (eds.), *Mediterranean Archaeological Landscapes: Current Issues*, 165–181. Philadelphia: University of Pennsylvania Museum of Archaeology and Anthropology.
Gkiasta, M. 2010. Social identities, materiality and connectivity in Early Bronze Age Crete. In: P. van Dommelen and B. A. Knapp (eds.), *Material Connections in the Ancient Mediterranean: Mobility, Materiality and Identity*, 85–105. London: Routledge.
Glennie, P. and N. Thrift. 2001. Time-geography. *International Encyclopedia of the Social and Behavioral Sciences:* 15692–15696.
Gosden, C. 1994. *Social Being and Time*. Oxford: Blackwell.
Gosden, C. 1999. *Anthropology and Archaeology: A Changing Relationship*. London: Routledge.
Gosden, C. and G. Lock. 1998. Prehistoric histories. *World Archaeology* 30: 2–12.
Gould, S. J. 1987. *Time's Arrow, Time's Cycle: Myth and Metaphor in the Discovery of Geological Time*. Cambridge, MA: Harvard University Press.
Gupta, A. and J. Ferguson. 1992. Beyond 'culture': space, identity and the politics of difference. *Cultural Anthropology* 7(1): 6–23.
Hadji, A. 2004. The construction of time in Aegean archaeology: archaeological time and third millennium BC Aegean chronology. Ph.D. dissertation, UC Berkeley, UMI pub. no. 3146870 (www.umi.com).
Hadji, A. 2009. The potential of Henri Lefebvre's work on space for archaeology. *Association of Social Anthropologists Annual Meeting*, Bristol, April 6–9, 2009.
Hadji, A. 2010a. Knowledge as heterotopy: academic practices and the creation of academic (phantasy) places. In: *Transformations and (Dis)continuities: Practices, Policies and Discourse on Urban Space*, 17–23. Athens: Alexandria Publications.
Hadji, A. 2010b. A theoretically informed study of urban contexts: the contribution of modern (social) urban theories. *TAG2010: 32nd Annual Meeting of the Theoretical Archaeology Group*, December 17–19, 2010, Bristol.
Hall, E. 1959. *The Silent Language*. Garden City, NY: Doubleday.
Hall, E. 1963. A system for the notation of proxemic behavior. *American Anthropologist* 65: 1003–1026.
Hall, E. 1966. *The Hidden Dimension*. Garden City, NY: Doubleday.
Hall, E. 1976. *Beyond Culture*. Garden City, NY: Anchor Press.
Harding, J. 2005. Rethinking the Great Divide: long-term structural history and the temporality of event. *Norwegian Archaeological Review* 38(2): 88–101.
Harris, W. V. 2006. The Mediterranean and ancient history. In: W. V. Harris (ed.), *Rethinking the Mediterranean*, 1–42. Oxford: Oxford University Press.
Hastrup, K. and K. F. Olwig. 1997. Introduction. In: K. F. Olwig and K. Hastrup (eds.), *Siting Culture: The Shifting Anthropological Object*, 1–15. London: Routledge.
Hawking, S. W. 1988. *A Brief History of Time*. Toronto: Bantam Books.
Heidegger, M. 1962 [1953]. *Being and Time*. Translated by J. Macquarrie and E. Robinson. New York: Harper and Row.
Heilbroner, R. L. 1980. *Marxism: For and Against*. New York and London: W. W. Norton.
Helms, M. W. 1978. Time, history, and the future of anthropology: observations on some unresolved issues. *Ethnohistory* 25(1): 1–13.

Herzfeld, M. 1984. The horns of the Mediterraneanist dilemma. *American Ethnologist* 11(3): 439–454.

Herzfeld, M. 2005. Practical Mediterraneanism: excuses for everything, from epistemology to eating. In: W. V. Harris (ed.), *Rethinking the Mediterranean,* 45–63. New York: Oxford University Press.

Herzfeld, M. 2009. Rhythm, tempo, and historical time: experiencing temporality in the neoliberal age. *Public Archaeology: Archaeological Ethnographies* 8(2–3): 108–123.

Hexter, J. H. 1972. Fernand Braudel and the monde Braudellien. *Journal of Modern History* 44: 480–539.

Hill, B. J. 2004. Time, scale and interpretation: 10,000 years of land use on the Transjordan plateau and multiple contexts of change. In: E. F. Athanassopoulos and L. Wandsnider (eds.), *Mediterranean Archaeological Landscapes: Current Issues,* 125–141. Philadelphia: University of Pennsylvania Museum of Archaeology and Anthropology.

Hodder, I. 2006. Memory. In I. Hodder (ed.), *Çatalhöyük Perspectives: Themes from the 1995–99 Seasons,* 183–195. Cambridge and London: McDonald Institute for Archaeological Research and the British Institute of Archaeology at Ankara Monograph.

Hodges, M. 2008. Rethinking time's arrow: Bergson, Deleuze and the anthropology of time. *Anthropological Theory* 8(4): 399–429.

Holdaway, S. and L. Wandsnider, eds. 2008. *Time in Archaeology: Time Perspectivism Revisited.* Salt Lake City: The University of Utah Press.

Horden, P. and N. Purcell. 2000. *The Corrupting Sea: A Study of Mediterranean History.* Oxford: Blackwell.

Horden, P. and N. Purcell. 2006. The Mediterranean and "the New Thalassology". *The American Historical Review* 111(3): 722–740.

Howe, L. 1981. The social determination of knowledge: Maurice Bloch and Balinese time. *Man* 16: 220–234.

Hoy Couzens, D. 2009. *The Time of Our Lives: A Critical History of Temporality.* Cambridge, MA: The MIT Press.

Hughes, D. O. and T. R. Trautmann, eds. 1995. *Time: Histories and Ethnologies.* Ann Arbor: University of Michigan Press.

Hurcombe, L. 2000. Time, skill and craft specialization as gender relations. In: *Gender and Material Culture in Archaeological Perspective,* 88–109. London: Palgrave Macmillan.

Husserl, E. 1991. *On the Phenomenology of the Consciousness of Internal Time* (1893–1917). Translated by J. B. Brough. Dordrecht: Kluwer Academic Publishers.

Ingold, T. 1993. The temporality of the landscape. *World Archaeology* 25(2): 152–174.

Janes, S. 2010. Negotiating island interactions: Cyprus, the Aegean and the Levant in the Late Bronze to Early Iron Ages. In: P. van Dommelen and B. A. Knapp (eds.), *Material Connections in the Ancient Mediterranean: Mobility, Materiality and Identity,* 127–146. London: Routledge.

Johnson, M. A. 1989. Conceptions of agency in archaeological interpretation. *Journal of Anthropological Archaeology* 8(2): 198–211.

Joyce, R. A. 2005. Archaeology of the body. *Annual Review of Anthropology* 34 (1): 139–158.

Joyce, R. A. and S. D. Gillespie, eds. 2000. *Beyond Kinship: Social and Material Reproduction in House Societies.* Philadelphia: University of Pennsylvania Press.

Joyce, R. A. and J. Lopiparo. 2005. Doing agency in archaeology. *Journal of Archaeological Method and Theory* 12: 365–374.

Karimali, E. 2005. Lithic technologies and use. In: E. Blake and B. A. Knapp (eds.), *The Archaeology of Mediterranean Prehistory,* 180–214. Oxford: Blackwell.

Karlsson, H. 2001. Time for an archaeological 'time-out'? In: H. Karlsson (ed.), *It's About Time: The Concept of Time in Archaeology*, 45–59. Göteborg: Bricoleur Press.

Keates, S. 2000. The ancestralization of the landscape. In: G. Nash (ed.), *Signifying Place and Space*, 83–102. Oxford: BAR International Series 902.

King, A. D. 1980. A time for space and a space for time. In: A.D. King (ed.), *Buildings and Society: Essays of the Social Development of the Built Environment*, 193–227. London: Routledge.

King, R., L. Proudfoot and B. Smith. 1997. *The Mediterranean: Environment and Society*. London: Arnold.

Knapp, A. B. 1992a. Archaeology and Annales: time, space, change. In: A. B. Knapp (ed.), *Annales, Archaeology and Ethnohistory*, 1–21. Cambridge: Cambridge University Press.

Knapp, A. B. 1992b. Bronze Age Mediterranean island cultures and the Ancient Near East. *The Biblical Archaeologist* 55(2): 52–72.

Knapp, A. B. 2003. The archaeology of community on Bronze Age Cyprus: Politiko 'Phorades' in context. *American Journal of Archaeology* 107(4): 559–580.

Knapp, A. B. 2007. Insularity and island identity in the prehistoric Mediterranean. In: S. Antoniadou and A. Pace (eds.), *Mediterranean Crossroads*, 37–62. Athens: Pierides Foundation.

Knapp, A. B. 2008. *Prehistoric and Protohistoric Cyprus: Identity, Insularity and Connectivity*. Oxford: Oxford University Press.

Knapp, A. B. 2009. Monumental architecture, identity and memory. *Proceedings of the Symposium: Bronze Age Architectural Traditions in the East Mediterranean: Diffusion and Diversity* (Gasteig, Munich, May 7–8, 2008), 47–59. Weilheim: Verein zur Förderung der Aufarbeitung der Hellenischen Geschichte e.V. 2009.

Knapp, A. B. 2012. Matter of fact: transcultural contacts in the Late Bronze Age eastern Mediterranean. In: J. Maran and P. W. Stockhammer (eds.), *Materiality and Social Practice: Transformative Capacities of Intercultural Encounters*, 32–50. Oxford: Oxbow.

Knapp, A. B. and E. Blake. 2005. Prehistory in the Mediterranean: the connecting and corrupting sea. In: E. Blake and B. A. Knapp (eds.), *The Archaeology of Mediterranean Prehistory*, 1–23. Oxford: Blackwell.

Knapp, A. B. and P. van Dommelen. 2010. Material connections: mobility, materiality and Mediterranean identities. In: P. van Dommelen and B. A. Knapp (eds.), *Material Connections in the Ancient Mediterranean: Mobility, Materiality and Identity*, 1–18. London: Routledge.

Kohring, S. E. and S. Wynne-Jones, eds. 2007. *Socialising Complexity: Structure, Interaction and Power in Archaeological Discourse*. Oxford: Oxbow.

Kolb, M. J. 2005. The genesis of monuments among the Mediterranean islands. In: E. Blake and B. A. Knapp (eds.), *The Archaeology of Mediterranean Prehistory*, 156–179. Oxford: Blackwell.

Kuijt, I., ed. 2000. *Life in Neolithic Farming Communities: Social Organization, Identity and Differentiation*. New York: Kluwer Academic/Plenum.

Leach, E. 1961. Two essays concerning the symbolic representation of time. In: E. Leach, *Rethinking Time*, 125–143. London: Athlone.

Lechtman, H. 1977. Style in technology: some early thoughts. In: H. Lechtman and R. Merrill (eds.), *Material Culture: Styles, Organization, and Dynamics of Technology*, 3–20. St. Paul, MN: West.

Lechtman, H. 1988. Traditions and styles in central Andean metalworking. In: R. Maddin (ed.), *The Beginning of the Use of Metals and Alloys*, 344–378. Cambridge, MA: MIT Press.

Lefebvre, H. 1991. *The Production of Space*. Translated by D. Nicholson-Smith. London: Blackwell.

Lefebvre, H. 2004 [1992]. *Rhythmanalysis: Space, Time and Everyday Life.* London: Continuum.
Lewis, J. D. and A. J. Weigert. 1981. Structures and meanings of social time. *Social Forces* 60: 433–462.
Lobao, L. M., G. Hooks and A. R. Tickamyer, eds. 2007. *The Sociology of Spatial Inequality.* Albany: State University of New York Press.
Low, S. M. and D. Lawrence-Zúñiga, eds. 2003a. *The Anthropology of Space and Place: Locating Culture.* Oxford and Malden, MA: Blackwell Publishing.
Low, S. M. and D. Lawrence-Zúñiga. 2003b. Locating culture. In: S. M. Low and D. Lawrence-Zúñiga (eds.), *The Anthropology of Space and Place: Locating Culture,* 1–47. Oxford and Malden, MA: Blackwell Publishing.
Lowenthal, D. 1975. Past time, present place: landscape and memory. *The Geographical Review* 65: 1–36.
Lucas, G. 2005. *The Archaeology of Time.* London and New York: Routledge.
Lüscher, K. K. 1974. Time: a much neglected dimension in social theory and research. *Sociological Analysis and Theory* 4: 101–117.
Manning, S. W. 1999. *A Test of Time: The Volcano of Thera and the Chronology and History of the Aegean and East Mediterranean in the Mid-second Millennium BC.* Oxford: Oxbow Books.
Manning, S. W. and L. Hulin. 2005. Maritime commerce and geographies of mobility in the Later Bronze Age of the eastern Mediterranean: problematizations. In: E. Blake and A. B. Knapp (eds.), *The Archaeology of Mediterranean Prehistory,* 270–302. Oxford: Blackwell.
Maran, J. and P. W. Stockhammer, eds. 2012. *Materiality and Social Practice: Transformative Capacities of Intercultural Encounters.* Oxford: Oxbow.
Marangou, C. and P. Della Casa. 2008. Islands in the Mediterranean: Introduction. *European Journal of Archaeology* 11: 171–177.
Maurice, A. 2001. De la méditerranée à l'asie: une comparaison nécessaire (commentaire). *Annales. Histoire, Sciences Sociales,* 56: 43–50.
McGlade, J. 1999. The times of history: archaeology, narrative, and non-linear causality. In: T. Murray (ed.), *Time and Archaeology,* 139–163. New York and London: Routledge.
McFayden, L. 2010. Spaces that were densely occupied: questioning 'ephemeral' evidence. In: D. Garrow and T. Yarrow (eds.), *Archaeology and Anthropology,* 40–52. Oxford: Oxbow Books.
Menin, S., ed. 2003. *Constructing Place: Mind and Matter.* London and New York: Routledge.
Merleau Ponty, M. 2012 [1945]. *Phenomenology of Perception.* Translated by D. A. Landes. London and New York: Routledge.
Meskell, L. and R. Joyce. 2003. *Embodied Lives: Figuring Ancient Maya and Egyptian Experience.* London and New York: Routledge.
Mizoguchi, K. 1993. Time in the reproduction of mortuary practices. *World Archaeology* 25: 223–235.
Montón Subías, S. 2007. Interpreting archaeological continuities: an approach to transversal equality in the Argaric Bronze Age of south-east Iberia. *World Archaeology* 39(2): 246–262.
Moore, J. D. 1996. The archaeology of plazas and the proxemics of ritual: three Andean traditions. *American Anthropologist* 98(4): 789–802.
Morris, I. 2003. Mediterraneanization. *Mediterranean Historical Review* 18(2): 30–55.
Movahedi, S. 1985. Cultural preconceptions of time: can we use operational time to meddle in God's time? *Comparative Studies in Society and History* 27: 385–400.
Munn, N. D. 1992. The cultural anthropology of time: a critical essay. *Annual Review of Anthropology* 21: 93–123.

Munn, N. 2003[1996]. Excluded spaces: the figure in the Australian Aboriginal landscape. In S. M. Low and D. Lawrence-Zúñiga (eds.), *The Anthropology of Space and Place: Locating Culture*, 92–109. Oxford / Malden: Blackwell Publishing.

Murray, T., ed. 1999. *Time and Archaeology*. London and New York: Routledge.

Nakassis, D., W. A. Parkinson and M. L. Galaty. 2011. Redistribution in Aegean palatial societies: redistributive economies from a theoretical and cross-cultural perspective. *American Journal of Archaeology* 115(2): 177–184.

Newton-Smith, W. H. 1980. *The Structure of Time*. London, Boston and Hekley: Routledge and Kegan Paul.

Olivier, L. 2011. *The Dark Abyss of Time: Archaeology and Memory*. Translated by Arthur Greenspan. Lanham, MD: Altamira Press.

Olsen, B. 2003. Material culture after text: re-membering things. *Norwegian Archaeological Review* 36(2): 87–104.

Olsen, B. 2007. Keeping things at arm's length: a genealogy of asymmetry *World Archaeology* 39(4): 579–588.

Panagiotopoulos, D. 2012. Encountering the foreign: (de)constructing alterity in the archaeologies of the Bronze Age Mediterranean. In: J. Maran and P. W. Stockhammer (eds.), *Materiality and Social Practice: Transformative Capacities of Intercultural Encounters*, 51–60. Oxford: Oxbow.

Parker B. J. and C. P. Foster, eds. 2012. *New Perspectives on Household Archaeology*. Winona Lake, Indiana: Eisenbrauns.

Papaconstantinou, D. 2006. *Identifying Domestic Space in the Neolithic Eastern Mediterranean: Method and Theory in Spatial Studies*. Oxford: British Archaeological Reports, Archaeopress.

Papageorgiou, D. 2008. The marine environment and its influence on seafaring and maritime routes in the prehistoric Aegean. *European Journal of Archaeology* 11: 199–222.

Parkinson, W. A. and M. L. Galaty, eds. 2010. *Archaic State Interaction: The Eastern Mediterranean in the Bronze Age*. Santa Fe, NM: School for Advanced Research (SAR) Press.

Patton, M. 1996. *Islands in Time: Island Sociogeography and Mediterranean Prehistory*. London: Routledge.

Peltenburg, E, 2007. East Mediterranean interactions in the 3rd millennium BC. In: S. Antoniadou and A. Pace (eds.), *Mediterranean Crossroads*, 141–161. Athens: Pierides Foundation.

Pepper, S. C. 1935. The order of time. In: *The Problem of Time*, 1–20. University of California Publications in Philosophy vol.18. Berkeley: University of California Press.

Perec, G. 1997. *Species of Spaces and Other Pieces*. London: Penguin Books.

Plog, S. 1983. Analysis of style in artifacts. *Annual Review of Anthropology* 12: 125–142.

Pullen, D. J. 2011. Redistribution in Aegean palatial societies before the palaces: redistribution and chiefdoms in mainland Greece. *American Journal of Archaeology* 115(2): 185–195.

Pullen, D. J. and T. F. Tartaron. 2007. Where is the palace? In: M. L. Galaty and W. A. Parkinson (eds.), *Rethinking Mycenaean Palaces II*, 146–158. Los Angeles: University of California, Cotsen Institute of Archaeology.

Purcell, N. 2003. The Boundless Sea of Unlikeness? On Defining the Mediterranean. *Mediterranean Historical Review* 18(2): 9–29.

Rabinow, P. 2007. *Marking Time: On the Anthropology of the Contemporary*. Princeton: Princeton University Press.

Rainbird, P. 2007. *The Archaeology of Islands*. Cambridge: Cambridge University Press.

Ramis, D. 2010. From colonization to habitation: early cultural adaptations in the Balearic Bronze Age. In: P. van Dommelen and B. A. Knapp (eds.), *Material*

Connections in the Ancient Mediterranean: Mobility, Materiality and Identity, 64–84. London: Routledge.

Richards, C. 1995. Knowing about the past. In: I. Hodder, M. Shanks, A. Alexandri, V. Buchli, J. Carman, J. Last and G. Lucas (eds.), *Interpreting Archaeology: Finding Meaning in the Past*, 216–219. London and New York: Routledge.

Robb, J. 2001. Island identities: ritual, travel and the creation of difference in Neolithic Malta. *European Journal of Archaeology* 4: 175–202.

Robb, J. 2007. *The Early Mediterranean Village: Agency, Material Culture, and Social Change in Neolithic Italy*. Cambridge: Cambridge University Press.

Robb, J. E. and H. R. Farr. 2005. Substances in motion: Neolithic Mediterranean "trade". In: E. Blake and A. B. Knapp (eds.), *The Archaeology of Mediterranean Prehistory*, 24–45. Oxford: Blackwell.

Rodman, M. C. 2003. Empowering place: multilocality and multivocality. In: S. M. Low and D. Lawrence-Zúñiga (eds.), *The Anthropology of Space and Place: Locating Culture*, 204–223. Oxford and Malden, MA: Blackwell Publishing.

Rossignol, J. and L. Wandsnider, eds. 1992. *Space, Time and Archaeological Landscapes*. New York: Plenum Press.

Rowlands, M. 1993. The role of memory in the transmission of culture. *World Archaeology* 25: 141–151.

Russel, A. 2010. Foreign materials, islander mobility and elite identity in Late Bronze Age Sardinia. In: P. van Dommelen and B. A. Knapp (eds.), *Material Connections in the Ancient Mediterranean: Mobility, Materiality and Identity*, 106–126. London: Routledge.

Saitta, D. S. 1994. Agency, class and archaeological interpretation. *Journal of Anthropological Archaeology* 13: 201–227.

Sato, M. 1991. Comparative ideas of chronology. *History and Theory* 30: 275–301.

Sato, M. 2001. Time, chronology and periodization in history. *International Encyclopedia of the Social and Behavioral Sciences:* 15686–15692.

Schoep, I., P. Tomkins and J. Driessen (eds.). 2011. *Back to the Beginning: Reassessing Social and Political Complexity on Crete during the Early and Middle Bronze Age*. Oxford: Oxbow Books.

Shankland, D., ed. 2012. *Archaeology and Anthropology Past, Present, and Future*. Association of Social Anthropologists Monographs number 48. London: Berg.

Shanks, M. and C. Tilley. 1987a. Abstract and substantial time. *Archaeological Review from Cambridge* 6: 32–41.

Shanks, M. and C. Tilley. 1987b. *Social Theory and Archaeology*. Cambridge: Polity Press.

Sherratt, A. 2012. The intercultural transformative capacities of irregularly appropriated goods. In: J. Maran and P. W. Stockhammer (eds.), *Materiality and Social Practice: Transformative Capacities of Intercultural Encounters*, 152–172. Oxford: Oxbow.

Silverman, E. K. 2001. Time, anthropology of. *International Encyclopedia of the Social and Behavioral Sciences:* 15683–15686.

Smith, M. E. 1992. Braudel's temporal rhythms and chronology theory in archaeology. In: A. B. Knapp (ed.), *Annales, Archaeology, and Ethnohistory*, 25–36. Cambridge and New York: Cambridge University Press.

Soja, E. W. 1989. *Postmodern Geographies: The Reassertion of Space in Critical Social Theory*. London: Verso.

Sollars, L. 2005. Settlement in the prehistoric Mediterranean. In: E. Blake and A. B. Knapp (eds.), *The Archaeology of Mediterranean Prehistory*, 252–269. Oxford: Blackwell.

Sorokin, P. A. and R. K. Merton. 1937. Social time: a methodological and functional analysis. *American Journal of Sociology* 42: 615–629.

Souvatzi, S. 2007. Social complexity is not the same as hierarchy. In: S. E. Kohring and S. Wynne-Jones (eds.), *Socializing Complexity: Structure, Interaction and Power in Archaeological Discourse*, 37–59. Oxford: Oxbow Books.

Souvatzi, S. 2008. *A Social Archaeology of Households in Neolithic Greece: An Anthropological Approach*. New York and Cambridge: Cambridge University Press.

Souvatzi, S. 2012a. Between the individual and the collective: household as a social process in Neolithic Greece. In: B. J. Parker and C. P. Foster (eds.), *New Perspectives on Household Archaeology*, 15–43. Winona Lake, IN: Eisenbrauns.

Souvatzi, S. 2012b. Space, place and architecture: a major meeting point between social archaeology and anthropology? In: D. Shankland (ed.), *Archaeology and Anthropology Past, Present, and Future*, 173–196. Association of Social Anthropologists Monographs number 48. London: Berg.

Souvatzi, S. 2013. Land tenure, social relations and social landscapes. In: M. Relaki and D. Catapoti (eds.), *An Archaeology of Land Ownership*, 21–45. London: Routledge.

Stevenson, D. 2003. *Cities and Urban Cultures*. Maidenhead, Berkshire: Open University Press.

Tartaron, T. F. 2008. Aegean prehistory as world archaeology: recent trends in the archaeology of Bronze Age Greece. *Journal of Archaeological Research* 16(2): 83–161.

Thomas, J. 1996. *Time, Culture and Identity: An Interpretative Archaeology*. London: Routledge.

Tringham, R., ed. 1973. *Territoriality and Proxemics: Archaeological and Ethnographic Evidence for the Use and Organization of Space*. Andover, MA: Warner Modular Publications.

Tringham, R. 2000. The continuous house: a view from the deep past. In S. Gillespie and R. Joyce (eds.), *Beyond Kinship: Social and Material Reproduction in House Societies*. Philadelphia: University of Pennsylvania Press.

Turnbull, D. 2002. Performance and narrative, bodies and movement in the construction of places and objects, spaces and knowledges: the case of the Maltese megaliths. *Theory Culture Society* 19: 125–143.

van Dommelen, P. 1997. Colonial constructs: colonialism and archaeology in the Mediterranean. *World Archaeology* 28(3): 305–323.

van Dommelen, P. and Knapp, A. B., eds. 2010. *Material Connections in the Ancient Mediterranean: Mobility, Materiality and Identity*. London: Routledge.

Vianello, A., ed. 2011. *Exotica in the Prehistoric Mediterranean*. Oxford: Oxbow.

Vis, B. N. 2009. *Built Environments, Constructed Societies: Inverting Spatial Analysis*. Leiden: Sidestone Press.

Wallerstein, I. 1998. The time of space and the space of time: the future of social science. *Political Geography* 17(1): 71–82.

Wandsnider, L. 2004 Artifact, landscape, and temporality in Mediterranean landscape studies. In: E. F. Athanassopoulos and L. Wandsnider (eds.), *Mediterranean Archaeological Landscapes: Current Issues*, 69–80. Philadelphia: University of Pennsylvania Museum of Archaeology and Anthropology.

Watkins, T. 2005. Architecture and 'theatres of memory' in the Neolithic of southwest Asia. In: E. DeMarrais, C. Gosden and C. Renfrew (eds.), *Rethinking Materiality: The Engagement of Mind with the Material World*, 97–106. Cambridge: McDonald Institute for Archaeological Research.

Witmore, C. 2007. Symmetrical archaeology: excerpts of a manifesto. *World Archaeology* 39(4): 546–562.

Zeder, M. A. 2008. Domestication and early agriculture in the Mediterranean basin: origins, diffusion, and impact. *Proceedings of the National Academy of Sciences of the United States of America* 105(33): 11597–11604.

Zerubavel, E. 1982. The standardization of time: a sociohistorical perspective. *American Journal of Sociology* 88: 1–23.

Zimmerman, V. 2007. *Excavating Victorians*. Albany, NY: SUNY Press.

2 Scales, Interaction, and Movement in Later Mediterranean Prehistory

Robert Chapman

INTRODUCTION

Some recent interpretations of later Mediterranean prehistory have prioritized the roles of cultural practices in the maintenance and change of social relations in time and space. There is talk of social identities, the importance of social memory, and the experience of landscapes in the daily lives of people (e.g., Van Dommelen and Knapp 2010). Time and space are seen less as archaeological frameworks than means by which people construct social relations and understand culturally the world in which they live and how they came to be where they are. These cultural practices are argued to have occurred at different scales, from days to centuries and from the household to the Mediterranean basin as a whole.

In this chapter my concern is with the main scales of time and space within which Mediterranean archaeologists have worked, how these scales have been constructed and are used by these archaeologists, what the deficiencies of such scales (as we know them) might be, and what theoretical assumptions (e.g., dynamic as opposed to static societies, isolated as opposed to connected social and political groups, top-down rather than bottom-up analyses of social reproduction) lie behind archaeological research. Given the size of the Mediterranean basin (22,000 miles from Jerusalem to Gibraltar, that is, three months' sailing before modern times), I am selective in the examples I use. The resolution of both spatial and temporal scales, as we currently know them, has implications for our understanding of social, political, and economic changes at critical points in the prehistory of the Mediterranean, from the adoption of agriculture to the emergence of institutionalized inequalities.

DYNAMIC AND STATIC SOCIETIES IN THE MEDITERRANEAN

Horden and Purcell (2000: 467) criticize historians and ethnographers for their emphasis on the 'timeless' nature of Mediterranean societies, by which rural societies are viewed as survivors from the past: static, traditional, and isolated. Mediterranean ethnographers have given the impression that at least some Mediterranean societies have been frozen in time and cut off from the wider world, and their study has emphasised "all that is archaic, culturally and economically primitive" (2000: 487). Such an approach creates a 'homogenised' (2000: 487) and 'Romantic' (2000: 28–29) view of Mediterranean societies, in which the past lives on in the present. Thus the historian Fernand Braudel (1972: 1239) wrote that "antiquity lives on round today's Mediterranean shores". This survivalism incorporated ahistorical stereotypes of 'traditional' rural economies (with transhumance and redistribution as necessary adaptations to environmental constants), along with dichotomies of agricultural/pastoral, mountain/lowland, and town/rural societies, as well as 'prime'/'marginal' environments.

Archaeological research on the later prehistoric Mediterranean in the first half of the twentieth century integrated this survivalism with a further dichotomy between the dynamic societies of the east (the sources of both agriculture and civilization) and the static societies of the west (which received economic and technological innovations through diffusion across a unifying sea). Individual Neolithic and Bronze Age artifact types, monument types, and symbols, as well as artifact assemblages, from 'nuclear areas' of the western Mediterranean were attributed to east Mediterranean 'colonists' or 'influences' (see Chapman 1990: 19–30 for examples from Iberia), and thereby the principles of cross-dating and association were used to extend the absolute chronology of the ancient Near East to Mediterranean and Temperate Europe. In many cases these artifact comparisons lacked ultimate conviction, whereas dating by the calibrated radiocarbon method established that specific artifacts and monuments in the west Mediterranean were earlier than their supposed prototypes in the east Mediterranean (Renfrew 1973).

As a result of the successful critique of diffusionism, greater emphasis was placed on autonomy in later prehistoric societies in all parts of the Mediterranean, but survivalism and the static/dynamic dichotomy continued to be in evidence. The Cambridge 'Palaeoeconomy' School emphasized long-term continuity in the relationship between people and Mediterranean environments, tracing seasonal movements between lowland and upland pastures in Greece, Italy, and Iberia from the Middle Palaeolithic to the present day (Higgs 1975). The influence of Braudel's (1972) *longue durée* was visible in Barker's (1981) research on Neolithic and later societies in central Italy, with its emphasis on slow, minimal change among peasant farmers. Malone (2003: 299) follows this approach for Neolithic Italy as a whole by proposing that "once established as farming communities, the societies appear to be conservative, and often rather isolated, except through the exchange of exotica". In other studies, major changes in macro-variables such as population pressure or the intensification of production were needed to jerk life into such static systems. The major change seemingly came in the Bronze Age, but while 'civilization' emerged in the East, stratification and 'complexity' have been attributed to a very limited number of areas in the west

Mediterranean before the development of the Etruscan state in the first millennium BC (see Chapman 2003 for discussion and examples). Mathers and Stoddart (1994: 16) even question the accuracy of the term 'stratification' for any pre–first millennium BC societies in the West.

FRAGMENTATION AND CONNECTIVITY

While the static/dynamic dichotomy has continued to shape archaeological assumptions on the Mediterranean past, the collapse of diffusionism encouraged archaeologists to explore both regional sequences of change and more localized networks of interaction over land and sea. One starting point was the Mediterranean itself.

There are a variety of criteria (e.g., geological, botanical, climatic, political) that are used to define the Mediterranean basin. The areas mapped by these criteria do not necessarily conform in their entirety, but the main object of debate concerns their boundaries. Horden and Purcell (2000: 80) criticize the notion that these boundaries are 'permanent' or 'uncrossable', referring to the Mediterranean as 'more softly edged'. In examining the geography of human history in the Mediterranean, these authors propose that "dense fragmentation complemented by a striving towards control of communications may be an apt summary of the Mediterranean past" (2000: 25). Although the fragmentary land masses and seas show a large degree of continuity through the postglacial past, they have been connected culturally in different ways at different moments in time. Degrees of isolation and connectivity have varied through time and were not subject to some form of constant environmental determinism.

This emphasis on autonomous development led archaeologists to explore regional sequences on land and the potential of islands to study how geographical isolation produced social and cultural idiosyncrasies. Such islands were viewed by some as closed 'laboratories' (Evans 1973), whereas others were more focused on how such islands became colonized by human populations (e.g., Cherry 1981) and what networks of interaction existed at different times between islands and between islands and mainlands, from the local and regional scales to the entire Mediterranean basin. This focus on connectivity required greater attention to the potential and constraints of both the sea (e.g., visibility, winds, currents) and the islands (e.g., distance from land, area, diversity), within the contexts of changing sea levels and available maritime technologies. In addition to Cherry's (1981) use of the theory of island biogeography to analyze island colonization across the Mediterranean, other authors have subsequently discussed the constraints of the sea on interaction, and the advantages of coastal navigation ('tramping') in areas like the west Mediterranean (Chapman 1990: 260–263), the Aegean (Broodbank 2000 on 'islandscapes'), the east Mediterranean (Manning and Hulin 2005), and right across the Mediterranean from east to west, as in the case of Phoenician colonization in the early first millennium BC (Aubet 2001).

In addition to the study of the potential for, and constraints on, interaction over the sea, others have focused on the characterization of raw materials to trace the production sources and spatial movement of artifacts. These range from the Neolithic of the Aegean (Perlès 1992) and the central Mediterranean (e.g., Malone 1985; Tykot 1996), to the trans-Mediterranean trade of the Mycenaeans (e.g., Harding 1984; Vianello 2005) in the late second millennium BC and the Phoenicians in the early first millennium BC (Aubet 2001). Interpretations of this interaction vary from localized exchange designed to establish and maintain social relations, through use of exotic materials/artifacts to gain political advantage (the model of Helms 1988), to the political strategies of expanding state societies. Such inferences face the challenges of moving from evidence of the exchange of individual materials/artifacts to the reconstruction of exchange networks, and of grasping the full scale of interaction, even among early states, given the biased nature of the archaeological record (e.g., Manning and Hulin 2005: 283–84 on the small scale of Late Bronze Age Aegean imports in the east Mediterranean).

Two examples illustrate how our ideas about insularity have been challenged. Robb (2001) argues that Malta's cultural isolation during the periods of temple building and rebuilding (c. 3600–2400 BC) has been overstated, given (a) the evidence for raw materials and goods (e.g., flint, obsidian, stone axes and amulets, lava, alabaster, red ochre, timber) introduced from Sicily (although see Malone and Stoddart 2004: 96), (b) the ability to cross from Sicily to Malta in one to three days in the summer, (c) the greater challenges of overland movement on Sicily as opposed to the sea crossing between southeast Sicily and Malta (see Robb 2001: Figure 1), (d) examples of cultural trends on Malta (e.g., decreased obsidian, local pottery styles) being shared by southern Italy and Sicily, and (e) cultural similarities (e.g., standing stone circles, rock-cut tombs, hypogea, and temples) between Malta, Sicily, southern Italy, and Sardinia. These observations contrast with the model of periods of interaction and contact alternating with those of ritual and isolation (e.g., Stoddart et al. 1993). Instead of Maltese temples being the peculiar, material outcome of isolation, they are argued to have symbolized "the construction of cultural difference" (Robb 2001: 192).

Lull et al. (2002: 118–119) argue against the determining nature of insularity per se and propose that "the social space of an island does not finish where its land ends, but integrates the surrounding sea as well". Thus "the perception of this space and the attitude of island societies toward the outside does not seem to be fixed but changes through time, due to different economic, social or political factors". Insularity is therefore 'most of all' a social rather than a natural construction, determined by both internal and external factors. This is shown by phases of 'increased interaction' (whether through population movement, exchange of material goods, or 'communication of information and values') on the Balearic islands (the

Early Beaker phase c.2500–2000 BC, the increase in new settlement and burial structures c.1600 BC, and changes in settlements, production, and ideology c.1200–850 BC). Interestingly, the period of talaiot construction on Majorca and Menorca (c.850–600/500 BC) shows little evidence of the overseas contacts and exchange that were an integral part of the Phoenician trade network that reached Ibiza by the mid-seventh century. The potential for extra-island contacts was not exploited by the main Balearic island populations. It is worth adding that, in the subsequent post-talaiotic period, which lasted until the Roman conquest in the late second century BC, the evidence for overseas interaction (e.g., Punic) increased again, so the construction of another, localized monument type, the taulas of Menorca, did not take place in a context of isolation, as would be predicted by the island laboratory argument.

The authors of these two case studies concur in the view that insularity is a human 'construction'. Insularity is also viewed as a relative rather than an absolute condition. Specific island sequences are studied in terms of the interaction of internal and external factors. From this perspective, the 'isolated' study of islands, especially in the context of the Mediterranean as opposed to the Pacific Ocean (the latter is seventy times the size of the former; Anderson 2011: 157), needs to be replaced by a larger-scale study of production and social reproduction over neighboring land and water masses (islands and mainlands). Socially and politically determined periods of greater 'insularity' and 'isolation' are also known from some mainland areas, such as is seen for the Argaric group of the earlier Bronze Age in southeast Spain, as compared with the preceding Chalcolithic (Risch 2002).

Even at a superficial level, the relatively small distances between Mediterranean mainlands and islands and the material evidence of interaction and colonization support the inference of social networks of different scales that made isolation a relative and socially contingent condition.

MEDITERRANEAN SOCIETIES AT THE LARGE SCALE

While a localized focus on social, economic, and political development, coupled with the tracing of exchange networks, marked out many areas of the Mediterranean in the 1970s and 1980s, a distinct approach has maintained the need for its study at the large scale. This owes its origin to world systems theory. In what follows, I will give a critical outline of this approach, coupled with examples of its application to the Mediterranean.

The development of world systems theory by Wallerstein (1974) was concerned with the historical study of the rise of capitalism as a world economy, the main dynamic being the asymmetric relationship between the providers of staple goods and raw materials in the peripheral regions (e.g., through trading outposts, colonies) and the dominant consumers of these goods and materials in core regions such as western Europe. Wallerstein's model did not include the study of noncapitalist and nonstate societies in their own terms. Such societies were not necessarily characterized by unequal exchange, or the movement of bulk goods, and Wallerstein's analysis of 'peripheral' societies denied their individual characteristics and histories, as well as any role in the development of 'cores'. These principal criticisms have led others, especially Chase-Dunn and Hall (for references, see Hall, Kardulias and Chase-Dunn 2011) to reframe 'world-systems theory' as 'world-systems analysis', an intellectual tool for a macro-scale study of how societies (and not just capitalist ones) interact with each other, and what the outcomes of that interaction might be. Peripheries are afforded a more active role, including resistance, in the development of the system, dominance by cores is not automatic (given technological capabilities and the extent of coercive powers), 'prestige' or 'luxury' goods are recognized as of greater importance than bulk goods for trade relations in noncapitalist societies, and there are cycles of expansion and contraction in the dominance of cores and the extent of world systems (e.g., Frank 1993).

Archaeologists have engaged both positively and critically with the application of world-systems approaches to the material remains of past societies. Stein (1999) criticizes the basic assumptions of Wallerstein's model (e.g., 'core dominance, asymmetric exchange, and long-distance exchange as the prime movers of social change') and takes Chase-Dunn and Hall to task that, "in attempting to generalize the applicability of the model while retaining its original terminology, they have eliminated its specificity as an explanatory construct and reduced the term 'world-system' into little more than shorthand for 'interregional interaction system'" (1999: 25), for which he proposes alternative models. McGuire (2002: 79) criticizes archaeological proponents of world-systems theory in North America for presenting "an economic reading of the old notion of interaction spheres". Among other criticisms, he notes that most archaeological applications of world-systems theory in north America have been "stripped of its Marxist foundation", interregional variations in 'cores' and 'peripheries' have been ignored, economic centrality may not equate with social or political centrality, and social change occurs at different scales and not just from the top down (2002: 137–138). Among Kohl's (2011) 'deficiencies' of the world-systems model are its 'terminological opacity'; the examples of multiple, rather than single, cores ("loosely integrated or overlapping") in contexts such as Sumer, Old Kingdom Egypt, and Shang China; "eminently transferable and widely shared" (rather than capitalism's "monopoly of force based on both organizational and technological superiority") Bronze Age technologies; and the permeable boundaries of Bronze Age societies, as opposed to modern nation-states. Kohl (2011: 83) accepts that there were "systemic, inter-regionally integrated networks of interaction" in the Bronze Age of western Eurasia, and changes in the boundaries of such interaction systems through time. Rather than stick to a rigid world-systems model, he advocates the study of societies and cultures within larger "social fields of interaction". Finally, McGuire

(1989: 40) reclaims the Marxist basis of the world-systems model by using the concepts of production, distribution, exchange, and consumption to analyze social reproduction within the context of larger scale systems of social, economic, and political relationships. Of particular interest are how relations of production, exchange, and distribution have led to interdependencies between human communities and larger-scale groupings.

Archaeological applications of world-systems models to the later prehistory of the Mediterranean fit more broadly into the context of a European 'periphery' on the edge of a Near Eastern urban 'core', as can be exemplified in Sherratt's (1993) presentation of a Bronze Age world system (see especially his figure 13, for a schematic mapping of four stages of the expansion of core and periphery c.3500–500 BC, with the "urbanised manufacturing zone with bulk transport and state organization" reaching the Aegean by 1500 BC and southern and eastern Spain by c.500 BC). A more detailed and theoretically explicit study is provided by Kristiansen (1998), who proposes the existence of a world-system of the Near East, the Mediterranean, and Europe from c.2000 BC, while recognizing that core-periphery relations differed between ancient and capitalist contexts. This "interpretive and explanatory sketch" (1998: 41), ultimately based on 1970s French structural Marxism, focuses on social reproduction operating at a larger, and more uncontrollable, scale (both temporal and spatial) than that of people's everyday lives, and foregrounds the study of the ways in which "the expansion of international exchange accelerated the pace of change in regional cultural traditions and—by the very nature of bronze technology—created a dependency in terms of supplies of metal and know-how between different regions that added a new dimension to change and tradition" (1998: 3). The Mycenaeans are accorded a central role as "transmitters and receivers of new influences between the east Mediterranean and central Europe" (1998: 361) and the establishment of trade contacts with the west Mediterranean.

Parkinson and Galaty (2009a) highlight the diversity of opinions on the utility of the world-system/core-periphery approach to the Bronze Age in the eastern Mediterranean, and particularly for our understanding of the emergence of Minoan and Mycenaean societies on Crete and the Greek mainland in the context of interaction with preexisting state and urban societies in the Near East. The editors (Parkinson and Galaty 2009b: 9) criticize world-systems analysis for its "broad brush lumping perspective that masks socio-cultural variability", arguing that the macro- and long-term scales of analysis do not account for the archaeological record of change as seen in medium and local/short-term scales of analysis. When it comes to the nature of the archaeological evidence for interaction between the Near East and the Aegean Bronze Age, debate centers on (a) the amounts of such evidence and its rates of deposition (e.g., Cline 2011 on uneven rates of deposition, as opposed to an averaging out of imports across six centuries) and (b) the weight that should be placed on what Sherratt (2011: 91) calls 'known unknowns', that is, goods and materials that survive poorly or not at all.

Clearly we need to examine critically the evidence for interaction that is known, as well as potentially unknown, but this still leaves us with an interpretive problem, namely, the role of this interaction in local social, political, and economic change. Thus from the Cretan evidence, Cherry (2011: 138) takes a 'minimalist view' of Near Eastern trade, that this interaction occurred, but did not *cause* the emergence of Minoan palaces and states.

The distinction between material evidence for interaction and the causes and effects of that interaction is also raised for the west Mediterranean by the finds of Mycenaean, Cypriot, and possibly Minoan imports (mostly pottery) from the seventeenth (Middle Helladic III) to the early eleventh (Late Helladic IIIC2) centuries BC. Blake (2008) and Vianello (2005) detail the nearly one thousand sherds from ninety-four sites from the (mainly southern and central) Italian peninsula, Sicily (and adjacent islands), Sardinia, and southern Spain, although pottery is not evenly distributed across this large distance throughout the at least six centuries of its production (whether Aegean or local), distribution, and consumption. A model of long-distance maritime trade has been preferred by some (e.g., Smith 1987; Kristiansen 1998), thus placing the west Mediterranean on the 'periphery' of the Aegean Bronze Age 'core', with the latter on the periphery of the Near Eastern core, and giving this interaction a primary role in the emergence of greater social complexity in the central Mediterranean.

This model is increasingly difficult to sustain, given analysis of the empirical record: there is no evidence for trading posts, nor for traders from the east Mediterranean; even allowing for the variable quality of the contextual and published evidence for pottery imports in the central Mediterranean, the majority of Italian Later Bronze Age sites in Italy lack imported Mycenaean pottery, and where it does occur, it is mostly in small numbers; and a majority of the 'Mycenaean' pottery was locally produced in Italy and Spain (Blake 2008). The numbers of copper oxhide ingots in Sardinia are only 10 percent of those found in the Uluburun wreck off the southern coast of Turkey, whereas there is no reason for Mycenaean metallurgical interest in southern Italy (where copper ores are absent) and no evidence for such interest in central Italy (where ores were present, but Mycenaean imports were scarce). There is no analytical evidence for west Mediterranean copper sources being used to produce Mycenaean metal objects, and the oxhide ingots on Sardinia, for all the debates about their analysis and attribution to a Cypriot source, have no compositional affinities with local sources of copper. Two further points are of key importance against a core/periphery interpretation: first, the mass production of exported Mycenaean pottery, coupled with contextual evidence for variable access (e.g., domestic/burial deposition, intrasite distributions), argue against any kind of 'luxury' or 'prestige' value; and second, production of the pottery appears to have been decentralized in both the Aegean and the west Mediterranean (Blake 2008).

The outcome of this careful evaluation of the current evidence for interaction between the east and west Mediterranean basins during the second millennium BC also supports a minimalist interpretation, rather than structured exploitation by an Aegean core of a west Mediterranean periphery. The scale of interaction between the east and west Mediterranean basins in the second millennium BC was small, irregular, and decentralized and cannot be accorded any primacy in subsequent local social, economic, and political changes. In all these characteristics, the interaction was incomparable to that seen from later Phoenician and Greek colonies (e.g., Aubet 2001).

While world-systems approaches follow a logical argument that change in human societies occurs at a variety of scales, they are illogical in the ways in which their proponents often fail to build up scales of change and interaction in time and space through the practices of production, distribution, exchange, and consumption that contribute to the reproduction of human societies. The west Mediterranean examples in particular show what can happen when what may seem satisfying and 'to be expected' at the large scale is confronted with an empirical record that offers it unconvincing support. We need to be cautious and critical of large-scale interpretations that present patterns in the material evidence in time and space without an equally critical study of the empirical record, evaluating the significance, specificity, and reliability of those patterns and the extent to which they articulate with local sequences of change. This is where we should be conducting a more detailed analysis of production and reproduction in time and space.

ARCHAEOLOGICAL STRUCTURING IN TIME AND SPACE

If we accept that there are multiple scales of analysis, then it follows that we need to structure our particular materials, and their contexts, in time and space and to understand the history and practice of such structuring. While there is a tradition of such practice in the Mediterranean since the late nineteenth century, it has been (and remains) demonstrably uneven in its resolution, and this has profound implications for our interpretations. There are variations in the nature and intensity of archaeological research across the Mediterranean basin. Within mainland areas such as Greece, Italy, and Spain, the intensity of fieldwork determines our uneven knowledge of site types and distributions, as well as regional sequences and knowledge of production, distribution, exchange, and consumption. Field survey was begun about two decades later in Spain (1970s) than in Italy and Greece (1950s). Bronze Age settlement in the Spanish interior has only been intensively studied since the 1970s. The Later Bronze Age of northern Greece has become the basis of more intensive fieldwork about a century after the classic Mycenaean sites were excavated in the Peloponnese. The human occupation of Cyprus has been extended backward by some four millennia during the last two decades. Early Neolithic open-air settlements in eastern Spain, representing the first two centuries of agricultural adoption, have only begun to be studied systematically in the last fifteen years. Ditched enclosures dating to the third millennium BC, in some cases (e.g., Valencina de la Concepción) of unexpectedly large sizes, have been the object of survey and excavation in southern Spain and Portugal during the last two decades, often discovered by a mixture of urban growth (leading to rescue excavations) and aerial photography. Flat Neolithic enclosures of the fourth millennium BC are also a new site type in Greece. In spite of archaeological fieldwork over nearly one hundred and fifty years, it appears as if we have barely touched the surface.

Relative chronologies of later prehistoric sites and materials in the Mediterranean basin began to be developed in areas such as Egypt (Flinders Petrie on the Predynastic sequence), the Aegean (Arthur Evans on Crete), and southeast Spain (the Siret brothers) in the late nineteenth and early twentieth centuries. Subsequent elaboration of the Cretan and mainland Greece sequences has produced "an incredibly rich, dense, *ad hoc*, and arcane system for describing and demarcating the time frame of the Aegean Bronze Age" (Manning 2010: 12), with variations between sites and regions, and in rates of change of different artifact types. The dating sequence for the second millennium BC in the Aegean exceeds any other area and period of the later prehistoric Mediterranean in its intensity and detail. There are many areas for which a relative chronology is still being developed and regions for which whole occupation periods are poorly known (e.g., the Neolithic of La Mancha, or the earlier Bronze Age of the southwestern central plateau of Spain; see Chapman 2008). Three-age periods such as the Chalcolithic in the west Mediterranean are still rarely divided beyond an early-middle-late sequence. At the other end of the scale, but still rare, are refined relative chronologies of human occupation, such as the daily, monthly, seasonal, or annual practices inferred through micromorphological study of the wall and floor deposits at Çatalhöyük (Matthews 2005).

An archaeological, absolute chronology for the later prehistoric Mediterranean (and indeed for central, western, and northern Europe) was devised in the early twentieth century by cross-dating from Egypt, using accepted imports or sometimes vaguer typological similarities of materials. The impact of radiocarbon dating in the last five decades has been dramatic, contested (e.g., the chronology for the eruption of the Santorini volcano), and uneven, in both time and space. High precision tree-ring dating is rare. Interregional variations in the intensity of carbon 14 dating are notable in every mainland: for example, four sites in the province of Almería (southeast Spain) have yielded more dates than all the sites together of the western Andalucían provinces of Cadiz, Seville, and Huelva (García Sanjuán and Odriozola Lloret 2010), with the absolute dating of the Argaric Bronze Age in Almería, Granada, Murcia, and Jaén contributing to the highest resolution second millennium BC sequence in Iberia (Castro et al. 1993–1994). Rates of agricultural colonization in the west Mediterranean, with their implications for our understanding of this process, depend on quality control of samples and their contexts (e.g., Zilhão 2011). Our understanding of the nature and rapidity of other major 'transitions' depends on such quality control, along with the detail and integrity of stratigraphic sequences. On Crete the structural remodelling of Knossos during the Protopalatial and Neopalatial occupations, coupled with the limited excavation areas and poor contextual information from 'transitional' phases (see Cherry 2011: 112) to Middle Minoan palatial society, make understanding of these phases and state origins a challenge. Critical assessment of radiocarbon dating for the transition from the Los Millares (Chalcolithic) to Argaric (Bronze Age) groups in southeast Spain (Lull et al. 2010) raises the possibility of a brief overlap, rather than a straight sequence. As in Crete, the resolution of the temporal sequence has implications for our understanding of degrees and rates of social, political, and economic change.

The structuring of sites and materials in space uses an archaeological record of variable quality and intensity, itself the outcome of a research and fieldwork infrastructure that lacks uniformity both locally and regionally. Since the late nineteenth century, the concept of stylistically coherent and bounded 'cultures' has been a basis for the spatial structuring of successive stages of regional sequences of European prehistory. Although there has been much debate over the meaning of these 'cultures' and their relationship to any kind of ethnic groupings, there is still an argument that stylistic variation measures the extent of social interaction, and that social boundaries vary in their fixity according to the nature of social integration (see Parkinson 2006). In the Mediterranean basin, there are examples of 'cultures' or styles that have internal stylistic homogeneity and spatial boundaries (e.g., the Argaric group in southeast Spain c.2250–1550 BC as opposed to the Valencian and Manchegan Bronze Age groups to the north and northwest), but such mutually exclusive distributions are rarer than less clearly demarcated cultures and pottery styles (e.g., the Italian Neolithic pottery styles whose edges are more blurred, or in some cases largely absent as regional or local markers; see Robb 2007: 181). Bernbeck (2012) roots the origins of the concept of bounded cultures within the context of European colonialism, challenges the idea of "internally coherent, externally clearly demarcated identities", and proposes that "we should rather think of the social as a potentially limitless network of groups and individuals", for which he adopts the term of a 'multitude', which is fluid in composition and extent as well as spatial definition. This fluidity and flexibility is argued to have been a major obstacle to the emergence of any long-term, unequal social relations, especially through mobility. This is an interesting proposal for thinking about Neolithic societies, their spatial patterning, and their material representation.

DIFFERENT SCALES OF MOVEMENT IN TIME AND SPACE

The colonization of Mediterranean islands, coupled with the introduction of material objects and domesticated plants and animals into empty or previously colonized islands, required labor organization and the maintenance of social networks to enable long-term reproductive viability. Coupled with the evidence of agricultural settlement and adoption on the Mediterranean mainlands of Greece, Italy, and Iberia, there is a strong argument for 'pioneer colonization' (e.g., Zilhão 2001), rather than continuous expansion in an east-west direction, with choice of the main settled areas related to basic, material conditions of existence (e.g., land cultivable with a digging stick technology, access to stone tools and raw materials for daily existence, and sufficient areas of pasture, as well as wild plant and animal resources). Where these conditions, recognized in cultural traditions, were not present in sufficient frequencies (e.g., the near absence of metamorphic rock for stone axe production in the Balearic islands, where settled agricultural colonization took place three millennia after that on the adjacent mainland of eastern Spain; Risch 2011), colonization was delayed. Even allowing for the time taken to expand and maintain social networks on mainlands and between mainlands and islands, as well as the organization of colonization in the contexts of local production and reproduction, the speed of movement was still five kilometers per year, over ten generations, from northern Italy to Portugal. Within and between these mainland and island areas, there was an increase in the movement of objects of social value, as seen through characterization studies.

In local areas, movement of people would have taken place on daily, weekly, seasonal, annual, and generational timescales. These movements would have varied in terms of involvement (which people?), distances moved, repetitiveness (always the same people?), and permanence. If we start with the agricultural system (and I do not imply that there was a uniform 'Mediterranean' system in the Neolithic), how intensive or extensive was it, and to what extent did it require mobility? What was the degree of this mobility, and how many people in the community were involved in it? What are the implications for the areas of land to which these early agricultural communities required inclusive or exclusive access? Where and how were the material objects required for daily existence produced in this space? How extensive were the networks of social interaction beyond the scale of individual communities (including periodic, short-term aggregations) that enabled the maintenance of social relationships and ultimately the reproduction of local societies?

These are all questions about landscapes of human movement and interaction. Although we have many characterization studies of objects of social value (although not yet much evidence for the scale of movement of livestock), we have yet to undertake sufficient isotopic analyses to assess lifetime human movement and its scale (e.g., unpublished studies from Spain and Portugal for the third millennium BC; Waterman, personal communication). For areas and periods where collective burial was practiced, the assumption of collective lives needs systematic, isotopic testing. Human osteological studies from Bronze Age southeast Spain suggest the hypothesis of matrilocal residence in the Vera Basin (Buikstra and Hoshower n.d.). Degrees of sedentism associated with Neolithic communities and agricultural practices have still to be explicitly evaluated in the archaeological record (e.g., discussion in Chapman 2008 on southern Spain). Still rarer are models of what Neolithic agricultural systems would have looked like in space: Robb (2007: 100) studies the space required for basic productive activities in Neolithic villages in southern Italy, concluding that "an intermarrying, stable demographic community would have utilized a minimum of between 80 and 200 km^2", and maybe up to 400 km^2, reconstructing some twenty to fifty tribal groups over the Italian peninsula, Sicily, and Malta (Robb 2007: 269).

We must be careful not to assume that these were stable, unchanging human landscapes. There were examples not only of the expansion and contraction (spatially and with altitude) of settled areas in the landscape, but also of settlement aggregation (e.g., the Neolithic ditched enclosures of the Tavoliere, and third millennium BC sites like Marroquiés Bajos and Valencina de la Concepción in southern Spain, which reached up to 400 hectares in size) and dispersal over periods of several hundred years.

FROM SMALL TO LARGE SCALES: A MATERIALIST APPROACH

Students of Mediterranean history and culture have struggled to escape dichotomous thought and to analyze their objects of study at different scales of time and space. Notions of static survivalism, especially in rural as opposed to urban areas across the basin, and of a dynamic east as opposed to a static west Mediterranean, still lie below the surface of archaeological thought. The radiocarbon revolution of the 1960s stimulated a greater emphasis on regional sequences of autonomous social, economic, and political development. Spatial connections were studied in terms of the characterization of objects of social value and island colonization as evidence of population movement: this was more multidirectional, rather than the unidirectional, east-west movement of diffusionism. By the 1980s, large-scale study of the later prehistory of the Mediterranean basin had reemerged in the form of world-systems theory/analysis. More recently, and following the stimulus of Horden and Purcell (2000), there has been greater emphasis on both fragmentation and connectivity in the Mediterranean, emphasizing 'insularity' as a social construction, whether we are talking about islands or social and political groupings.

Archaeologists work with material evidence of past communities that varies in terms of its preservation and intensity of study, and we group this material (whether it be objects, sites, etc.) together in temporal units (e.g., periods, relative and absolute chronologies, different scales from the seasonal and annual through several generations) and spatial units (e.g., houses, communities, 'cultures', regions, landmasses). These groupings are our constructions, but they are based on an empirical reality. Their advantage is that they enable us to work at different scales. Their disadvantage is that they break up what is often a continuity of human action and movement in time and space. Diffusionism overemphasized the degree, continuity, and scale of this movement, whereas critics of this approach stressed fragmentation at the expense of connectivity.

Whereas much current discussion in archaeology focuses on the lives of our prehistoric ancestors as culturally constituted (and I do not deny the role of the cultural and symbolic in everyday lives and across generations), the emphasis in this chapter has been on how archaeology works. Using our groupings of material evidence in time and space, our first priority should be to study the production and reproduction of social life. It is one thing to make assertions of large-scale systems of such reproduction, but it is another to demonstrate the scale and extent of such systems through empirical analysis from the bottom up. We must also be wary of focusing on the cultural construction of memory, tradition, and identities when we do not understand the structures in which people lived. According to this materialist argument, the much-vaunted human agents produce and reproduce social life on the basis of their structural capacities, rather than negotiating their identities and daily lives in a structureless void. To paraphrase Marx and Engels, history cannot be created and culture cannot be lived without satisfying the basic material needs of life. These needs are met through mental and physical labor in the course of daily social practices and cultural life. Such practices and lives are best studied locally and contextually (e.g., What are the conflicting demands of different forms of production, and how is labor divided?), working from the local to the regional level, and beyond.

ACKNOWLEDGMENTS

I would like to thank Stella Souvatzi and Athena Hadji for their invitation to write this overview and for their advice on its content. I also thank Anna Waterman for discussing details of isotopic analyses of Copper Age burials in Spain and Portugal.

REFERENCES

Anderson, A. 2011. Islands from the south: an Oceanic perspective on island colonisation. In: N. Phoca-Cosmetatou (ed.), *The First Mediterranean Islanders: Initial Occupation and Survival Strategies*, 157–171. Oxford: University of Oxford School of Archaeology.

Aubet, M. E. 2001. *The Phoenicians and the West: Politics, Colonies and Trade*. 2nd ed. Cambridge: Cambridge University Press.

Barker, G. 1981. *Landscape and Society: Prehistoric Central Italy*. London: Academic Press.

Bernbeck, R. 2012. Multitudes before sovereignty: theoretical reflections and a Late Neolithic case. In: T. L. Kienlin and A. Zimmerman (eds.), *Beyond Elites: Alternatives to Hierarchical Systems in Modelling Social Formations. Volume 1*, 147–168. Bonn: Verlag Dr. Rudolf Habelt GmbH.

Blake, E. 2008. The Mycenaeans in Italy: a minimalist position. *Papers of the British School at Rome* 76: 1–34.
Braudel, F. 1972. *The Mediterranean and the Mediterranean World in the Age of Phillip II*. London: Collins.
Broodbank, C. 2000. *An Island Archaeology of the Early Cyclades*. Cambridge: Cambridge University Press.
Buikstra, J. E. and L. Hoshower. n.d. Análisis de los restos humanos de la necrópolis de Gatas.
Castro, P.V., R. W. Chapman, S. Gili, V. Lull, R. Micó, C. Rihuete, R. Risch and M. E. Sanahuja. 1993–1994. Tiempos sociales de los contextos funerarios argáricos. *Anales de Prehistoria y Arqueología* 9–10: 77–105.
Chapman, R. 1990. *Emerging Complexity: The Later Prehistory of South-east Spain, Iberia and the West Mediterranean*. Cambridge: Cambridge University Press.
Chapman, R. 2003. *Archaeologies of Complexity*. London: Routledge.
Chapman, R. 2008. Producing inequalities: regional sequences in later prehistoric Southern Spain. *Journal of World Prehistory* 21: 195–260.
Cherry, J. F. 1981. Pattern and process in the earliest colonisation of the Mediterranean islands. *Proceedings of the Prehistoric Society* 47: 41–68.
Cherry, J. F. 2011. Sorting out Crete's Prepalatial off-island interactions. In W. A. Parkinson and M. L. Galaty (eds.), *Ancient State Interaction: The Eastern Mediterranean in the Bronze Age*, 107–140. Santa Fe: School for Advanced Research Press.
Cline, E. H. 2011. Bronze Age interactions between the Aegean and the Eastern Mediterranean revisited: mainstream, periphery or margin? In: W. A. Parkinson and M. L. Galaty (eds.), *Ancient State Interaction: The Eastern Mediterranean in the Bronze Age*,161–180. Santa Fe: School for Advanced Research Press.
Evans, J. D. 1973. Islands as laboratories for the study of cultural process. In: C. Renfrew (ed.), *The Explanation of Culture Change: Models in Prehistory*, 517–520. London: Duckworth.
Frank, A. G. 1993. The Bronze Age world system and its cycles. *Current Anthropology* 34: 383–413.
García Sanjuán, L. and C. Odriozola Lloret. 2010. La cronología radiocarbónica de la Edad del Bronce (ca. 2200–865 cal ANE) en el Suroeste de la Península Ibérica. In: J. Jiménez Ávila (ed.), *Actas de la Reunión Sidereum Ana II: El Río Guadiana en el Bronce Final (Mérida, Mayo de 2008)*. Mérida: CSIC.
Hall, T. D., P. N. Kardulias and C. Chase-Dunn. 2011. World-systems analysis and archaeology: continuing the dialogue. *Journal of Archaeological Research* 19: 233–279.
Harding, A. F. 1984. *The Mycenaeans and Europe*. London: Academic Press.
Helms, M. W. 1988. *Ulysses' Sail: An Ethnographic Odyssey of Power, Knowledge and Geographical Distance*. Princeton: Princeton University Press.
Higgs, E., ed. 1975. *Palaeoeconomy*. Cambridge: Cambridge University Press.
Horden, P. and N. Purcell. 2000. *The Corrupting Sea: A Study of Mediterranean History*. Oxford: Blackwell.
Kohl, P. L. 2011. World-systems and modelling macro-historical processes in later prehistory: an examination of old and a search for new perspectives. In: T. C. Wilkinson, S. Sherratt and J. Bennet (eds.), *Interweaving Worlds: Systemic Interactions in Eurasia, c.7th to 1st Millennia BC*, 77–86. Oxford: Oxbow Books.
Kristiansen, K. 1998. *Europe Before History*. Cambridge: Cambridge University Press.
Lull, V., R. Micó, C. Rihuete and R. Risch. 2002. Social and ideological changes in the Balearic islands during later prehistory. In: W. H. Waldren and J. A. Ensenyat (eds.), *World Islands in Prehistory*, 117–126. Oxford: British Archaeological Reports International Series 1095.

Lull, V., R. Micó, C. Rihuete and R. Risch. 2010. Límites históricos y limitaciones del conocimiento arqueológico: la transicón entre los grupos arqueológicos de Los Millares y El Argar. In: P. Bueno, A. Gilman, C. Martín Morales and F. J. Sánchez Palencia (eds.), *Arqueología, Sociedad, Territorio y Paisaje,* 75–94. Madrid: Consejo Superior de Investigaciones Científicas.

Malone, C. 1985. Pots, prestige and ritual in Neolithic Southern Italy. In C. Malone and S. Stoddart (eds.), *Papers in Italian Archaeology IV: The Cambridge Conference,* 118–151. Oxford: British Archaeological Reports.

Malone, C. 2003. The Italian Neolithic: a synthesis of research. *Journal of World Prehistory* 17: 235–312.

Malone, C. and S. Stoddart. 2004. Towards an island of mind. In: J. F. Cherry, C. Scarre and S. Shennan (eds.), *Explaining Social Change,* 93–102. Cambridge: McDonald Institute for Archaeological Research.

Manning, S. W. 2010. Chronology and terminology. In: E. H. Cline (ed.), *The Aegean Bronze Age (ca. 3000–1000 BC),* 11–28. Oxford: Oxford University Press.

Manning, S. W. and L. Hulin. 2005. Maritime commerce and geographies of mobility in the Later Bronze Age of the eastern Mediterranean: problematizations. In: E. Blake and A. B. Knapp (eds.), *The Archaeology of Mediterranean Prehistory,* 270–302. Oxford: Blackwell.

Mathers, C. and S. Stoddart, eds. 1994. *Development and Decline in the Mediterranean Bronze Age.* Sheffield: J. R. Collis Publications.

Matthews, W. 2005. Life-cycle and life-course of buildings. In: I. Hodder (ed.), *Catalhöyük perspectives: Reports from the 1995–99 Seasons,* 125–149. Cambridge: McDonald Institute for Archaeological Research/British Institute at Ankara.

McGuire, R. H. 1989. The greater Southwest as a periphery of Mesoamerica. In: T.C. Champion (ed.), *Centre and Periphery: Comparative Studies in Archaeology,* 40–66. London: Routledge.

McGuire, R. H. 2002. *A Marxist Archaeology.* New York: Percheron Press.

Parkinson, W. A. 2006. Tribal boundaries: stylistic variability and social boundary maintenance during the transition to the Copper Age on the Great Hungarian Plain. *Journal of Anthropological Archaeology* 25: 33–58.

Parkinson, W. A. and M. L. Galaty, eds. 2009a. *Ancient State Interaction: The Eastern Mediterranean in the Bronze Age.* Santa Fe: School for Advanced Research Press.

Parkinson, W. A. and M. L. Galaty. 2009b. Introduction: interaction and ancient societies. In: W. A. Parkinson and M. L. Galaty (eds.), *Ancient State Interaction: The Eastern Mediterranean in the Bronze Age,* 3–28. Santa Fe: School for Advanced Research Press.

Perlès, C. 1992. Systems of exchange and organization of production in Neolithic Greece. *Journal of Mediterranean Archaeology* 5: 115–164.

Renfrew, C. 1973. *Before Civilisation.* London: Jonathan Cape.

Risch, R. 2002. *Recursos naturales, medios de producción y explotación social.* Mainz am Rhein: Verlag Philipp von Zabern.

Risch, R. 2011. Social and economic organisation of stone axe production and distribution in the western Mediterranean. In: V. Davis and M. Edmonds (eds.), *Stone Axe Studies III,* 99–118. Oxford: Oxbow Books.

Robb, J. 2001. Island identities: ritual, travel and the creation of difference in Neolithic Malta. *European Journal of Archaeology* 4(2): 175–202.

Robb, J. 2007. *The Early Mediterranean Village: Agency, Material Culture and Social Change in Neolithic Italy.* Cambridge: Cambridge University Press.

Sherratt, A. 1993. What would a Bronze Age world system look like? Relations between temperate Europe and the Mediterranean in later prehistory. *Journal of European Archaeology* 1(2): 1–58.

Sherratt, S. 2011. The Aegean and the wider world: some thoughts on a world-systems perspective. In: W. A. Parkinson and M. L. Galaty (eds.), *Ancient State*

Interaction: The Eastern Mediterranean in the Bronze Age, 81–106. Santa Fe: School for Advanced Research Press.
Smith, T. R. 1987. *Mycenaean Trade and Interaction in the West Central Mediterranean 1600–1000 BC.* Oxford: British Archaeological Reports International Series 371.
Stein, G. J. 1999. *Rethinking World-Systems: Diasporas, Colonies and Interaction in Uruk, Mesopotamia.* Tucson: University of Arizona Press.
Stoddart, S., A. Bonanno, T. Gouder, C. Malone and D. Trump. 1993. Cult in an island society: prehistoric Malta in the Tarxien period. *Cambridge Archaeological Journal* 3: 3–19.
Tykot, R. 1996. Obsidian procurement and distribution in the central and western Mediterranean. *Journal of Mediterranean Archaeology* 9(1): 39–82.
Van Dommelen, P. and A. B. Knapp, eds. 2010. *Material Connections in the Ancient Mediterranean: Mobility, Materiality and Identity.* London: Routledge.
Vianello, A. 2005. *Late Bronze Age Mycenaean and Italic Products in the West Mediterranean: A Social and Economic Analysis.* Oxford: British Archaeological Reports International Series 1439.
Wallerstein, I. 1974. *The Modern World-System: Capitalist Agriculture and the Origins of European World-Economy in the Sixteenth Century.* New York: Academic Press.
Zilhão, J. 2001. Radiocarbon evidence for maritime pioneer colonization at the origins of farming in west Mediterranean Europe. *Proceedings of the National Academy of Science USA* 98: 14180–14185.
Zilhão, J. 2011. Time is on my side . . . In: A. Hadjikoumis, E. Robinson and S. Viner (eds.), *Dynamics of Neolithisation in Europe: Studies in Honour of Andrew Sherratt,* 46–65. Oxford: Oxbow Books.

3 What Does a Settlement's Layout Show about the Society That Inhabits It? On the Importance of Thinking-Through-Images

Stavros Stavrides

Archaeologists often try to infer from the layout of settlements and cities defining characteristics of the societies that inhabited them. Observing forms of spatial enclosure, arrangements of differentiated spaces, and patterns of spatial elements, archaeologists attempt to understand social relationships. What is presupposed in these interpretive practices often is that spatial arrangements correspond through specific and distinct ways to 'social arrangements', to the layout, so to speak, of an ordered and stable society.

Especially during the 1960s and 1970s, archaeological research, deeply influenced by the dominant paradigms of functionalism and positivism, was in search of "the causal factors that determine the relationship between social organizations and spatial patterning" (Souvatzi 2012). This has strengthened an almost naturalized view that space is a product and a reflection of society, so that by examining space we can discover a society's organization, logic, and functioning. An epistemological shift in the 1980s and 1990s has radically questioned such research trajectories. The postmodernist critique has fundamentally shaken any "assumed isomorphism of space, place and culture" (ibid.), suspending, at the same time, any certainty about the power of inhabited spaces to reveal the social relations that inhabit them.

This kind of critique has opened the ground for the construction of dynamic models concerning the space-society relationship. A plethora of social actors has indeed been recognized and studied, leading to a multiplicity of interpretations. However, a renewed interest in phenomena of cultural symbolism has encouraged research on the representational role of socially produced space. In this prospect, archaeologists can assume that cities and settlements not only provide clues on social organization but, perhaps, also show how these spatial arrangements were purposefully created by their inhabitants to represent their society's view of itself. A society, in this reasoning, is not simply revealed by the spaces that it inhabits but is explicitly represented in these spaces. In this case we can speak of a city as an image of the society that inhabits it.

Here of course problems of a different nature arise. As any form of representation contains distortions, deviations, and illusions, how can we, in a way, 'correct' the fallacies of the image, if we are to discover the characteristics of a specific society? An alternative research agenda could of course attempt to locate not a society's characteristics as such but only the characteristics of the image this society has created for itself. Such an agenda could easily be misunderstood as a kind of phenomenology of the city-as-image. There is, however, an obvious counterargument to this: can we really even describe a city-image without knowing (or discovering) the ways people actually lived and collectively produced the beliefs that were necessary for the creation of this image? In other words, we should not assume that we can interpret the correspondence of spatial layouts to specific societies without investigating first the ways in which inhabited space is connected to social life, as the latter develops in space and time.

One possible way of reformulating the problem of the city-society relationship, whether this is conceived as a correspondence problem or as a representation problem, is to start from an approach to inhabited space as performed space. If people relate to space not simply because they live in it, but because they actually 'mold' it, then they create in and through it not only what they need in order to survive, but also the meaning of their actions. Through space (and time) people connect to other people, they measure the distances that separate them from others as meaningful distances, as distances that mold the potentialities and evaluations of their actions. Space, thus, cannot be considered as prior to or 'outside' of social actions (a kind of container), but as a necessary component and defining characteristic of any such action.

What follows is an attempt to rethink the city-society relationship considered as a relationship made meaningful in the lives of a city's inhabitants. Cities, settlements, and spatial arrangements, in this prospect, can be perceived through socially constructed images, which can present societies to their members as coherent and recognizable totalities. In such images, people recognize patterns of behavior, represented values, and ordered social positions. But people do something more than recognize or "see" images; people think-through-images. Employing the concept of "thought-image", this chapter will tackle with the problems archaeological (as well as social and spatial) theory has to face in studying cities that are constructed (or interpreted) as images of the corresponding societies.

Inhabitants and researchers alike tend to ignore the fact that through images of space we learn to think about our society, whether to reproduce dominant values and social relations or to question or even defy them. What will be termed "diagrammatic thinking" will be shown to be one form of thinking-through-images that projects to patterns of spatial arrangements legitimated social hierarchies and social role taxonomies.

This chapter seeks to contribute to the overall project of this book by problematizing the concept of space. Connecting space to practices of thinking-through-images, which deeply influence the social meaning of inhabiting acts, means understanding space as a process rather than as an entity or a perceptual category. Reformulating the problem of the space-society relationship as a problem that does not simply involve the tracing out of correspondences or representational links can perhaps help archaeology avoid various forms of hasty determinism: space neither simply shows (represents, reflects, etc.) society nor simply hides it (obscuring, distorting, etc., its functioning).

APPROACHING INHABITED SPACE AS A FOCUS OF THOUGHT-IMAGES

Returning to the problem of the city-as-image, we can reconsider its formulation; we can assume that people produce their space through their actions, which are necessarily spatially defined, and that they 'see' themselves as actors in the image of spaces they produce.

However, this process is not smooth, repetitive, and peaceful. In his investigation of the ways in which people project their identities to "their space" (be it a national region or a group's shelter), Bourdieu (1991: 221) refers to an unavoidable "struggle over representations in the sense of mental images". This struggle takes obviously different forms in different historical contexts, but we cannot and should not forget that people attach themselves to specific representations of the society they participate in through the necessary mediation of power relations.

Should we, then, not seek to devise new tools in the research on the city-society relationship that will take into account the form-giving dynamics of a latent or explicit struggle over representations? As Smith (2007) has shown, ancient cities cannot be divided into planned and unplanned ones, and it is more appropriate to observe various degrees of planning. Collectively or through the decisions and acts of a ruling elite, cities assume a form that is not simply the result of some kind of socially defined survival efforts but the outcome of precisely the struggle over representations. Cities were produced through practices that were expressing not only needs and interests but also values and meaning. Cities, thus, are material and cultural artifacts at the same time.

It is not, however, the production of cities that reveals more clearly the fact that they are equally constituted by meaning and matter. It is habitation as a process of ongoing production. By inhabiting cities, people transform them. What they transform is actually the social meaning of space they inhabit because they continuously create and re-create spatial relations. Space produced and made meaningful through, and because of, habitation is space constituted as a network of spatial relations. It is in this network, constantly 'performed' and refashioned, that people recognize meaningful constellations of spatial elements that they understand as images.

What is or, rather, what can be a thought-image? At first, thought-images were a specific kind of concise literary text written by theorists linked to the Frankfurt School (W. Benjamin, S. Kracauer, and E. Bloch). As Richter (2007: 7) describes it, "The *Denkbild* [. . .] is a brief aphoristic prose text [. . .] that both illuminates and explodes the conventional distinctions among literature, philosophy, journalistic intervention and cultural critique". He discerns in these texts a "fragmentary, explosive, and decentering force" (ibid.: 8). This force gives to thought-images the power to discover and reveal in the mundane and seemingly insignificant aspects of everyday life clues that activate new interpretations of the prevailing culture.

It is not a mere coincidence that the *Denkbild* writers focus on images of the city to draw on unexpected comparisons and to turn these images to thoughts about the city considered as the locus and representation of culture. City images thus become something like hieroglyphic fragments of culture; they are not simply recognized as life pictures but are being treated as potentially decipherable riddles. *Denkbilder* are constructed as a peculiar mixture of abstract thoughts and images that does not have the definitive meaning of a literary metaphor. *Denkbilder,* thus, attempt to present thoughts that are created through images rather than simply explain these thoughts by illustrating them.

Richter (2007: 12) believes that thought-images actually evoke what cannot be said in words. One would say that *Denkbilder* hint toward ways of understanding social reality that need to combine concreteness and abstraction. The specificity of an image, which represents some kind of recognizable reality, is made indicative of a general process or an abstract thought. What is important is that the specificity of the image is not considered as a vehicle for abstract generalizations but, rather, constitutes a necessary part of a dialectic movement between perception and thought.

A crucial characteristic of *Denkbilder* is that they are explicitly aimed at destroying habitual understanding of both city life and the culture that inhabits and sustains it. As Szondi (1991: 28) writes about Benjamin's *Denkbilder,* "The language of images makes it possible to understand unfamiliar things without their ceasing to be unfamiliar". Although Szondi still characterizes this use of images as metaphoric, he nevertheless rightly locates in the power of thought-images the ability to preserve a feeling of unfamiliarity and strangeness, a feeling of experiencing something "for the first time" (ibid.: 22). This feeling is often equated in Benjamin's work with a sought-for reinvention of the revealing astonishment a child feels even when exploring his or her own city. "Habit has not yet done its work" (Benjamin 1985c: 78). For example, the Marseilles cathedral, immense and empty, appears to Benjamin as a bizarre "gigantic railway station that could never be opened to traffic" (1985c: 82). This image develops to a chain of comparisons that triggers thoughts about religion and vanity. "This is the Marseilles religion station. Sleeping cars to eternity depart from here at Mass times" (ibid.).

So far it seems that thought-images are necessarily written and that their power to reveal new constellations in the interpretation of culture is based on bold experimentations on the limits of language's representational power.

Surely, for Benjamin and Kracauer this is fundamentally true. However, especially Benjamin's philosophy seems to hint toward a more ambitious prospect: the ability to think in images. Weigel (1996: 53) insists that Benjamin's "thinking-in-images constitutes his specific and characteristic way of theorizing, of philosophizing and of writing". According to Weigel (ibid.: x), the specificity of Benjamin's theory lies in this form of thinking that examines the ways "the idea of reality is formed and the images of history are handed down". Thus, "images are not the object, but rather the matrix and medium of his theoretical work" (ibid.).

Let us take this last phrase as a starting point for a potential theoretical 'usurpation' of Benjamin's theoretic-political project. Matrix and medium: this most likely means that thinking in images is not simply an inventive interpretive strategy but a potentially generative process. Thinking-in-images can be something more than an idiosyncratic philosophical strategy. It can rather describe a more general attitude that might acquire an anthropological horizon, as it possibly characterizes the way people treat their experiences and perceptions during their everyday praxis. People seem to think through the things they perceive by using images of the world around them as means to generalize, to compare, and even to form abstract concepts. This is not to reiterate the obvious idea that people develop the means to deal with their everyday needs by evaluating concrete realities. People transcend through praxis the alleged necessities of praxis. People recognize and give meaning to their acts and to the acts of others by thinking through images, by thinking in images. This kind of thinking can combine the concreteness and singularity of individual experiences with the process that gives to these experiences a social meaning, aligning these experiences to experiences collectively formed.

The role of the city, and of inhabited space in general, is very important in this process. What kind of more abundant and more appropriate resource can one imagine to be used as a fuel for this thinking-in-images than the always proliferating images of the inhabited environment? Inhabiting practices involve a socially accentuated ability to recognize spatial relations, to evaluate distances, and to identify crucial social meanings in the arrangement of spatial elements. Spatial relations, made socially meaningful, condense in images. Such images take shape in literary texts, in everyday discussions, in paintings and photographs, in personal letters, and in publicity campaigns, in maps, and in address sketch notes too. In other words, those images permeate any act of converting individual experiences of and in space to experiences recognized as common and potentially shared. We should not reduce this process to a mere conformity to stereotyped and reproduced iconic messages, though. Even such a reductionist view still has to explain the construction of iconic messages through the necessary intervention of socially recognized abstractions and generalizations (human encoding procedures essentially involve them). If, on the other hand, we see in this process more than blind and automatic conformity, we should recognize the constructive role of thinking through images. People create from their perceptual inputs images that are already 'contaminated' with meaning. Images, thus, are a kind of selective and purposeful articulation of spatial elements aimed at capturing and representing an important evaluation of perceived inhabited environment. This evaluation potentially opens to every possible social orientation of meaning: it can connect images of the inhabited environment with specific and more general interests, it can legitimate their value by comparing them to an external referent (be it religious or secular), and it can multiply their revealing power by connecting them to realities beyond perceived reality as the 'universe', 'nature', 'society as a whole', 'humankind', and so forth.

We can extract a definable and analyzable array of its characteristics from this somewhat speculative description of thinking-in-images. First of all, thinking-in-images involves an approach to the human built environment considered as a dense network of spatial relations. Images emerge as people temporarily and selectively fix these relations to meaningful representations of human relations. Of course, that gives to concrete, historically specific dominant powers the means to control and 'educate' people. Images created through the directive mediation of a socially powerful authority become, thus, a means to reduce the interpretive-productive dynamics of inhabiting. Many societies have used this power to control the meaning of images of inhabited space through acts of planning the city-as-image. For example, Sennett (1994: 89) explicitly speaks about an obsessive and characteristic practice during the period of Roman Empire which "had made order and imperial power inseparable". This practice was focused, according to him, to the production of recognizable forms of "visual order" in cities and space, which were guided by the principles of a "reassuring geometry" (ibid. 90; see also Carl 2000: 330). New Roman towns were "impressed onto the conquered territory" (ibid. 90) through the establishment of a clearly recognizable plan based on simple geometrical relations. Destroying preexistent streets, shrines, or buildings thus meant effacing traces of a different preexisting spatial order that could threaten imperial power.

In a different context, the Angkorian cities in Southeast Asia were "designed to recreate the world of the gods" through an architectural representation of the holy mountain (Mount Meru), with its peaks symbolized in the towering pyramid-like constructions of the temples (Higham 2000: 357). The "ultimate purpose of this quest for perfection may be found in the divine nature of kingship" (ibid.). The explicit orientation toward a shaping of cities in the image of a well-ordered universe ("cosmos"), or any other kind of socially meaningful spatial symbolization, seems to be just one of the ways that the generalized human ability to think-in-images by using inputs from spatial experience develops.

Second, thinking-in-images, apart from being necessarily relational, is also performative. One does not think in images without being involved in acts of inhabiting. One has to 'see' the city from the angles offered or discovered during one's life as a member of a specific society. Angles of view produce possibilities of combinations and comparisons, and they enhance and limit selectivity at the same time. This is why dominant institutions or rulers attempt to fix angles of seeing and to direct practices of seeing. For example, "linear perspective came to symbolize a harmony between the mathematical regularities in optics and God's will" (Jay 1988: 5–6) and were used to depict a well-ordered, knowable, and controllable world. The fixed eye of linear perspective was essentially considered as the center of a sovereign gaze, resembling that of a ruler surveying his dominion.

John Berger, to name just one among many inventive theorists, has shown that our appreciation and evaluation of our social world is directed by socially controlled and reproduced "ways of seeing" (Berger 1972; also Jay 1988).

Third, if thinking in images is a relational and performative process of generating images, and thoughts through images, then, in principle, this is an open process. Societies attempt to control, regulate, and standardize it, but people can and do find ways to improvise and invent. Thinking-in-images can be a dissident practice. Benjamin (1992) has developed a theorizing stance that praised and performed human emancipation. His theorizing, thus, emphasized the power of this kind of thinking to retain unfamiliarity with the city as an environment full of unrealized possibilities. Seeking to redeem the liberating program of modernism, Benjamin developed a thought immersed in images that were combined to create new, unexpected constellations of meaning. For him, thinking-in-images was a practice of discovery. And when he used this thinking to contemplate history, discovery meant for him unearthing latent promises buried in the past. Images of the past, in this context, meant images of personal and collective memory, that were given new meaning as they were re-created in the present. In his famous Arcades Project, Benjamin seems to incorporate the following in his thinking on images: "The past has left images in literary texts, images comparable to those which are imprinted by light on a photo sensitive plate. The future alone possesses developers active enough to scan such surfaces perfectly" (Benjamin 1999: 482, [N 15a, 1]). In this formulation, images in texts acquire the dynamics of potential representations of the past that can only be activated fully by the power of the future. It is as if images can be considered more than a representation (whether in words or not) because they possess an inherent potentiality that is based on their comparability to the present of their "recognizability" (ibid. 463, [N3, 1]).

Can we not imagine people in different societies, ancient ones included, being able to think in images and thus being able to rethink their life and society? This does not mean that we should project our own, contemporary reflective thinking to past societies. However, anthropological evidence does seem to show that even though abstract reasoning has its specific historic forms connected to intracultural genealogies, thinking with the use of images, with the mediation of images, is a universal practice. At least after

Levi-Strauss we know that generalizing thought developed from an acute awareness of perceivable differences emerging from the world of experience. "Savage thought", thus, is a thought through images in its own distinct way: "The savage mind deepens its knowledge with the help of *imagines mundi.* It builds mental structures which facilitate an understanding of the world in as much as they resemble it. In this sense savage thought can be defined as analogical thought . . . savage thought [is] a system of concepts embedded in images" (Lévi-Strauss 1966: 263).

UNDERSTANDING CITIES AS DIAGRAMS: THINKING THROUGH IMAGES?

In his book *Discipline and Punish: The Birth of the Prison,* Foucault (1995: 200–210) has described at length a kind of spatial mechanism that he associated with societies of surveillance. As we know, this mechanism, 'Panopticon', was composed of a series of spaces (or cells) arranged around an observation tower through which a single unseen person could see what everyone in the prison cells was doing. Panopticon as a machine of surveillance was historically embedded in specific spatial arrangements and buildings, primarily in prisons but also in asylums, schools, or army barracks. However, Panopticon was not simply a spatial arrangement. It also was not a specific image of a surveillance mechanism. It was a diagram. Foucault did not try to analyze a specific instance of a surveillance arrangement, he did not try to illustrate the logic of social surveillance by using a concrete example of a surveillance space, and he did not use Panopticon as a metaphor for surveillance society either. Probably he did something that contained all these possible gestures but also went beyond them: he attempted to think through the image of Panopticon about the set of power relations that characterize a specific society. In this sense, Panopticon "is the diagram of a mechanism of power reduced to its ideal form" (Elden 2001: 149). Power relations, in Foucault's reasoning, can be analyzed through their impact on the configuration of social space and social time, but they are not reducible to those results. There is an inherent potentiality in power relations, a potentiality that creates both practices of domination and practices of resistance.

A "disciplinary diagram" (Foucault 1995: 198, translation modified in Elden ibid. 145) is a diagram of forces that does not simply contain a concrete arrangement or balance of forces but a field of potential force relations that correspond to the dynamics of a specific society. "The diagram or abstract machine is the map of relations between forces", in Deleuze's understanding of Foucault's reasoning (Deleuze 1988: 36). Like a map, which, although it appears as a finalized definition of locations and distances, can nevertheless contain an immense yet not unlimited amount of possible routes, so does a diagram contain a field of possible acts and spatiotemporal arrangements, which can be produced by a specific set of power relations. A diagram may thus be depicted or made presentable through specific images (with varying degrees of abstraction), but it is essentially meant to confer an understanding of relations rather than an appreciation of forms. In any diagrammatic understanding, forms mediate comparisons. To put it in another way, images understood as diagrams are images meant to be reflected upon as indicative of abstract, symbolized relations. This is how a city-image (or a map for that matter, viewed as a city-image) can be taken as a diagram of recognized (or recognizable) social relations (understood as power relations) or as a diagram of a certain universal order (which often legitimates a certain social order).

Characteristically, Benjamin himself ventured to reflect on diagrammatic thinking by describing an autobiographical incident. He was "struck", as he writes (Benjamin 1985a: 318–319), by the idea of drawing the diagram of his life. And this diagram looked more like a "series of family trees" producing the image of a labyrinth. "Installed in the chamber at its enigmatic centre [was] ego or fate", and all the entrances leading into the interior were recognized as "primal acquaintances". This diagram is an image, created out of spatial patterns and elements, that seeks to depict a dense network of relationships between specific persons and Benjamin himself. A spatial image, therefore, becomes the vehicle to explore one's life and with it to investigate patterns of potential human relations ("passageways . . . always in the most diverse periods of life, guide us to the friend, the betrayer, the beloved, the pupil or the master" [ibid. 319]). Through such a thought-image, one can rethink one's life as revealing the dreams and nightmares of a troubled epoch.

Do people possess this power of diagrammatic thinking? We have evidence to assume that people were addressed by rulers or dominant elites as though they indeed possessed this power. For example, it seems that people in late traditional Chinese cities were familiar with the "cosmographical conceptions articulated by the cultured elite" (Henderson 1994: 215), and those conceptions had influenced greatly Chinese domestic architecture. Moreover, in many ancient cities, ranging from Teotihuacan to Beijing, people probably would not miss the monumentality of the general urban layout, even when they were not informed about the exact cosmologic meanings of those sovereign city-planning gestures (Smith 2007: 35). True, in certain societies only the elite (religious, administrative, or intellectual) can appreciate diagrammatic reasoning. This is the case, for example, for Egyptian necropolises. Only the priests probably could read in spatial arrangements cosmologic symbolizing geometries and correspondences. Also, in Kandy (Sri Lanka's last precolonial capital), cosmological planning "was evident to the ruler of the city and his Buddhist monks and scribes [while] others were unaware" (Coningham 2000: 353). But many forms of city planning seem to address a presupposed generalized ability to perceive and understand inhabited space diagrammatically. The city of New Delhi offers a very good example. As is known, it was created near the historic city of Delhi out of a plan commissioned by the British colonial administration and King George V himself. Modeled as a "theater of colonial domination" (Jyoti 1992: 83), the city was designed following the spatial arrangement and hierarchical layout of a specific ritual celebration of Mughal India called 'durbar'. In durbars, "relative placement of people and objects was used to symbolize their relationship to the ruler. Proximity indicated status" (Jyoti 1992: 87). As a kind of durbar built on stone and bricks, the imperial 'New Delhi' accomplished a correspondence of place and status for its inhabitants, arranging their residences and public buildings in a hierarchical layout, which culminated in the placement of Viceregal's palace at the end of a dominating central axis avenue. Thus, the "layout of the imperial city . . . became an experience in complete domination for indigenous visitors" (ibid. 96).

New Delhi was designed to represent and enforce a 'legitimated' imposed authority by using spatial relations to express and strengthen social relations. Clearly an authority wanted to instruct a subordinate population how to behave and think about their rulers. Is this not a case in which a planning authority relies on the ability of people to understand and inhabit space through a diagrammatic thinking-in-images? Were the inhabitants of a city built ex nihilo not treated as people who should adjust their lives to a city-image, a city not simply depicting an imperialist project but providing the means to be appropriated as a naturalized diagram of power relations? As in many other cases of cities planned to show diagrammatically a dominating central authority, it is important to observe that the diagrammatic understanding is based on the acts of inhabiting space. People 'perform' the diagram as well as think through it about their place in their society.

Perhaps it is more appropriate to think of cities built with an overarching intent similar to that of New Delhi not as "cities built as images" (Smith 2007: 30) but as cities built to be understood as diagrams. Accordingly, various efforts to discover an overarching meaning connecting every part of the city to a representation of society or cosmos—using, for example, Eliade's (1959) ideas on the "cosmological significance of settlements" or Lynch's (1981) "theory of magical correspondences" (both mentioned in Smith 2007)—could refashion and reevaluate their inferences through consideration of the performative and relational character of thinking through space images. "Cosmograms" (Carl 2000: 330) are just one version of spatially constructed diagrams, and they were meant to provide thought-images to those who inhabited them.

What remains crucially important, in this context, is the actual temporality of the process of thinking-in-images while inhabiting space. Sometimes we cannot be sure about when this process was triggered, or meant to be triggered, or if this process was explicitly encouraged by city planning.

We may have to prepare ourselves for findings and interpretations that show that utilitarian purposes produce city layouts that appear as diagrammatically meaningful but were in reality invested only after the city's construction with diagrammatic intentions. As Smith (2007: 38) suggests for the case of the Aztec capital of Tenochitlán, "the spatial structure of Tenochitlán was determined by the generally orthogonal courses of canals and causeways. Once the city had developed an orthogonal layout around these linear features, the city's plan could have been given a cosmological interpretation by its rulers, priests, and planners".

In a different but relevant context, 'interpretations' of ancient spatial arrangements that continue to puzzle archaeologists could profit from the hypothesis of inhabiting space through thinking-in-images. For example, Turnbull's (2002: 137) hypothesis that the Neolithic Maltese megaliths were arranged as social "theatres of knowledge" is based on the idea that "space is performative" and that "Maltese knowledge traditions were performed" (ibid.). These monuments, then, were, according to Turnbull, arrangements that directed movement and regulated the transmission of knowledge through the so-called oracle holes. Regardless of whether this theory is plausible or not, one can observe again a presupposition that effectively hints toward a diagrammatic appreciation of inhabited space. People learn to move through space, but they also learn to evaluate this movement, to recognize possible trajectories, and to connect movement to meaning (Massey 2005: 13, 117–125).

Placing the emphasis on the experience of lived space, Thomas (2001: 173) chooses to understand landscape as a "network of related places". "While a represented landscape is an object or entity, a lived landscape is a set of relationships" (ibid.). People, however, interpret, that is, attribute meaning to inhabited landscapes, by converting their relations to other people, places, and things to representations that generalize those relations. It is exactly this generalizing process that explains the connection between lived and thought space and gives to thought-images their power to carry and transfer social meaning. Thinking in images involves practices that perform space by activating potentialities created by spatial arrangements. Crucial symbolic intentions are recognized as worldviews are expressed in spatial relations.

In both social anthropology and archaeology we can trace the uses of concepts that actually have the characteristics of thought-images, whether they are supposed to be representing the inhabitants' view of their space or a model constructed by researchers in order to understand the inhabitants' view. Without oversimplifying rich discussions and theoretical interpretations, we could say that both the idea of 'house' and the idea of 'threshold', which were employed as social artifacts with rich cultural meaning for the understanding of the societies that used them, are actually thought-images: they can combine descriptions of specific spatial elements with the symbolic meaning attributed to them (abstract and general but based on those concrete images). For example, thresholds, as van Gennep's (1960) work proves, were spatial artifacts connected to various "rites of passage". Thus, "the passage from one social position to another is identified with a territorial passage" (ibid. 192). Societies think through the image of passage about the ways in which their members are expected to cross thresholds separating different social roles and positions (Stavrides 2010: 16–17).

There are strong indications that at least during the Neolithic period in the Mediterranean, the image of the house, for example, as condensed in miniature clay house models, was more than a mere description of a spatial element. Findings hint toward the possibility of these models being used symbolically to express the identity of individual groups as well as being activated in collective rituals as regenerators of a dominant ethos of collectivity (Souvatzi 2008: 197–198, 217–220, 234; also Borić 2007).

Ingold's (1993) famous theoretical proposal about 'the temporality of the landscape' can be placed within a similar interpretive framework. If landscape is, as Ingold (ibid. 162) argues, a "work in progress", a work created by the multifarious and multirhythmic layers of human activity, interpreting landscape means a lot more than just finding a key or a code of correspondences. This is the case in the example of the Apache that Ingold (ibid. 171) mentions, who intend through their stories "to allow listeners to place themselves in relation to specific features of the landscape". Thus, interpreting a landscape means participating in an effort to connect, combine, place, and compare spatial elements that temporarily become organized in images. Those images are, to recall Weigel's remark (1996: x), both the matrix and the medium of a mode of thinking that opens hermeneutic trajectories by combining concrete experiences with abstract thoughts.

CONCLUSION

Thinking-in-images as a generalized ability of humans can and is manipulated by different power regimes to direct the attention of people and to mold their way of thinking. People learn to think through images in ways that reaffirm dominant world views and legitimize social relations. The city can easily and productively become the locus in which these image-thoughts are generated and activated.

It seems that people use their ability to observe, recognize, and inhabit spatial relations in order to create images of their society and decipher the meaning of the parts of social life in which they participate. To the inhabitants of socially produced space (understood as a network of meaningful spatial relations), images of space can become the matrix and the medium of a thinking that develops by bridging the distance between concrete experience and abstract thought. Images of space become thought-images, images that support a generative understanding of society immersed in culturally specific ideas, mythologies, and values.

Can this mode of thinking shed new light on questions focused on the meaning of the layout of settlements and cities, given that this layout is usually taken as a strong indication of the corresponding society's organization and culture? Perhaps yes, if we can develop an understanding of the mechanism of thinking-in-images that gives people the power to think-through-inhabiting and to recognize social values and socially meaningful ideas in spatially constructed images. Maybe this prospect gives us the opportunity to think of the members of a society not simply as conforming subjects, but as potentially inventive human actors. Constellations of thoughts and images, as Benjamin (1992) has urged us to think, can produce unexpected combinations, revealing comparisons, promising new ideas. Benjamin's own work has explicitly focused on late nineteenth and early twentieth century cities and attempted to reveal the buried promises of modern city life through the thought-images he constructed. His famous city portraits were not idiosyncratic poetic descriptions of big modern cities (Berlin, Paris, and Moscow but also Marseilles and Naples) but attempts to discover in city images emblematic representations of hidden potentialities and misrecognized practices (see, for example, Benjamin 1985b, in which Naples is approached through thought-images revolving around the concept of 'porosity'). By retaining the inventive and astonished look of children, Benjamin sought to capture in cities the powers, ambiguities, promises, and nightmares of modern culture. And he imagined the *flâneur,* drawing from Baudelaire, as the thinker-in-city-images par excellence, an emblematic hero of modernity who used the power of thought images to rediscover the city as a constellation of spaces and times compared in inventive, creative, and liberating ways.

Can we learn from this project something about the power of city images not simply to instruct but also to inspire people? Can we rethink the great projects of various ruling elites of antiquity who used the city as an image in order to impose dominant values, worldviews, and representations of society, knowing that people can think through images in not always predictable ways?

What kind of history can we write if we take seriously this generalized ability to think-in-images, beyond dominant thinking and beyond dominant images? Did not the people of Çatalhöyük think through the images of their town about their relations with their neighbors in different ways than, for example, the people of Mycenae? Was it not important for people to diagrammatically be aware of the specific egalitarian or hierarchical characteristics of the corresponding built-up areas? Among the findings in Neolithic Çatalhöyük is an amazing "cartographic representation of cityspace, stunningly detailed yet abstractly perceived" (Soja 2000: 40–41). This almost diagrammatic presentation of a 'city' layout possibly shows that inhabitants could recognize the logic of the spatial patterns that corresponded to the "relatively egalitarian society" (Hodder and Pels 2010: 179) in which they lived.

Referring to the possible social meaning of differentiations between specific types of houses they have excavated, Hodder and Pels (2010: 164) propose to name exceptional "elaborate" houses as "history houses", houses, that is, in which "more transcendent knowledge and symbolic capital than in others" was accumulated. One can guess that spatial differences are made socially meaningful through a generalized ability of inhabitants to think through images, to distill general ideas out of images. Bloch (2010: 152, 154–159) proposes that we take Neolithic Çatalhöyük as an example of "house based societies", in which the house is not only a specific material entity, the module out of which society's space is created, but also a "concept" through which society is represented as a recognizable and stable network of human relations.

Perhaps, then, we need to ask: How did people, who were experiencing major changes in their society and city-culture, use the image of their city to attempt to think about the different society that was emerging? Did they try to imagine a different future by thinking-through the images of their present? After all, it is not by mere chance that most of the world's reformers and visionaries, but also most of the inspired commoners, visualized a different world by thinking through images of imagined ideal cities. One has to remember Thomas More's Utopia, the Ideal Cities of the Renaissance (Filarete, Da Vinci, and others), those of the neoclassical eighteenth century (Boulee and Ledoux), the utopias of the so-called utopian socialists (e.g., Fourier's Phalanstère and Godin's Familistère), but also those liberal utopian cities about which Jules Vernes and Jonathan Swift have imaginatively written.

Thought-images seem to possess enormous power: not only can they contribute to the reproduction of dominant social relations; they can also support the dreams and thoughts that transgress them.

REFERENCES

Benjamin, W. 1985a. A Berlin chronicle. In: W. Benjamin, *One Way Street and Other Writings,* 293–346. London: Verso.

Benjamin, W. 1985b. Naples. In: W. Benjamin, *One Way Street and Other Writings,* 167–176 . London: Verso.

Benjamin, W. 1985c. One way street. In: W. Benjamin, *One Way Street and Other Writings,* 45–104. London: Verso.

Benjamin, W. 1992. Theses on the philosophy of history. In: W. Benjamin, *Illuminations,* 245–256. London: Fontana Press.

Benjamin, W. 1999. *The Arcades Project.* Cambridge, MA: The Belknap Press.

Berger, J. 1972. *Ways of Seeing.* London: Penguin Books and BBC.

Bloch, M. 2010. Is there religion at Çatalhöyük or are there just houses? In: I. Hodder (ed.), *Religion in the Emergence of Civilization: Çatalhöyük as a Case Study,* 146–162. New York and Cambridge: Cambridge University Press.

Borić, D. 2007. The house between grand narrative and microhistory: a house society in the Balkans. In: R. Beck, Jr. (ed.), *The Durable House: House Society Models in Archaeology,* 97–129. Center for Archaeological Investigations, Occasional Paper No. 35. Southern Illinois University.

Bourdieu, P. 1991. *Language and Symbolic Power.* Cambridge: Polity Press.

Buck-Morss, S. 1991. *The Dialectics of Seeing: Walter Benjamin and the Arcades Project.* Cambridge, MA: MIT Press.

Carl, P. 2000. City-image versus topography of praxis. *Cambridge Archaeological Journal* 10(2): 328–335.
Coningham, R. 2000. Contestatory urban texts or were cities in South Asia built as images? *Cambridge Archaeological Journal* 10(2): 348–354.
Deleuze, G. 1988. *Foucault.* Minneapolis: University of Minnesota Press.
Elden, S. 2001. *Mapping the Present: Heidegger, Foucault and the Project of Spatial History.* London: Continuum.
Eliade, M. 1959. *Cosmos and History: The Myth of the Eternal Return.* New York: Harper and Row.
Foucault, M. 1995. *Discipline and Punish.* New York: Random House.
Henderson, J. B. 1994. Chinese cosmographical thought: the high intellectual tradition. In: J. B. Harley and D. Woodward (eds.), *The History of Cartography,* vol. 20: *Cartography in Traditional East and Southeast Asian Societies,* 203–227. Chicago: University of Chicago Press.
Higham, C. 2000. The symbolism of the Angkorian City. *Cambridge Archaeological Journal* 10(2): 355–357.
Hodder, I. and P. Pels. 2010. History houses: a new interpretation of architectural elaboration at Çatalhöyük. In: I. Hodder (ed.), *Religion in the Emergence of Civilization: Çatalhöyük as a Case Study,* 163–186. New York and Cambridge: Cambridge University Press.
Ingold, T. 1993. The temporality of landscape. *World Archaeology* 25(2): 152–174.
Jay, M. 1988. Scopic regimes of modernity. In: H. Foster (ed.), *Vision and Visuality,* 3–23. Seattle: Bay Press.
Jyoti, H. 1992. City as Durbar: theater and power in Imperial Delhi. In: N. Alsayyad (ed.), *Forms of Dominance,* 83–105. Aldershot: Avebury Ashgate.
Lévi-Strauss, C. 1966. *The Savage Mind.* London: George Wiedenfeld and Nicholson Ltd.
Lynch, K. 1981. *A Theory of Good City Form.* Cambridge, MA: MIT Press.
Massey, D. 2005. *For Space.* London: Sage.
Richter, G. 2007. *Thought-Images: Frankfurt School Writers' Reflections from Damaged Life.* Stanford, CA: Stanford University Press.
Sennett, R. 1994. *Flesh and Stone: The Body and the City in Western Civilization.* London: Faber and Faber.
Smith, M. E. 2007. Form and meaning in the earliest cities: a new approach to ancient urban planning. *Journal of Planning History* 6(1): 3–47.
Soja, E. 2000. *Postmetropolis. Critical Sudies of Cities and Regions*, London: Wiley.
Souvatzi, S. 2008. *A Social Archaeology of Households in Neolithic Greece: An Anthropological Approach.* New York and Cambridge: Cambridge University Press.
Souvatzi, S. 2012. Space, place, architecture: a major meeting point between social archaeology and anthropology? In: D. Shankland (ed.), *Archaeology and Anthropology: Past, Present and Future,* 173–196. Association of Social Anthropologists Monographs number 48. London: Berg.
Stavrides, S. 2010. *Towards the City of Thresholds.* Trento: Professional dreamers.
Szondi, P. 1991. Walter Benjamin's city portraits. In: G. Smith (ed.), *On Walter Benjamin,* 18–32. Cambridge, MA: MIT Press.
Thomas, J. 2001. Archaeologies of place and landscape. In: I. Hodder (ed.), *Archaeological Theory Today,* 165–186. Cambridge: Polity Press.
Turnbull, D. 2002. Performance and narrative, bodies and movement in the construction of places and objects, spaces and knowledges: the case of the Maltese megaliths. *Theory Culture and Society* 19(5–6): 125–143.
Van Gennep, A. 1960. *The Rites of Passage.* London: Routledge and Kegan Paul.
Weigel, S. 1996. *Body- and Image-Space: Re-reading Walter Benjamin.* London: Routledge.

4 On Stone Houses and the Co-creation of Worlds and Selves

Rachel Harkness[1]

It was raining heavily when we approached the stone house in the wood. Our sandaled feet squelched in muddy grass and the path made by footsteps of others before us led through trees, then around towards the building. Hurrying, head-down, I was hoping for some relief from the downpour so didn't stop long to consider the rectangular building of stone, with its pitched roof of slate. I did see that it was all the silky colours of soaked sealskin: somehow strange and promising.

A plain wooden door was in the middle of the far gable end. Relieved, at last, also to notice a small square window high up, I opened this door and entered. There was hesitation then, in that moment, for as the wash of rains drummed behind me, I stepped into blackness, so into another sensory realm. Before me, just beyond the threshold on which I stood, the earth opened up, and in the relative darkness of the interior I struggled to comprehend. Reeling a little, I adjusted to the feeling that here the ground was at once both falling away and rising up. Slowly, I saw the craggy magnificence of a rock-face revealed where a floor might have been, and, in a moment of suspension, of silence, I felt the presence of the stone—old and terrific, it seemed—that occupied every inch of that house.[2]

PATTERNS OF MOTION, PATTERNINGS OF ACTION

The paragraphs above are an autobiographical account of a visit I made to the house in the wood in the summer of 2011. Since that visit I have returned only once to the site, again to experience the landscape and the form of the house through summer showers. Even with such brief visits, however, the house—of which I hope this description gives a good sense—forms one of two case studies of sorts for this chapter. Whereas the stone house is a very specific and localized phenomenon, briefly encountered, the second case study is quite different in that it is a phenomenon that I researched for more than five years during the course of my doctorate in social anthropology. This second phenomenon is a type of ecological or environmentally friendly dwelling that is built by its owner-dwellers, is off-grid, and goes by the name of the Earthship. Earthships, having originated with builders in the southwestern United States in the early 1970s, now pepper the globe, and there must be at least a few hundred of these types of homes now in existence. From the spring of 2005 to the autumn of 2006, I worked alongside the builders of more than fifty Earthships in both Scotland and New Mexico in the United States of America, and using ethnographic research methods and my own apprenticeship, I investigated their culture of building, their practice as an eco-architectural 'movement', and the wider questions of how to dwell ecologically in today's world.

I was interested in learning about what Earthship building and living taught of the world, of social life, and of humanity's relationship with the nonhuman environment. I hoped that concentrating on the art of making things would allow me to further thinking on the relationship between theory and practice, design and sustainability, on labor and power and their connection to hope for the future. The Earthship proved to be a wonderfully rich phenomenon to study, not least for its builders' belief in the democracy of low-tech self-build and its aesthetic blend of vernacular style and salvaged materials. In its forming, the Earthship made manifest the dwellers' interwoven temporal and spatial understandings and experiences, concerns, and dreams.

In this chapter, I draw on the work I did with the Anglo-American and Scottish-based women and men who built one another's Earthship homes, from mud and earth, reclaimed wood, and old bottles and tires. I learned much from them, including a great appreciation of the material nature of our shared, interdependent, and limited world, the importance of sensitivity to the local environment and to the vernacular, and the significant role of craft and handiwork in the forging of just, anti-alienating, and sustainable futures. Their stories and practices and creations underpin my own, and their way of dwelling heavily informs the interpretation of space and time that follows. Before we hear more of this, however, let us first return to that damp wood in central Scotland for a time.

The experience of encountering the house in the woods was powerful: it played with my assumptions and expectations of what a stone structure such as this would ordinarily hold or be, and I perceived it acutely through my sensing, moving, critical body. Though I was challenged to make sense of it, I felt time and space, self and world, concentrated within that event that was perhaps at its most powerful at the moment of the darkness's reveal. It was an experience that was visceral and moving (in the myriad senses of this term), for being in that house was both an immediate experience in that the building and its inhabitant—if they were not the same entity—pressed themselves upon me as a visitor, yet simultaneously they allowed me, perhaps even invited me, to leave that particular place and be elsewhere in other times.

This stone house, its engaging and transcendent quality, and its affective presence will be my recurring touchstone in this essay, as—in keeping with the volume's aim to foreground the temporal and spatial dimensions of society—I endeavor to consider the experiential interrelation of space and time by way of drawing together broadly anthropological reflections on the nature of architecture, the environment, and building. Intentionally, I will continue to use language that brings together often seemingly opposing concepts or notions, in order to suggest that the dichotomous dualism to which we are accustomed in the so-called West (e.g., in terms of nature as other to and distinct from culture, female from male, body from mind, etc.) are, as the writer and geographer Doreen Massey (1993: 147) has put it, "at heart problematical and a hindrance to either understanding or changing the world". Instead, I want to think in a way that acknowledges the dialectical, the relational, and always ongoing character of life and living.

In addition to the stone house being an edifice the color of midwinter skies and it being the experience of the familiar ground falling away, it is also a piece of work by the acclaimed British sculptor, photographer, and environmentalist Andy Goldsworthy.[3] In this guise, it is called 'Stone House—Bonnington'. Goldsworthy (2009) has described his stone houses (for there are different incarnations of stone houses in his work) as becoming "forum(s) in which the nature of the place and human nature meet". I take from this the idea of the meeting place—with its temporal, spatial, and social connotations—and weave it in and out of my discussion. Furthermore, this notion that different natures or characters meet in the Stone House is helpful, I think, in that it immediately suggests a kinship between human beings and the nonhuman (which in this case is the rock standing in for the nature of the place); both may have varying natures, and in their meeting they unfold, they reveal, they are shaped by one another. So, while thinking on the notions of meeting, the nexus, the intersections and minglings, I am drawing on a variety of thinkers and writers who have theorized time and space and their interrelation, or have cast light upon the subject.

The position from which I write is one that is Marxist influenced and phenomenological, and it has grown out of my research interests in contemporary ecological-building practices. It reflects something of what the anthropologist David Graeber has called the 'Heracleitian' tradition for the way in which it sees, as he puts it (Graeber 2001: xii), "what seem to us to be fixed objects as patterns of motion, and what seem to be fixed 'social structures' as patternings of action". I am interested in these 'patternings of action', in power, or agency, in making or production, and it is this sense of motion or movement that lies at the core of my approach.

This motion might be thought of as process or the processual, but it is not to be mistaken for continuous high-speed change. I am not suggesting that life everywhere is rapidly uprooting. Instead, I am suggesting that continuous movement encompasses phases of rest, pauses, hesitations, and faltering increases; it is creeping *and* galloping, the meager as well as the more generous. What I am describing here is a *both/and* approach (not a stark either/or). This is an approach that evokes the notion of a sliding scale moved along between extremes of magnitude, say, or up and down various gradations that become more or less, relationally speaking.

It is perhaps obvious now that in discussing an evocative contemporary artwork encountered in temperate Scottish climes and eco-homes of radical form on either side of the Atlantic within a book concerned with the Mediterranean prehistoric, I hope to be illustrating the power and importance of the multiscalar approach to space and time. I am arguing, then, *by way of* encouraging the reader to leap and to move with me from damp summer woodland to stone quarry, from cultural notions of the comforting, sheltering home, to alienation brought about by certain architectures, from the long geological times of stone's formation to the changing traditions of working with certain materials.

It is by this course that I mean to suggest that it is in our *movement* (including that movement that is imagined, intended, and remembered) through a fundamentally relational, material, and ever-changing world that we come to better know and understand it, and the others (human and nonhuman) that constitute it with and in relation to ourselves. We see that the spatial and the temporal are intrinsically linked—their firm tie revealed and re-bound in the making of movements-between and the contemplation or recognition of relations-between and their quality.[4] Taking the phenomenological approach, experience—the social and material as well as the individual—and a belief that there is simultaneous construction of self and world, are at the heart of this analysis.

EXCAVATING

Look in to the land—not out of the building.

Goldsworthy (2009)

Goldsworthy had waited long to find a place in which the bedrock lay close to the surface of the ground yet did not puncture it. At Bonnington Wood, he found such a place. Here, the bedrock's undulating face was revealed through careful excavation of only a small amount of topsoil. As Goldsworthy (2009) points out, "it is difficult to believe that something so strong and powerful can be so hidden whilst being so close", and it is this closeness in the confines of the almost windowless house that encourages the temporary human dwellers to engage with the stone through all of their senses: the stone touches as it invites touch, its heavy taste is breathed in on the damp air, and in the dim light it activates looking, evoking a fearful mix of awe and curiosity. In the form of the walls and in the rising ground the stone is the still-changing manifestation of formations and contortions that have been happening for longer than people have lived on Earth, and its presence here is undeniable. As the anthropologist Kathleen Stewart (2007: 84) has noted, "power is a thing of the senses". In the stone house, the air bristles with the thing-power of the bedrock laid bare.

On the layers of earth that have been carefully removed, Goldsworthy (2009) remarks that they "are evidence of time and change," and that their peeling-back brings an awareness of what lies beneath the feet. Revealing the bedrock in excavation at Bonnington is a purposeful act of art-making, then, but one that follows conventions in the reading of the passage of time in matter. This practice perhaps began in earnest with James Hutton and his observations of the rocks at Siccar Point, southeastern Scotland—not far from Bonnington, in fact. Here, Hutton observed that the Earth was vastly ancient, and as Steven Baxter (2003: 2) puts it in his biography of Hutton, he saw that "every particle of its surface had undergone countless cycles of erosion and uplift". In its exposing of rocks separated in their creation by eighty million years, in what came to be called its 'radical unconformity', Siccar Point showed Hutton and others that land is crumpled and it rises up, tectonic plates spasm and strata break and fold. "Great forces within the Earth distort the neat layers" of consolidated debris (Baxter 2003: 3): processes we can see if we look with deep geological time in mind. With this change in perception, argues Baxter (2003: 230), we could see "mountains bloom and fade, as transient as flowers".

So, layers of earth do not always roll back neatly to reveal the record of an earlier epoch. Instead, material accumulations and entanglements have built upon one another, intermingled, and supported, furnished, and been utilized in human dwelling practices in various ways. Excavations reveal the material traces of these entanglements of ongoing lifeways, elements and forces (nonhuman as well as human). As such, and to quote the archaeologist Lesley McFadyen (2007: 117), space is revealed "as mobile and rendered through force". Furthermore, in order to complicate overly simplistic correlations of excavating and traveling back in time (tropes all too common in campaigns to 'modernize' troglodytic dwellings, for instance), we might add that because we dig in order to build, so we dig into the future as much as we do the past (with our foundations, and dwellings and gardening practices); likewise, we build as much into the past as we do the future (with our memorials to events, to celebrated men and women, to eras past).

Of course, turning back to the Stone House and its disclosure, some visitors there might be unsurprised to see the rock beneath their feet: those who study the changing, continuously forming landscapes of our planet, for instance, may well perceive their environments with this version of the so-called internal view that Goldsworthy hopes to provoke. However, for many visitors, the concealment and revelation, staged so effectively in the Stone House, will direct their attention anew and provide them with a contrasting view of the landscapes surrounding them (Figure 4.1). Such concealments and revelations also highlight the importance of ambiguity. In the dark interior of the building at Bonnington, the excavated ground may have revealed the surprising closeness of the lurking mass of rock, and it does play on

Figure 4.1 The woodland setting and the exterior of the 'Stone House—Bonnington', by Andy Goldsworthy, in Bonnington Wood, Edinburgh, Scotland. Photo taken by the author, 2012.

our ideas of being able to read the past in the material present, but there is something more than a textual temporality here, something more than the palimpsest (Figure 4.2). For as we know, the possibility of surprise depends on unknowns, and the potential for change on the conditions of possibility,

Figure 4.2 The bedrock, inside the Stone House. 'Stone House—Bonnington', by Andy Goldsworthy, Bonnington Wood, Edinburgh, Scotland. Photo taken by the author, 2012.

and such conditions do not thrive in mapped-out territories or harshly lit rooms. The past can only be interpreted from the matter and its rich murkiness, and just as it has played a role in the making of the environments in which we currently live, so too will the rock play a role in the processes of future-makings. In our interpretations of environments and materials, of constructions and places of habitation, then, as Caitlin DeSilvey (2006: 320) puts it when referring to working with that which is decaying, we must work with "sensations of ambiguity".

FOLLOWING THE THREADS

In a piece of beautifully illustrated writing about the art of quarrying stone, Joel Fischer (2011) describes the global crafts and skills of splitting, cutting, and extracting stone from the earth. Fischer introduces the reader to a world of carving, of the gradual cutting away of and into the earth, the finding and widening of seams and cracks and fault lines, the sawing of stone with sand and with diamonds. To read these tales, then, is to see the labor that has sculpted these wall stones, the hands that have deciphered, prised, fired, and drilled rocks across the ages.

The stones that compose the walls of the Stone House have come from quarries such as these. They have been the focus of attention while huge swathes of landscape have been cut and great caverns formed. Turning to them again now, we find that like the bedrock, they also have much to reveal. Following the stones takes us to where they have been and to processes of excavation and extraction on an industrial scale.

Describing quarries of Italian marble, pink Kemnay granite, and Irish limestone, Fischer notes their columns and archways, their cavernous qualities, quarries that "can have cathedral-like interiors" (2011: 59), and remarks that the often "attractive physical space of a quarry is created as a by-product of paying attention to something else" (ibid.).[5] Further on in his survey, Fischer elaborates on this relationship between excavation and construction, describing the quarry—its clusters of chambers "retreating deeper and deeper into the earth" (Fischer 2011: 64)—as almost the inverse of the city's elevations above. Thus, an understanding of building with stone is emerging that is one of both excavations and elevations. It is one of feeling in the darkness, of reveling in the creations of human hands and great forces of pressure and heat, and of extending skyward. As Juhani Pallasmaa (2005: 67) puts it, "at the same time as making us aware of the depth of the earth, it makes us dream of levitation and flight".

My own work with builders of the ecological homes called Earthships also takes me, takes us, down into the earth and up into the skies. Here too, there is evidence that building can be seen, felt, and thought in terms of excavations and elevations. For these contemporary builders dig down into plains and into hillsides to create curved recesses within which to build and dwell. The homes that they fashion are long and low or terraced houses that gather sunlight through south-facing glass walls and use the dense mass of rammed-earth walls (compacted by the energy of many laboring together) along with the enveloping landscape to regulate the temperature of the home (see Figure 4.3).

Figure 4.3 An exterior shot of an earthship in central Scotland. The photo shows the south-facing, semi-subterranean position of the building, which is fashioned from materials such as tires rammed with compacted earth, heated by way of its passive solar design and powered by renewable technologies. Photo taken by the author, 2005.

Encouraged by the makers of these solar edifices, one could see their building as a process of moving into the earth and out again: a constant interweaving of people, materials, light, energies. Furthermore, their semi-subterranean eco-homes are not boxlike constructions that sit upon the earth, either literally or metaphorically, but rather are understood by builders to be *of the earth* and *from it* and built *into it*. Processes of excavating, of digging, of planting, of reaching solid ground, of cutting away, of shaping-out, of moving mass from one location to another, are all fundamental parts of building and dwelling that come more forcibly into view having taken this ecological approach.

During my research with the Earthship builders, I once saw a photographer's time-lapse depiction of the period of a home's construction that afforded a similar sort of perspective, though, in this case, by speeding up the replay of action. It struck me, by way of its long view: here were flows of people, timber, glass frames, spurts of action and calm between them, flurries and tides, and the weather's flickering but steady change through the seasons. Furthermore, retaining this temporal perspective provides a view of these homes as conduits: conduits for water, which falls as rain, is filtered for usage by the dwellers and is used up by interior gardens; for air, as the lively breeze passively ventilates the house, turns the wind turbine, and is cleaned by the garden plants; and for solar energy, which ebbs in and out of the house daily and across the course of the year retreats in the summer and makes welcome advance in the winter months.

These warm, bright eco-houses are made of earth, and wind and light and water flow in and out of them, their flows orchestrated by the builder in time with wider, greater rhythms. They make apparent an environment-world that is extending in all directions around us, not divided into sky above and earth below with us living on a thin surface between, but as Tim Ingold (2011: 119) argues, sky and earth as "manifolds of movement directly implicated in one another". Living, then, is 'living through' and is as much a process of digging and climbing, stooping and reaching as it is moving along.[6]

QUALITIES OF DYNAMISM

This quality of dynamism that I describe is perhaps akin to what Doreen Massey (2005) is referring to when she encourages an understanding of space as dynamic and political. She speaks of space in terms of the relations between nonessentialized entities, in terms of simultaneous coexistence, and in terms of the radical openness of the future. These are ways of understanding space that I agree with. I also see them reflected in the creativity and imaginations of environmental builders: where things and people are not fixed objects but are changing, growing, learning; where life is lived in relation to the other vital constituents of an interdependent world in a way that is nonlinear; where futures, always in-the-making, should be imagined and hoped for, but never fully determined. Furthermore, I am keen to agree again with Massey (1993) that space and time are not opposites of one another. Instead, I would contend that they are seldom experienced or imagined by people either in isolated singularity or as abstract notions. Their qualities are known as part of living in the world, and very rarely is one understood without the other.

For instance, in changing their architecture and how their dwelling impacted in the world (through chains of material suppliers to places such as forests and quarries) and upon people (from the health of family members to that of the distant dweller affected by dam building or global warming), the eco-dwellers with whom I worked changed how they and others might live now and in the future. By doing what they called 'self-building', the builders fundamentally altered the relationship between time and labor so entrenched since the rise of capitalism, and they took control of the building of their own home, with their own manual labor.[7] In shifting their calendars and units of measurement to those of the patterns and rhythms of plants, the daylight, the seasons, of aging and of growth, and in acknowledging what we might call the temporal values of their building's energy sources and materials (for example, renewables' cycling longevity), they were creating a way of existing that was impacting upon and relating to the world differently, that was conscious *for* the coexisting other.[8]

The central position of the person to this activity is crucial here: eco-building is (at its best) anti-alienating. It is intentional in the case of the self-build that the laboring done lessens people's experience of alienation in and from the world and from one another (an alienation believed by the builders I worked with to be exacerbated by so much of today's architecture). To have a hand in the construction and cultivation of one's own dwelling place is invaluable to the eco-builder. In the case of the Stone House, as a visitor, you experience, through the soles of your feet and up into the whole of your sensing body, the stone-rock. Being able to feel, for yourself, is important and key to the experience, as is being able to communicate this to and appreciate it in others, and therefore, in that way, to transcend your own self. What is happening here is not a changing of time into space or an understanding of one through the other but rather a feeling of the ongoing processes of which you are part, an enfolding of yourself into the fabric of life.

REINVENTING HOUSING

> *Containing the bedrock inside a house accentuates the sense of revelation of the stone floor. Houses are usually places of security, shelter and comfort for people. There is something unnerving about entering a building in which nature is the occupant.*
>
> Goldsworthy (2009)

Cultural assumptions about the house being a place for people that is apart from the natural abound in the contemporary West. Arguably, very practical concerns about shelter underlie these assumptions, but even more significant are our cultural understandings of how humanity relates and should relate to nonhuman nature. We might say that, generally speaking, 'to house' is to differentiate, to set apart from something we call other-nature.[9] Our buildings tend to exclude other-nature at all costs.

However, just as for decades now, thinkers have been successfully destabilizing, complicating, and undoing the so-often revived dichotomous dualism of Western nature and culture (e.g., Cronon 1996; Descola and Palsson 1996), so too have groups of people in the so-called West been building homes that consciously house 'within' the environment; houses that are of the world rather than against it! Like the Stone House, these eco-constructions embody and express the value of being able to play with cultural assumptions and cemented forms, as they challenge, mix, explore, and contest. They ask questions of what fits in a house and of what belongs there. In the Stone House this is done with the presence of the bedrock; in the eco-house it is with the inclusion of the indoor gardens that process sewage or earth-plastered walls or the use of reclaimed, previously 'rubbished', materials. Such challenging is crucial to the builder, but also more widely, to society, for playful disruptions of the social ordering are potentially part of the essential—and so quite serious—reinventions and revitalizations of the way that we build and dwell.

Another cultural assumption to dispel is the idea that architecture is static. In the form of the house this assumption is accentuated by the fact that notions of security, shelter, and comfort have become associated with the static and stationary: a castellated model of shelter, if you will; one that is about building tall, solid, rigid walls around oneself; ignoring the ways in which flexible structures often survive the storms that topple the rigid. In fact, architecture is always moving, it is always in constant, if relative, motion, not as a succinct object that is moved across a landscape such as is the locomotive carriage, but as if it were a birds' nest in a tree: some living branches intertwined with fallen leaves, broken twigs, feathers, and moss found in the maker's way and incorporated in the act of nest-making.

Thus, in the buildings around us today we can see that people build extensions and convert attic spaces; the birth of children and their eventual leaving of the nest influences the fabric of the building. So too do fashions and fads in design: the conservatory or the veranda springs up and lean-tos and prefabs are pulled down. Elsewhere, houses subside, floods engulf settlements, and damage needs mending. There is movement also in the way that builders and designers, working over a time, adapt, refine, and improve their designs and hone their skills. On smaller scales still, the plaster that these craftspeople apply with the repeating waves of arching arms takes a time to dry properly between coats, and by the season window shutters and screens are added to dwellings at the appropriate time of year, then put away again when the seasons turn. Deserted buildings begin to peel and flake away at the edges, weeds lodge in the gutters, and grass cracks through the pavings. Architecture, it seems, is something in flux, along with and part of the flow of human social life, and processes of decay, erosion, and growth.[10]

From this perspective, a perspective that provides a good view of patternings of motion (Graeber 2011), the thing that is the house can no longer be comfortably categorized as a distinct object. This emphasis on patterns of motion also chimes with Lesley McFadyen's (2006) depiction of the different paces of Neolithic architecture as a choreography of slow phases, lapses in momentum and pauses, against quick and rapid moments. She paints a picture of architecture as full of rhythm and movement, and understands it as a matter of the creative junctions of forces and people. In this light, the posthole, for instance, is the crush of bodies heaving together, the thuds of weight shouldered against wooden post or stone. In the wake of this image, I can see the stones of the house at Bonnington rising in conjuncture with its human builders; I furnish the image with the labor of lifting stones and the rasping mix of mortar that I know of from eco-building, and with the stories from builders of how their hands learned to find the fit between stone placed upon stone or of how their bodies learned the rhythm of moving with stony weight.

Just as the dance or the infectious beat can sweep one away, architecture is moving in this sense too; it is a shaping and affecting force. Experiencing architecture is to be influenced by it, for architectural spaces are indeed produced or culturally created (Lefebvre 1991), but they are also producers. Our built environments help to shape our bodies, determine our comportment, and color and inform our values and our actions. In shaping our environments, we shape ourselves, and our future selves. We do not always do it consciously, and as the eco-builders that I worked with on Earthships would argue (as they were very much united in their negative critique of contemporary mass housing), we do not always do it for the better. However, it is this potential for better ways of living for future and other beings, for the planet at large, that partly attracts people to self-building and the eco-house: the builders see the promise of architecture.

THE MICROCOSM

In his book *The Poetics of Space,* Gaston Bachelard's (1994) phenomenological study of the house or the image of the house, habitation, and dreaming are intertwined. Many lives center in a thing or a place such as the house: "the house is not experienced from day to day only, on the thread of a narrative, or in the telling of our own story", says Bachelard (1994: 5). "Through dreams," he suggests (ibid.), "the various dwelling places in our lives co-penetrate and retain the treasures of former days". House images "are in us as much as we are in them" (Bachelard 1994: xxxvii). Moreover, the house and universe are dialectically related and the house can be a world in itself (1994: 37). The poetics of the house depend on containment and secrecy, on the dialectics of the miniature and of immensity, of open and closed, within and without.

I am struck by "the power of [this] poetic logic" (Pallasmaa 2007: 19). According to this phenomenological approach, architecture is necessarily tasked with mediating and reconciling seemingly different and disparate elements and forces; building is a practice of negotiating, of working out relations in and with the world. As Juhani Pallasmaa (2007: 19) describes it, architecture "integrates irreconcilable opposites". In the house, the everyday dwelling place of the common person and perhaps the most ubiquitous architectural form in the West, we see "synthesis and polyvalency" (Pallasmaa 2007: 19). Stone House, poetic image, house of the rock, exhibits this polyvalency, this ability to bind with many things, and it allows us the chance to hold a myriad of senses, places, images, meanings, and feelings, *simultaneously.* Like the eco-house, many of these co-present senses, these connections that it makes, binding things and beings, events and places together temporarily, are not those that might traditionally be found in the house as 'cultural container', 'society's mirror', or 'exclusion zone'.

Furthermore, while it might seem that the Stone House is unique in this way, this poetic quality of architecture is most definitely more widespread.[11] In fact, some of the most engaging recent works on affect, things, and assemblage are those by Jane Bennett (2010) and Kathleen Stewart (2007), who both consider how life is made up of everyday moments, events of varying magnitudes that sweep us away, or gnaw at our senses, or arrest us momentarily. They describe how the agencies of things implicate us in often the most mundane ways (such as via the food we ingest or the paths we take along pavements (Bennett 2010), and they highlight the 'potential that animates the ordinary', those multiple moments of fascination where the world pulls us into strange daydream, or hits us, sharp, in the gut (Stewart 2007). Thus, in the everyday of stone houses and eco-houses there is the revolutionary; in the materials made ordinary through common use and in the habitual movements of dwelling there are powerful forces at work, mingling and shaping each other.

EMBODIMENT: BODY, STONE, AND MAKING

In his book on architecture and the senses, *The Eyes of the Skin,* Pallasmaa (2005) adds more weight to our arguments when he contends that in the process of making, "movement, balance and scale are felt unconsciously through the body as tensions in the muscular system and in the positions of the skeleton and inner organs" (2005: 67), and these bodily sensations are "mirrored" in those that encounter the building, even if, as he notes, the experience is centuries later (ibid.). Once more we return to the important centrality of the body, as Pallasmaa encourages us to understand architectural spaces as experienced through what he calls the "mimetic capacity of the body" (ibid.); for him, this feature of building allows person to know person, and it allows communication through the building and through the ground of the body.

In the Stone House it is not just the architecture that speaks through the body: the senses are quickened as the watery darkness unveils the rough compactness of the bedrock, its reaches and its crevices, its uneven levels underfoot, and the changing textures, ridges, and seams of grain that speak of ages of pressure and heat in their formation. The connections between body and stone spark all sorts of considerations of the active relationship between our fleshy human bodies and what the sculptor Jessica Harrison (2011: 183) has noted is too often considered as the solid and inert material of stone.

Following a route similar to Harrison, who explores the movement of stone intertwined with the sculptor, the 'in-between body and stone'

(Harrison, 2011: 185), we might consider the encounters between stone and human body, and the comparisons these encounters generate too: those of firmness and strength, solidity and sound, texture and temperature. Our perspective here is furnished by the experience of the Stone House, with its melding of the ordinary and extraordinary affects of stoniness (Stewart 2007). Harrison's perspective is that of someone working with stone, with hand tools and power tools, and so when she speaks of how stone outlives us by thousands of years, she does so as someone who also knows stone to move, to change and breathe, to grow hot and cold, and to 'be affected by us and by the world around us' (2011: 185).

The cultural geographer Tim Edensor agrees, pointing out in his account of a sandstone church called St Ann's that "the constitution of [. . .] stone is [. . .] affected by the agencies that flow around and nestle in the building in which it is a constituent" (2011: 241). In his study Edensor reveals the "vital, mutable and stabilizing qualities of stony matter, and the circulating, contingent and changing agencies of the non-humans" that affect it (2011: 243).[12] Harrison's 'in-between' can thus be extended or multiplied to the many and various in-betweens of bodies that together form buildings-in-the-world such as the house or the church.

Returning to that threshold in the Stone House, balancing there revealed the kinesthetic, synesthetic, empathetic ground of our being; perception, it seems, is always strivingly multisensory and mobile in its multiperspectival nature. Thus, we sense from here to here, from that to that, between this and this. We orientate ourselves toward things, toward one another, in the world, this way, and the forms and movements of living, decaying, growing, and transforming are multiple, curvilinear, entangled, unpredictable. They might be thought of as lifelines, or as forces that stir, prompt, and engage the imagination, or as the reverberations and infectious vibrations of lively materials.

For the body is the nexus for our experiencing of the world: it is a synthesizer extraordinaire; it is the center from which simultaneous coexistence is sensed, understood, generated. Furthermore, if our orientation in the world is toward things and forces that we understand as being fundamentally related to us, then the potential ethical outcome of this seems promising. However, our mindful bodies are also assemblages in their own right, and we move amidst a world of powerful agencies and shifting temporary assemblages that include the nonhuman. The in-between of the stone-rock and (human) body at Bonnington is a token of that dynamic world of nonhuman as well as human forces.

EXTENDING RELATIONS

Whether we consider St. Ann's Church with Tim Edensor, the Stone House with Goldsworthy, or the semi-subterranean Earthships with their eco-builders, we might see the building as something that "incorporates materials that have life histories of their own and that may have served time in previous structures, living and non-living" (Ingold 2004: 240). As such, buildings embody their relations with their surroundings (ibid.), relations that extend in all manner of directions, to all manner of scales.

Eco-builders tend to see the incorporation of certain materials into their buildings as resulting in homes that embody their values and their aspirations for the future. Thus, the use of nontoxic paints, of straw-bale insulation, of beams cut from tree trunks hewn in nearby forests: all tell stories of pasts experienced and futures imagined. Furthermore, the builders with whom I worked have expanded their sense of materials to include discarded post-consumer and post-industrial materials ('wastes' to some). The houses made of such materials looked both to that which has happened and that which is still to unfold while being situated in the present of making. The materials carry in them stories of development and comfort, of resource exploitation and petroleum oil, of the organic and the industrial.

These eco-homes were positioned at the nexus of flows of stuff, stories, lifelines, and their incorporations of these nonconventional materials—from the rubber tires or glass bottles to the wooden offcuts and quarry rubble—meant that the buildings manifest what Shannon Lee Dawdy (2010) calls complex 'refoldings of time'. Here, the builders' practice was to purposefully enfold and entangle any clean lines of supposed 'progress', or any idea that the material impact of ways of life and consumption enjoyed could be quietly and easily dealt with elsewhere (and by others). We might understand these forms, these refoldings, as practical critiques of a Western environmentalism (widely construed) that shirks its responsibilities (its impacts) and seeks virgin ground on which to grow 'the good life', denying the limits of the Earth. They are also practicing critique through a creative stitching-together, always partial and in-the-making, that is akin perhaps to montage or bricolage (Lévi-Strauss 1996 [1962]).

Eco-building therefore teaches us to follow the materials, to think of the 'embodied energies' invested in a thing up to the point at which we meet it or the 'carbon footprints' it will make as we set it out into the world.[13] It does so by focusing on making, craft, building: assembling a wall, mixing an earth-plaster mix, cutting joints for the pieces of a wooden cabinet, laying down a mosaic tile floor, raising a rafter, digging a trench for a pipe. These are the means with which to know self and world—through thoughtful, social, practical engagement with materials.

For as McFadyen (2007: 121) describes, "all practices of making extended out to other materials and elsewhere (so from working flint, to the processing of plant remains, to the gathering of plants, to the movements of animals and their eating habitats)". In the eco-home, all practices of making extended out to other people including the craftspeople, laborers, and technicians who might have created any preconstructed elements of the house such as pipes or light fittings or photovoltaic panels. Similarly, they extended to those people living elsewhere and in other times who would be affected by the build now or in the future. As Edensor (2011: 239) puts it, the building, as assemblage, "highlights the multi-scalar origins of the heterogeneous elements that constitute it, refuting notions that posit reified scales at local, regional, national and global level".

The mindfulness or what some called 'consciousness' of the builders extended to other parts of the world and to other beings; and they extended out in time to lives lived and passed, and those only dreamed of. They were spoken about in terms of where materials had come from, the people that were entangled in the materials' tale, the children of the future, and the possibilities of their lives on the Earth. Just as the slow voice of the Stone House's inhabitant gives rise in its visitor to an awareness of humankind's rapid flash of movement across the Earth, so the use of a material in the eco-home challenged the dwellers to consider their implication in its life so far and life to come; and its in theirs.

SYNTHESIS (WITH ALTERITY)

This emphasis on the material is now recognizable as a recurring trope, and it directs us once again to its synthesizing quality. For in the lively material of stones, earth, woods, and our human bodies, of lichens' growth, masons' tools, or diggers' blades, we find temporary meetings of stories and lifelines, the meetings of space-times and scales that might have been called modern or prehistoric, natural and cultural, organic and nonorganic, global and local, rooted and transitory, ephemeral and lasting.

This approach of mine is an imaginative and experimental one. It may ally with many of those who have been calling for the "collapsing [of] archaeological and ethnographic time" (Dawdy 2010: 762) in order (for anthropology in particular) to escape from the clutches of what is dubbed 'progressive time', to escape from a tyrannical Modernity that claims a radical break with a now obsolescent past (Dawdy 2010; Latour 1993). Without a notion of progressive time chapping at our heels and claims of clean breaks and the dawns of new eras separating us irreconcilably from days past, a focus on the life and vitality experienced by the constituents of the world brings a different way of considering things. A multiperspectival stance on material continuity, in the minerals, in the elements that circulate through the world, now part of water, now of cloud, now of stone, and now of plant or flesh, reveals alternative views.

If we are to attempt to synthesize these varying perspectives, if we were to imagine waste materials as building materials or landscapes as shifting plates and the traces or promise of violent eruptions, then we would have before us, through us, a dance of elements, encompassing yet always becoming: one of forces, whirls of organisms, tides of species, a rich tapestry of clouds, caverns, and molten movements deep within the earth.

We can see both Stone House and the eco-houses as careful interweavings of materials and imagery. They are earthly, polyvalent entities accessible to us at a human scale, yet also open out to others. In a world of motion that is unfolding, open, and generative, these houses have the resultant power to engage thinking, feeling bodies in their ongoing stories. As such, they are celebrations of creativity and power, and they gather interdependencies of the elements, forces, and beings that together constitute the world.

Drawing on material from both an affective experience within a stone dwelling and much longer-term research on environmental architecture and building practices, this chapter offers one interdisciplinary and somewhat different perspective to this volume, albeit one that shares the aims of stressing the importance of integrated and multiscalar approaches to space and time. It is a broadly anthropological perspective—one with contemporary cultures of environmental art and building and an emphasis on the experiential at its core—which demonstrates how sites of construction might be creatively and usefully considered dynamic and polyvalent meeting places or forums. Furthermore, it demonstrates how these forums or entanglements are more than human (in that they involve the material, the animal, the imagined, the geological, etc.) and that they are extensive, powerful, and enfolding in that they reach out in time and space in all sorts of directions (up and down, far and wide, to people, places, times, forces, events). They reach out and draw in, weaving relations-between, forming and consisting of patterns and patternings of motion and action. By providing a different vantage point, I hope to have been able to show (not least in the movement the chapter embodies and encourages) that it is in the interweaving, in the moving (in its broadest sense), in the relations between and the bound temporarily together, that we see self and world as part of the same undulating fabric, in constant formation, and space and time as fundamentally related.

NOTES

1. My warm thanks to the editors of this volume, Stella Souvatzi and Athena Hadji, for their constructive comments and very helpful advice on drafts of this chapter. Thanks are also due to participants in the Fellow's Writing-Up Seminar held at the University of Edinburgh's Institute for Advanced Studies in the Humanities in the early spring of 2012, as I received wonderfully thorough and thoughtful feedback from them when presenting an earlier version of this work.
2. The description that opens this chapter is an autobiographical account of an experience I had in the summer of 2011. I describe it here, in the hope that I have captured in the personal and poetic something of the rich sensory engagement with space and time that I felt it afforded me, and also that you, the reader, will allow me to pursue in this chapter a form of writing that reflects my interdisciplinary standpoint.
3. While my interpretation here does not try to do justice to the work's relationship with/in the world of environmental or land art to which Goldsworthy's

pieces are so often attributed, nor does it refer to other scholarship on the piece, I treat the work respectfully while using it to think upon space-time.

4. Moving between and relations-between are both subject to differing qualities: that is, a connection might be missed or exploitative, a relation strained or severed, etc.
5. This is a quality worth highlighting: this noticing of by-products, this creation of negative images from efforts to produce the positive, this attention to the excavated stuff as well as the emptied spaces, this peripheral vision or indirect glance.
6. 'Living through' translates into *erleben* in German, which also means experience.
7. See Harkness (2009) for more on this.
8. A consciousness for, not of, as this is an open caring or empathy, not a determined knowing.
9. By this 'other-nature' I mean the nonhuman orders, patterns, organisms, and forces of the world.
10. See Stewart Brand (1994) for more on this quality of architecture.
11. We have already seen it in our glimpse of quarrying, of course.
12. Ecologist Aldo Leopold wrote of the need to develop a land ethic that pivoted on a sense of community that was inclusive of the nonhuman (1949). With an ethic such as this, these nonhuman agencies are viewed as co-constituent and co-creator of an ever-shifting assemblage, of which the stone was a part.
13. For social theory that has been urging likewise in one way or another, see the Actor Network Theory of Bruno Latour (2005), the case for multisited ethnography with which to trace mobile things and flows by Marcus (1995), and the 'Social Life of Things' highlighted by Appadurai (1986).

REFERENCES

Appadurai, A. 1986. Introduction: commodities and the politics of value. In A. Appadurai (ed.), *The Social Life of Things: Commodities in Cultural Perspective,* 3–63. Cambridge: Cambridge University Press.

Bachelard, G. 1994. *The Poetics of Space: The Classic Look at How We Experience Intimate Places.* Boston: Beacon Press.

Baxter, S. 2003. *Revolutions in the Earth: James Hutton and the True Age of the World.* London: Phoenix.

Bennett, J. 2010. *Vibrant Matter: A Political Ecology of Things.* Durham and London: Duke University Press.

Brand, S. 1994. *How Buildings Learn: What Happens after They're Built.* New York and London: Penguin.

Cronon, W., ed. 1996. *Uncommon Ground: Toward Reinventing Nature.* New York: W.W. Norton.

Descola, P. and G. Palsson. 1996. *Nature and Society: Anthropological Perspectives.* London: Routledge.

DeSilvey, C. 2006. Observed decay: telling stories with mutable things. *Journal of Material Culture* 11(3): 318–338.

Dawdy, Shannon Lee. 2010. Clockpunk anthropology and the ruins of modernity. *Current Anthropology* 51(6): 761–793.

Edensor, T. 2011. Entangled agencies, material networks and repair in a building assemblage: the mutable stone of St Ann's Church, Manchester. *Transactions of the Institute of British Geographers* 36(2): 238–252.

Fischer, J. 2011. The quarry. In: Andrew Patrizio (ed.), *Stone: A Legacy and Inspiration for Art,* 28–69. London: Black Dog Publishing.

Goldsworthy, Andy. 2009. *Artists Statement for 'Stone House—Bonnington' http://www.jupiterartland.org/the_works/the_artists/andy_goldsworthy/* accessed December 9, 2011.

Graeber, D. 2001. *Toward an Anthropological Theory of Value: The False Coin of Our Own Dreams.* New York and Hampshire, England: Palgrave.

Harkness, R. 2009. *Thinking Building Dwelling: Examining Earthships in Taos and Fife.* Unpublished thesis awarded by the University of Aberdeen, Scotland.

Harrison, J. 2011. Body and stone. In: Andrew Patrizio (ed.), *Stone: A Legacy and Inspiration for Art,* 180–205. London: Black Dog Publishing.

Ingold, T. 2004. Buildings. In: S. Harrison S. Pile and N. Thrift (eds.), *Patterned Ground: Entanglements of Nature and Culture,* 238–240. London: Reaktion Books.

Ingold, T. 2011. *Being Alive: Essays on Movement, Knowledge and Description.* London and New York: Routledge.

Latour, B. 1993. *We Have Never Been Modern.* Cambridge, MA: Harvard University Press.

Latour, B. 2005. *Reassembling the Social: An Introduction to Actor-Network-Theory.* Oxford: Oxford University Press.

Lefebvre, H. 1991. *The Production of Space.* Oxford and Cambridge, MA: Blackwell.

Leopold, A. 1949. *A Sand County Almanac and Sketches Here and There.* New York: Oxford University Press.

Lévi-Strauss, C. 1996 [1962]. *The Savage Mind.* Oxford: Oxford University Press.

Marcus, G. 1995. Ethnography in/of the world system: the emergence of multi-sited ethnography'. *Annual Review of Anthropology* 24: 95–117.

Massey, D. 1993. Politics and space/time. In: M. Keith and S. Pile (eds.), *Place and the Politics of Identity,* 141–161. London: Routledge.

Massey, D. 2005. *For Space.* London; Thousand Oaks; New Delhi: Sage Publications.

McFadyen, L. 2006. Building technologies: quick and slow architectures and early Neolithic long barrow sites in southern Britain. *Archaeological Review from Cambridge* 21(1): 117–134.

McFadyen, L. 2007. Mobile spaces of Mesolithic Britain. *Home Cultures* 4(2): 117–128.

Pallasmaa, J. 2005. *The Eyes of the Skin: Architecture and the Senses.* West Sussex, England: John Wiley and Sons.

Pallasmaa, J. 2007. New architectural horizons. *Architectural Design* 77: 16–23.

Stewart, K. 2007. *Ordinary Affects.* Durham and London: Duke University Press.

5 Bodies of Evidence? Re-imagining a Phenomenological Approach to Space and Time in Prehistoric Malta

Eimear Meegan

INTRODUCTION

The use of phenomenological theory has become somewhat of a taboo within archaeological research. Once a common characteristic of post-processual approaches to space, and to a much lesser extent time, the last number of years have seen its application grind almost to a halt, amidst increasing criticism regarding what is perceived to varying degrees as its incompatibility with archaeological inquiry (Brück 1998; 2005; Fleming 1999; 2005). Taking this criticism as a point of departure, this chapter asks whether in abandoning phenomenological principles altogether we would, perhaps, be throwing the proverbial baby out with the bath water. Can phenomenology in fact play a significant and fruitful role in the multiscalar approach to space and time that is advocated in this volume? In exploring this possibility, I briefly consider what drew archaeologists to this particular strand of philosophy in the first instance and how it has typically been employed in relation to spatial and temporal aspects of material culture. Referring back to the primary philosophical texts, it becomes apparent that phenomenological archaeology, as we know it, is a less than accurate reflection of the underlying principles of the philosophy. Indeed, it is argued here that so-called landscape phenomenology, rather than illustrating the potential value of phenomenologically driven investigations into space and time, has worked to obscure it. To remedy this, I suggest a reimagining of our conventional understanding of phenomenology, one that I feel shows a greater appreciation of what Thomas (2006: 43) has described as "the subtlety of phenomenological thought"; and one that, I propose, has much to offer as a means of engaging with, and thinking through, the convergence or interweaving of space, time, and action in the past; and the many scales at which this interplay worked. Taking monumental construction in Late Neolithic Malta as a case study, this proposition is put to the test as the spatiality and temporality of temple building are viewed through a recalibrated phenomenological lens.

PHENOMENOLOGY AND THE POST-PROCESSUAL MOVEMENT

During the 1980s and 1990s, as more and more archaeologists began to challenge the customary treatment of time, and particularly space, as mere 'containers for action' (e.g., Barrett 1994; Bender 1992; Hodder 1982; Shanks and Tilley 1987; 1988), phenomenology's appeal was its opposition to Descartes's subject/object division, and the theorization of time and space that had come to dominate modern thinking as a consequence. The Cartesian thesis is of course famously built on the premise that 'I think therefore I am'. This exaltation of the thinking subject inevitably resulted in the corresponding objectification of the physical world and all its attributes, and thus goes a long way in accounting for the traditional archaeological portrayal of space and time as neutral and divorced from the meaning-laden activities of the human subject. The aim of phenomenologists such as Heidegger and Merleau-Ponty was to overcome this dichotomy, which they saw as running totally contrary to the reality of the human condition (Heidegger 1962 [1927]: 73). As they observed, we simply cannot abstract ourselves from the material world. We are "world involved" (Moran 2000: 233), immutably enmeshed through our bodies in space and in time, which, as a consequence, are never neutral and can never be disengaged from our actions. Accordingly, phenomenology sought to put the subject back into the material world, to situate the body both temporally and spatially (Husserl 1970 [1900]; Merleau-Ponty 2002 [1945]: xii–xiii). Only by doing this, it was argued, would it be possible to explore the co-constitutive discourse that exists between time, space, and action, and which is the essence of existence. Given this reconceptualization of space and time as playing an active and essential role in the making of meaning, together with the foregrounding of material engagement, the potential value of phenomenological theory as a means of redressing the objectification of traditional Cartesian approaches is undeniable. Its widespread influence on post-processual thought is thus easily understood. Helped largely by Foucault's (1977) seminal work on the inherent power of structured space, this influence translated largely in spatial terms, specifically as explorations into the way in which the experience of archaeological landscapes, or simply monuments, was conditioned or determined by the very form they took (Bradley 1993; Richards and Parker Pearson 1994; Thomas 1991; Tilley 1993). The significance of this particular approach lay in the light that it was believed to shed on the motivations and ideology of those responsible for the structuring of the space in question. As Thomas (2006: 43) has observed, though, not all of those who have drawn on phenomenology in this way have been strictly explicit in doing so. There are, however, a number of people who have been most unambiguous in their advocacy of it. Among these, Christopher Tilley (1994; 2004; 2008) has undoubtedly been the most vocal.

Published in 1994, Tilley's *A Phenomenology of Landscape: Places, Paths and Monuments* was the first full volume to explicitly use phenomenological theory as the basis of its investigation into archaeological space. In this hugely influential and pioneering study, the author sets out his own reading of the primary philosophical texts, and on the basis of this understanding, as is well known, he advocates the use of the contemporary body as a means of exploring past experience of prehistoric spaces. Drawing heavily on the work of both Heidegger and Merleau-Ponty, he recognizes space as inescapably humanized and dialectical, constituted and constitutive by virtue of our bodily insertion in the world (Tilley 1994: 17). Framing these phenomenological insights within the context of archaeological inquiry, he then goes on to argue that the "bones of the land" have remained largely unchanged since Mesolithic times, as has the body, which continues to be the "privileged vantage point from which we apprehend the world" (Tilley 1994: 73–74). By this reckoning, therefore, we already have at our disposal the means to explore the prehistoric dialectic between space and action. In fact, according to Tilley, we are the means.

There was, as Thomas (2006: 55) puts it, "a fresh and compelling character" to this use of personal bodily experience as a means of "thinking through the evidence encountered in the field" (Tilley 1994: 71); and the ease with which this kind of an approach could be implemented made it all the more appealing to others. Indeed, in the years following the publication of his work, Tilley's embodied narrative inspired numerous archaeologists to carry out very similar studies on a wide range of prehistoric landscapes, mainly within Britain. As Thomas has previously highlighted, though, not all of these have "the philosophical grounding of the original" (Thomas 2006: 55). In fact, I would go so far as to say that most seem to have accepted, without question, Tilley's reading of original phenomenological texts, and offer little more than a cursory nod in their direction (Brophy 1999; Cummings 2002; Fraser 1998; Watson 2001). *A Phenomenology of Landscape* thus became almost a handbook of what was perceived as the phenomenological method; and to this day, the power of this seminal work is such that, within archaeology, the use of phenomenological theory is still equated with an embodied methodology.

THE DECLINE AND FALL OF THE EMBODIED NARRATIVE

The many and varied problems inherent in this kind of narrative are well rehearsed (e.g., Brück 1998; 2005; Fleming 1999; 2005), so it is unnecessary to relay them here in any detail. Suffice to say, the overriding concerns that have been raised are in relation to the concepts of a universal body and a timeless, unchanging landscape, both of which are implied by the entirely subjective nature of the methodology described. These concerns, despite having been raised more than a decade ago, have yet to be satisfactorily addressed, and consequently seem to have been taken by many as proof that phenomenology and archaeology are simply, and irreconcilably, incompatible. The approach's ailing popularity is thus to be expected, but perhaps it is not beyond revival; perhaps the problem lies not with phenomenology itself but with the manner in which it has tended to be used or, as John Barrett and Ilhong Ko (2009) have suggested, misused by archaeologists.

For Barrett and Ko (2009), this misuse stems from what they perceive to be a contradictory interpretation of Heidegger's concept of 'Being-in-the-world', which is the central premise of Tilley's embodied narrative. As they have observed, this concept is described by Tilley (1994: 12) as a process of objectification in which people "objectify the world by setting themselves apart from it". It is this notion of *Being* as characterized by a distancing from the world that Barrett and Ko see as the root of the underlying confusion. This interpretation, they have argued, approaches *Being* specifically from the point of view of what Heidegger referred to as '*Being* present at hand' and thus fundamentally misconstrues his philosophy (Barrett and Ko 2009: 281–282).

'Being present at hand', as Relph has explained, is a mode of existence that implies a distance between humans and nature—"[a]n element of self-awareness inserts itself between me and my world" (Relph 1985: 18). It is a reflective or theoretical way of *Being* that takes a step back from the immanence of our day-to-day practical encounter with the world—the so-called 'Being ready to hand'. It is the former, reflective state of *Being* that traditional Cartesian narratives take as their point of departure, and that Heidegger saw as the source of the subject-object division. 'Being present at hand' is thus precisely the state his hermeneutical phenomenology was intended to overcome. Unsurprisingly, therefore, the focus of his work is the quotidian, prereflective activity of the 'ready to hand' that he saw as the source of the theoretical and the fundamental basis of 'Being-in-the-world' (Heidegger 1962 [1927]: 61); it is the way we "reach for an object to act as a hammer, we see a tree as a source of wood or shelter from the rain, and so on" (Moran 2000: 233). In view of this, Barrett and Ko (2009: 281) have interpreted Tilley's emphasis of objectification and distancing as breaking with Heidegger's philosophy rather than adhering to it; and it is this apparent misinterpretation, and the objectification it implies, that they regard as mistakenly legitimizing an embodied approach to space or, as they put it, a 'subjective epistemology' (ibid.: 279).

Undeniably, the inherent subjectivity of Tilley's methodology suggests a spatial objectification that seems altogether at odds with phenomenological theory. However, on further reflection, and contrary to Barrett and Ko's argument, misinterpretation of Heidegger's 'Being-in-the-world' does not appear to account for this contradiction. 'Being-in-the-world' does indeed describe the process by which we step back from the prereflective nature of day-to-day praxis, and thus come to live our lives in a contemplative state, holding the world, objectively, at arms length. The source of the problem does not, therefore, lie in any misapprehension of the structure of Being-in-the-world. Rather, it seems to come from the specious inference that it is possible to deconstruct *Being* by working backwards from the reflective to the prereflective, from an objective space to a contingent one.

This is clearly implied in Tilley's identification of a number of spatial classes, namely somatic, perceptual, existential, architectural, and cognitive. The first two of these categories describe the space of "habitual and unselfconscious" (Tilley 1994: 16), day-to-day activity; while existential and architectural spaces are bounded or constructed, "made tangible, visible and sensible" (Tilley 1994: 17). Finally, cognitive space "provides a basis for reflection and theorization with regard to understanding the others" (Tilley 1994: 17); it is, Tilley claims (1994: 17), the space of "discussion and analysis" in which steps can literally be retraced, revealing the motivations of those whose habitual movements made prehistoric landscapes. According to Heidegger's philosophy, though, it is not possible to move backwards from, or undo, reflection in this way. As Barrett and Ko (2009: 288) put it, "time's arrow is not reversible". Thus, Heidegger famously advocated the 'destruction' or 'dismantling' of the philosophical tradition in order that we might overcome theorization and engage directly with the habitual, taken-for-granted actions of 'Being-present-at-hand'—"those primordial experiences in which we achieved our first ways of determining the nature of Being" (Heidegger 1962 [1927]: 44). It is Tilley's proposed regression, therefore, that jars with Heidegger's philosophy; but what is the driving force behind this backtracking? The answer to this perhaps lies in the nature of the question phenomenological reasoning has tended to be employed to address.

THE ANSWER IS IN THE QUESTION

As previously mentioned, phenomenology has generally been used to explore how the physical form of monuments or landscapes impacts or determines the manner in which they can be physically experienced and understood, and what this might mean in terms of social hierarchy and the distribution of power. This kind of question in itself objectifies space and simply perpetuates the division of subject and object, making it fundamentally inappropriate to a phenomenological inquiry. It is a line of thinking that, as noted above, was heavily influenced by Foucault's (1977) seminal work on power and discipline, and seemingly it had started to seep into the archaeological psyche before any serious or concerted effort had been made to engage analytically with the details of phenomenological theory (e.g., Bradley 1993; Miller and Tilley 1984; Thomas 1990; Tilley 1993), and before Tilley had fully established the theoretical foundations of his embodied 'landscape phenomenology' (1994). So, it is certainly plausible to suggest that his approach was skewed by his continued use of a wholly unsuitable line of questioning that left him with little option but to work backwards from an objectified, timeless landscape, using the only thing that links past experience of place with the present: the body. The centrality of the body within phenomenological theory would only have served to reinforce the logic of this embodied narrative. The end result is that rather than providing genuine phenomenological insight, this approach actually obscures what phenomenology really has to offer archaeological research, that is, the light it sheds on the manner in which the possibilities of being, and thus the possibilities of space and of time, are disclosed to us through our everyday, taken-for-granted encounter with the world.

Though embodied approaches continue to be held up as the hallmark of phenomenological archaeology, some alternative suggestions have in fact been made as to how else phenomenological theory might be used to inform material culture studies; and these suggestions reflect much more closely the ideas underpinning the philosophy. In keeping with the interpretation set out above, Thomas (2006: 53) and Barrett and Ko (2009: 288) have recognized that it is only by engaging with material culture as process rather than product that we can begin to approach a truly phenomenological archaeology. They cite the work of Ingold (2000) and MacFadyen (2006; 2007), respectively, which, though not explicitly phenomenological, inadvertently provide a taste of how phenomenology can and should be used within archaeology. There is, however, a lot more to be said in terms of teasing out a clearly developed phenomenological archaeology that utilizes the principles of the philosophy to their fullest effects, particularly when it comes to aspects of space and time. To do this, I will now turn my attention to the process of megalithic construction in Late Neolithic Malta.

MONUMENTAL MALTA

Rising from the depths of the western Mediterranean Sea, the Maltese archipelago comprises three rough-hewn limestone islands, where widespread and differential outcropping of coralline and globigerina limestone has caused deep faults or fissures, making for a very distinctive landscape indeed. The remains of some thirty stone-built temples are scattered throughout the larger islands of Malta and Gozo and bear witness to the efflorescence of a distinct cultural group between the mid fourth and mid third millennia BC, when it appears to have come to quite an abrupt end (Figure 5.1).

Though these impressively monumental buildings are by no means identical, they do display a considerable degree of uniformity in terms of their layout, the techniques and materials used in their construction, and their decoration and furnishings. As such, they are thought to have housed relatively standardized ritual practices, most likely associated with the living as contemporary funerary activity took the form of collective inhumation, either in rock-cut tombs or in much more elaborate hypogea (Evans 1959).

The earliest temples are roughly trefoil in plan, as at Ta' Ħaġrat, where a short passageway leads into a central space that is flanked by a pair of lateral

Figure 5.1 The entrance to one of three temple buildings at Mnajdra.

apses and terminates in a third one (Evans 1959: 85–86). Later structures are considerably more complex in form and indicate a gradual elaboration of the standard trefoil arrangement that culminated in the construction of the confusing, poly-lobed temples at Tarxien and Ħaġar Qim (Evans 1959: 85–86). Here, up to three pairs of apses can be identified within a single structure (Figure 5.2). Within these more labyrinthine temples, access to certain spaces appears to have been increasingly restricted, with V- and U-shaped "rope-holes" often found in the jambstones of later structures suggesting the addition of a screen or a door across particular thresholds to allow the goings on within to be shielded from prying eyes (Evans 1959: 136).

Though temple layout seems to have become increasingly convoluted and restricting, the materials used remained the same throughout. Coarse, durable coralline limestone was used, where available, for the bulk of the building, with the softer and more malleable globigerina often used for doorjambs or decorative elements such as altars (Clark 2004: 368). The appearance of such sophisticated freestanding architecture at such an early date has long intrigued prehistorians, thus yielding a wealth of research into so-called temple society. This research, however, has tended to approach the temples from an entirely objective standpoint, as finished products representing singular expressions of space and time. Even when details of construction have been considered, it is the practical and functional aspects of the architecture and engineering that dominate the discussion (Clark 1998; 2004; Torpiano 2004). Little or no thought has been given to the flow and dynamics of construction as a habitual process, and what this may have

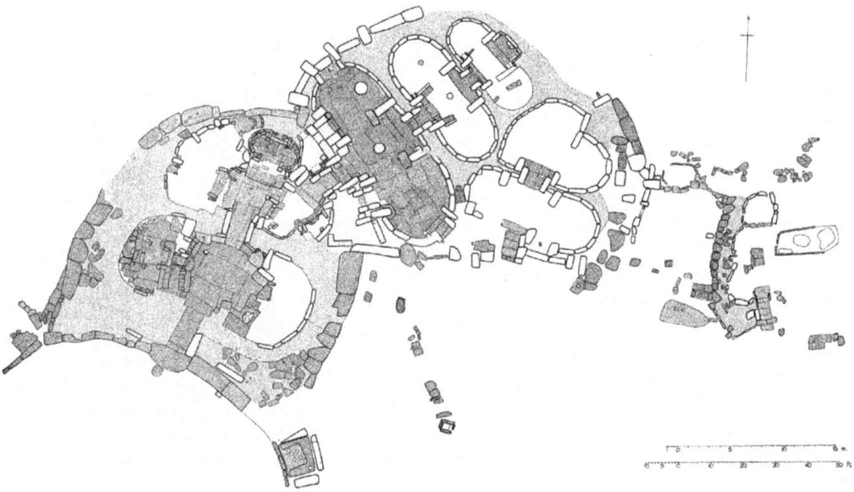

Figure 5.2 The complex ground plan of the Tarxien temples. After Evans 1971, plan 30A.

meant in terms of shifting scales of space and time. This is not to say that the latter approach should be seen as an alternative to earlier more empirical work. Rather, it should be seen as building on the invaluable evidence it provides us with. When all is said and done, it is only after the practical details of building have been discussed and debated that can we even begin to think about what it might have been like to live through them. For this reason, they are summarized below.

Unsurprisingly, the stone used in temple building appears, for the most part, to have been sourced close to the sites of construction (Clark 1998: 143). Given the nature of Maltese geology, this was not particularly restricting. In Malta, you are never very far from a coralline or a globigerina outcrop. The two temples at Ġgantija, perched on top of the Xagħra plateau on Gozo, are noteworthy in that the globigerina blocks used here are likely to have been sourced on the outskirts of Xewkija, about 1.2 kilometers away from the site (Clark 1998: 153). The bulk of the material at all sites is thought to have been quarried by wetting wooden wedges that had been inserted into specially cut slots, causing the wood to expand and split the block from its parent outcrop (Clark 1998: 162–164) (Figure 5.3).

Once freed from its source, the block could then be levered out and brought to the site. At Ġgantija, most of the blocks used weighed between 0.5 and 2 metric tons, though there are a considerable number that tip the scales well beyond this. Clark has estimated that while blocks under 2 metric tons may have been carried by groups of up to thirty-six people, anything heavier than this would have had to be moved along the ground. This, he suggests, was executed either by means of wooden levers and stone rollers,

Figure 5.3 A decorated block from one of the Ħaġar Qim temples.

or by means of a wooden sledge that ran along specially laid tracks (Clark 1998: 168–170).

While the very hard nature of the coralline limestone seems to have precluded any kind of working (Clark 1998: 126), the much softer globigerina blocks seem to have been extensively shaped and often decorated before being moved into position (Clark 1998: 181). At Ġgantija and Skorba, the remains of limestone plaster were found still adhering to some of the undressed coralline walls, suggesting that the rough exterior of the stone had been smoothed over. Known locally as torba, this plaster has a long history of use in Malta. It was made by grinding globigerina into a powder and adding water to form a paste that was applied in layers, becoming very hard as it dried (Clark 1998: 184). Though some doubts have been expressed as to how often torba was used on the walls, there is extensive evidence to support its very frequent use on temple floors. It has also been suggested that, once prepared, the plastered walls and floors may have been painted using imported red ochre, traces of which were found at Ġgantija, Skorba, and Tarxien (Clark 1998: 126–130). This idea is given further credence by extensive use of ochre to decorate the ceiling of the Hal Saflieni hypogeum.

Though no indisputable evidence of roofing survives, the pristine condition of the decorated stones excavated at Tarxien (the first of only a handful of temples to be excavated scientifically) does suggest that they were in fact covered over. The discovery of a miniature model of a seemingly stone-roofed temple during the excavations at Ta' Ħaġrat lends further support to this idea, as does the depiction of a roofed temple etched into a globigerina upright at Mnajdra. Given the lack of direct material evidence, though, it is not surprising that the questions of how these sometimes vast, lobed spaces were spanned, and with what materials, have been debated at length for more than seventy years. The most recent hypothesis has been put forward by Clark (1998: 120–126). While he concedes that a type of stone slab corbelling may have been used in smaller, earlier structures, he argues that the use of pitched timber beams covered with brush is a much more likely scenario for the larger, later temples. The rather extensive evidence of burning on the walls of the central temple at Tarxien, as well as Trump's (1966: 6–7) interpretation of a "sterile layer of crumbly orange clay" as "presumably the decayed remains of a roof of clay on brush", seem to fit well with this hypothesis. However, Clark (1998: 123) suggests that the brush roofs may well have been sealed with torba rather than clay, as it would have been a far more effective means of waterproofing, and would also have allowed for the addition of guttering that, he argues, would have been essential given that Malta is prone to torrential downpours.

Faced with such advanced techniques at such an early stage in Mediterranean prehistory, it is easy to get caught up in the mechanics of the tasks described above and to forget that these activities are brought into being by material bodies moving through a material world; and that they are rendered routine and unconscious by their repetition time after time, day in and day out. Phenomenology reminds us of this by putting the body back into the world and into motion. There are no subjects or objects where this unconscious movement is concerned, there is simply an intertwining of what Merleau-Ponty (2004 [1968]: 255) describes as the flesh of the body and the flesh of the world; and this intertwining or so-called chiasm "is not limited to submitting passively to space and time, it actively assumes them, it takes them up in their basic significance which is obscured by the commonplaceness of established situations" (Merleau-Ponty 1962 [1945]: 117). The sundry nature of the physical phrasing demanded by the various construction activities described thus suggests that temple building brought several configurations of space and time into play; but it is only by engaging with the movement of construction as the fleshy interdigitation of the builders with the material world that we can begin to shed any light on the temporal and spatial flux implied.

BODIES OF FLESH AND STONE: THE INTERTWINING OF CONSTRUCTION

If we reconsider the practical and functional details of the building process in light of a material intertwining, it is the sheer level and intimacy of contact between the builders and the stone they were using that is particularly striking. Where once we saw the abstract execution of mechanical tasks, we now see a prolonged meeting of flesh and stone. Right from the outset, the builders' movement or body schema was entirely caught up in stone. Even the selection of a suitable outcrop to quarry meant them moving in toward and around the stone, running their hands over and back across its surface to familiarize themselves with it. This physical encounter implies a kind of friction or tension between the two, which would have become increasingly acute as the construction process unfolded. Whether they were quarrying and maneuvering a block of coralline, shaping or decorating globigerina, or in fact grinding lumps of globigerina to make torba, the builders' entire being was directed toward and consumed by stone. Even when lines of direct contact were broken by the use of wooden levers and perhaps ropes, their moving bodies continued to be guided and defined by their relationship with stone. As well as this, limestone is of course known for its very chalky nature and the copious amount of dust that is thrown up as is it extracted and worked. Thus, whether or not the builders were actually touching the stone, they were undoubtedly still connected with it through the ether. As they moved around the site, the limestone dust lingering in the air would have settled on their skin, in their hair, and under their nails. Indeed, the simple and essential act of breathing would have drawn this residue into their lungs and into their bloodstream. This last observation captures, most effectively perhaps, the transformative nature of this intertwining. As explained above, there are no subjects or objects at this level of existence and so, locked together in the movement of construction, the builders and the stone they were using came to sustain each other from within. In this way, the boundaries between them would have become blurred, making it difficult to tell where flesh ended and stone began.

This blurring of boundaries has previously been explored by Lesley McFadyen (2006; 2007), whose work on Long Barrow construction in Late Neolithic Britain, as lauded by Barrett and Ko (2009), advocates an approach that takes architecture as action rather than object. Describing the building of the structure at Ascott-under-Wychwood, she explains how the setting of stones on end within the mound transformed this material from structurally sound and stable to precarious and in need of support, which then came in the form of dumps of material. Building in this way meant that these stones would have to be propped up and held in position as the supporting materials were rapidly arranged around them. The speed at which this would have happened has led McFadyen (2007: 353) to christen this type of building 'quick architecture', and she provides a very visceral description of how this 'quick architecture' may have played out within society, discussing the way in which subject and object might have become interdependent, morphing fluidly into each other and opening people up to the possibilities of being.

> Imagine the intensity and entwined movements of people and things propping each other and everything else up in close proximity. Bodies building picked up the soil and dust of the worked earth. There was an

> intermingling of people, grass,wood, rubble, marl, clay and stone. In this space, bodies were made to matter through a negotiation of junctions with other materials or living things. (McFadyen 2007: 353).

The notion of the intermingling of materials implies a relationship based on exchange, in which the materials involved bleed into each other or come to assume each other's properties or qualities. This kind of reciprocal relationship between people and the things they come into contact with has famously been explored by Marilyn Strathern (1988) in her groundbreaking work on the Kula ring in Melanesia. A renowned and powerful exchange network, the Kula ring is based on the circulation of a very specific type of shell necklaces and armbands by the men of the Trobriand Islands. The significance of these so-called Kula valuables, as Strathern explains, can be understood in terms of both metaphor and, crucially, metonymy (Strathern 1988: 196). They not only stand for or represent the men who have made them and those who have exchanged them, but they are in fact considered to *be* parts of those individuals, accumulating and carrying with them something of the essence of every man they encounter as they move within the ring of exchange. These men are in turn transformed by their encounters with Kula objects and accrue to themselves something of the character of every necklace and armband that passes through their hands. Admittedly, exchange within the Kula network is inherently political. It proceeds in a carefully considered manner and is thus far removed from the unconscious intermingling described by McFadyen. However, its mutually transformative nature is similarly based on the blurring of boundaries between man and materials and should perhaps be understood as the residue of a less reflective type of being. By this token, the intertwining of flesh and stone in prehistoric Malta can perhaps be understood in terms of the temple builders taking on the material essence of Maltese limestone and vice versa; but what implications does this have with regard to the spatial and temporal configuration brought to bear through the construction process?

MAKING SPACE FOR US AND TIME FOR ME: SCALES OF SPACE AND TIME IN TEMPLE BUILDING SOCIETY

Though there is some debate surrounding the extent to which Malta's geology was covered by now eroded soil during the temple period, there can be little doubt in our minds that, for the builders, stone was a defining feature of their island home, if not *the* defining feature. Everywhere they turned they were faced with stone—along the bare, rocky coastlines, in the deep fissures that cut through and traversed islands, and in the caves that were home to early settlers. Thus, the material intertwining of construction possibly implies not just a dialectic between flesh and stone but a dialectic between the islanders and their home. In building the temples, therefore, they were forging an intimate relationship with the landscape, and in the process they were forging a space and time of belonging, locality, and commonality. If we also consider the very close physical quarters at which many of the builders must undoubtedly have been working, it is likely that they were as in tune with each other's movements as they were with their own, perhaps to the extent that they saw each other as extensions of themselves. Thus, the blurring of the boundaries between stone and flesh may well have extended to the boundaries between the builders themselves, creating a space and time of community and kinship, which could only have served to strengthen that of belonging referred to above. These different but simultaneous spatial and temporal configurations would inevitably have made for a very strongly rooted and widely shared island identity. Having said this, though, if, as Clark (1998: 129–130) has suggested, the plastered walls and floors of the temples were painted with ground ochre from Sicily, the builders intertwining with this foreign substance would have created a space and time that extended beyond the local and the community and embraced the wider world. This was perhaps a space and time of memory and the ancestral past as the temple builders' ancestors are believed to have migrated from Sicily in the mid sixth millennium BC. The acts of grinding and of painting, moreover, introduce yet another kind of intertwining that differs from the communal sort described above.

There were in fact numerous tasks that would have been performed single-handedly or perhaps in small, loosely formed groups, rather than in the large teams it would have taken to quarry and maneuver the stone uprights. Like the grinding of ochre, the processing of globigerina for torba would also have involved working as an individual, most likely using quern stones, examples of which have been found associated with the temples at Skorba (Trump 1966: 15) and possibly Kordin (Evans 1971: 23). This is not to say, however, that the processing of these materials was necessarily an entirely solitary task. Based on evidence of thickness, Clark has suggested that a total of thirty-one cubic meters of torba plaster was required to cover the walls and floors of the two adjacent temples at Ġgantija (1998: 131). This, he estimates, would have taken twenty person days per metric ton to grind (Clark 1998: 185). Thus, it is certainly possible, if not likely, that the grinding of limestone and ochre was carried out by a considerable number of people at any given site. This being the case, they would surely have worked close by each other, perhaps in a designated area, which would inevitably have fostered communication and a sense of fellowship among them. It is certainly hard to imagine that they would have knelt or sat in total silence, working their way through the day without uttering a word to each other. As O'Connor (2009: 154), drawing on Watts (forthcoming), has observed, the social significance of grinding stone "in many contemporary societies is a context for learning and singing". Ultimately, though, it is a very personal act in which, as O'Connor (2009: 158) puts it, "the mind is free to wander and travel, to mull over and work through ideas, thoughts and memories";

and consequently, it constitutes a spatial and temporal configuration of individual introspection and contemplation. This space and time of individual intertwining with the materials of construction was perhaps further exaggerated or intensified through the decorative carving of blocks and also the painting of walls and floors. It is very difficult to understand the patterns we see today as anything other than the discrete work of individuals, whether painted on the ceiling of Hal Saflieni or etched into stone at one of the temples. There is a compositional coherence in each instance that would seem to preclude group work and points to the existence of a number of, for want of a better word, artists who worked alone.

The complex material intertwining of Maltese temple building thus hints at a variety of spatial and temporal scales, operating simultaneously and ranging from the local and the communal, to the exotic and the individual. Inevitably, though, the balance struck between these was a dynamic one, with different configurations coming to the fore at different times. In the early stages of the temple period, it was perhaps the local and the communal that dominated. That this was the case is suggested by the large numbers of quite formulaic temples scattered across the two main islands. As building continued, though, and the temples became more and more convoluted and inaccessible to all but very small groups of people, all the signs point to an increasingly hierarchical or bureaucratic social structure and the rise of an elite (Bonanno et al. 1990). Communal space and time thus seems to have been gradually deposed and replaced by a configuration that allowed, more and more, for the loosening of community ties and for progressively smaller groups of individuals to come to the fore. It would seem, then, that the more individually focused intertwining of construction gradually came to overpower that which had fostered the growth of a tightly knit island identity, and it was perhaps this that began to undermine the integrity of society, leading ultimately to the implosion of temple building society in the mid third millennium.

CONCLUSION

In the past, phenomenological approaches to space have been on the receiving end of some very harsh but largely deserved criticism. Here, however, I have argued that the many issues flagged by past critiques can be accounted for by the use of an embodied methodology that is entirely at odds with the underlying principles of the philosophy itself. Returning to the primary texts, we have seen that it is only by engaging with habitual material practice as it unfolds day-to-day that we can lay any real claim to following a genuine phenomenological course. We have also seen that the value of this approach, and in particular Merleau-Ponty's concept of practice as an unconscious intertwining of flesh, is the insight it can provide into the many scales and permutations of space and time that pervaded and defined life in the past.

If this capacity is to be fully recognized, though, and if phenomenological theory is to have any future within archaeology, we must, once and for all, put paid to the idea of phenomenology as a distinct methodology and start to consider it instead as a way of thinking about the dynamic interplay of space, time, and action. In doing this, however, we must not forget that it is Tilley's pioneering work that, in spite of its shortcomings, has paved the way for us.

ACKNOWLEDGMENTS

My thanks to the editors for their patience and comments, and also to Gabriel Cooney and Joanna Brück for their comments on an earlier draft of this chapter. My research is kindly supported by the UCD John Hume Institute for Global Irish Studies.

REFERENCES

Barrett, J. 1994. *Fragments from Antiquity: An Archaeology of Social Life in Britain 2900–1200 BC*. Oxford: Blackwell.

Barrett, J. and I. Ko. 2009. A phenomenology of landscape: a crisis in British archaeology? *Journal of Social Archaeology* 9(3): 275–294.

Bender, B. 1992. Theorizing landscapes, and the prehistoric landscapes of Stonehenge. *Man* 27: 735–755.

Bender, B. 1998. *Stonehenge: Making Space*. Oxford: Berg.

Bonanno, A., T. Gouder, C. Malone and S. Stoddart. 1990. Monuments in an island society: the Maltese context. *World Archaeology* 22(2): 190–205.

Bradley, R. 1993. *Altering the Earth: The Origins of Monuments in Britain and Continental Europe*. Edinburgh: Society of Antiquaries of Scotland.

Brophy, K. 1999. Seeing the cursus as a symbolic river. *British Archaeology* 44: 6–7.

Brück, J. 1998. In the footsteps of the ancestors: a review of Christopher Tilley's 'A Phenomenology of Landscape'. *Archaeological Review from Cambridge* 15: 23–36.

Brück, J. 2005. Experiencing the past? The development of a phenomenological archaeology in British prehistory. *Archaeological Dialogues* 12: 45–72.

Clark, D. 1998. *Insular Monument Building: A Cause of Social Stress? The Case of Prehistoric Malta*. Ph.D. thesis, University of Bristol.

Clark, D. 2004. Construction techniques. In: D. Cilia (ed.), *Malta before History*, 367–377. Miranda: Sliema.

Cummings, V. 2002. Between mountains and sea: a reconsideration of the Neolithic monument of south-west Scotland. *Proceedings of the Prehistoric Society* 68: 125–146.

Evans, J. D. 1959. *Malta*. London: Thames and Hudson.

Evans, J. D. 1971. *The Prehistoric Antiquities of the Maltese Islands: A Survey*. London: Athlone Press.

Fraser, S. M. 1998. The public forum and the space between: the materiality of social strategy in the Irish Neolithic. *Proceedings of the Prehistoric Society* 64: 203–224.

Fleming, A. 1999. Phenomenology and the megaliths of Wales: a dreaming too far. *Oxford Journal of Archaeology* 18: 119–125.

Fleming, A. 2005. Megaliths and post-modernism: the case of Wales. *Antiquity* 79: 921–932.

Foucault, M. 1977. *Discipline and Punish: The Birth of the Prison*. London: Penguin.

Heidegger, M. 1962[1927]. *Being and Time*. Oxford and Cambridge (U.S.): Blackwell.

Hodder, I. 1982. *Symbols in Action*. Cambridge: Cambridge University Press.

Husserl, E. 1970 [1900]. *Logical Investigations*. Translated by J. N. Findlay. New York: Humanities Press.

Ingold, T. 2000. *The Perception of the Environment: Essays in Livelihood, Dwelling and Skill*. London: Routledge.

McFadyen, L. 2006. Building technologies, quick architecture and Early Neolithic Long Barrow sites in southern Britain. *Archaeological Review from Cambridge* 21(1): 117–134.

McFadyen, L. 2007. Making architecture. In: D. Benson and A. Whittle (eds.), Building Memories: The Neolithic Cotswold Long Barrow at Ascott-under-Wychwood, Oxfordshire, 348–354. Oxford: Oxbow.

Merleau-Ponty, M. 2002 [1945]. *A Phenomenology of Perception*. London and New York: Routledge.

Merleau-Ponty, M. 2004 [1968]. The intertwining—the chiasm: selection from the visible and the invisible. In T. Baldwin (ed.), *Maurice Merleau Ponty: Basic Writings*, 247–272.

Miller, D. and C. Tilley. 1984. *Ideology, Power and Prehistory*. Cambridge: Cambridge University Press.

Moran, D. 2000. *An Introduction to Phenomenology*. Routledge: London and New York.

O'Connor, B. 2009. Re-collected objects: carved worked and unworked stone in Bronze Age funerary monuments. In: B. O'Connor, G. Cooney and J. Chapman (eds.), *Materialitas: Working Stone, Carving Identity*, 147–161. Oxford: The Prehistoric Society and Oxbow Books.

Relph, E. 1985. Geographical experiences and being-in-the-world. In: D. Seamon and R. Mugeraur (eds.), *Dwelling, Place and Environment: Towards a Phenomenology of Person and World*, 15–31. Dordrecht, Boston and Lancaster: Nijhoff.

Richards, C. and M. Parker Pearson. 1994. *Architecture and Order: Approaches to Social Space*. London: Routledge.

Shanks, M. and C. Tilley. 1987. *Re-constructing Archaeology: Theory and Practice*. Cambridge: Cambridge University Press.

Shanks, M. and C. Tilley. 1988. *Social Theory and Archaeology*. Cambridge: Polity in association with Blackwell.

Strathern, M. 1988. *The Gender of the Gift*. Berkley and London: University of California Press.

Thomas, J. 1990. Monuments from the inside: the case of the Irish megalithic tombs. *World Archaeology* 22(2): 168–178.

Thomas, J. 1991. Reading the Neolithic. *Anthropology Today* 7: 9–11.

Thomas, J. 2006. Phenomenology and material culture. In: C. Tilley, W. Keane, S. Kuechler, M. Rowlands and P. Spyer (eds.), *Handbook of Material Culture*, 43–59. London, California and New Delhi: Sage.

Tilley, C. 1993. Art, architecture, landscape [Neolithic Sweden]. In: B. Bender (ed.), *Landscape, Politics and Perspectives*, 49–84. Oxford: Berg.

Tilley, C. 1994. *A Phenomenology of Landscape: Place, Paths and Monuments*. Oxford: Berg.

Tilley, C. 2004. *The Materiality of Stone: Explorations in Landscape Phenomenology.* Oxford: Berg.
Tilley, C. 2008. *Body and Image: Explorations in Landscape Phenomenology* 2. California: Left Coast Press.
Torpiano, A. 2004. The construction of the megalithic temples. In: D. Cilia (ed.), *Malta before History,* 347–365. Sliema: Miranda.
Trump, D. 1966. *Skorba.* Oxford: The Society of Antiquaries.
Watson, A. 2001. Composing Avebury. *World Archaeology* 33: 296–314.
Watts, S. forthcoming. *Contexts of Deposition: The Structured Deposition of Querns in the South West of England in the Prehistoric Period.* Ph.D. thesis, University of Exeter.

6 Time and Place, Memory, and Identity in the Early Neolithic of Southwest Asia

Trevor Watkins

Between about 22,000 and 10,000 years ago in southwest Asia, the age-old fission-fusion mode of social organization of mobile hunter-gatherer bands was transformed. Groups relied increasingly on stored harvests of cereals and pulses, becoming more and more tied to the place where those resources were available, forming permanent settlements where they stored their food. They used the framework of sedentism and storage as the basis for building larger, more cohesive permanent communities that numbered hundreds, even thousands, of people. From at least 12,000 years ago, these earliest sedentary communities were cultivating selected wild species, leading toward their full domestication, and engaging in herd management strategies with the animals that later became fully domesticated. The continuing growth in size of these large, permanent communities was sustained by the artificial intensification of the food supply, that is, by farming crops and herding animals (Watkins 2012). But the process of creating and sustaining large, permanent communities required more highly developed cognitive and cultural skills (Watkins 2008; 2010). This chapter explores the novel symbolic means that created a sense of belonging and place and sustained the collective memory that was essential to a sense of community identity. Living in these new, large, permanent communities fundamentally required, among other things, that their members felt they belonged in that place, in that house, or in that community; and the sense of belonging in a place was based in community identity, collective memory, and history. Thus, the Neolithic of southwest Asia saw not only new forms of society but also new ways in which humans related to time and space.

We need to be able to support the assertions at the end of the previous paragraph and make a reasoned explanation of the changes that we see in the Epi-palaeolithic and early Neolithic periods. It has become the standard view that changes in subsistence strategies in response to either population pressure or climatic and environmental pressures drove the social and cultural changes that we see in the archaeological record. I seek to set these phenomena within a context of human cognitive and cultural evolution. In common with Cauvin (1994; 2000), I see the changes as being led by cognitive and cultural innovations (Cauvin called them a 'psycho-cultural' revolution), rather than by the shift toward farming and the domestication of plants and animals. Cauvin pointed in particular to the small, female human figurines and the various representations of bulls as the material representations of a pair of divine beings. I have been impressed by the monumental architecture and extraordinary sculpture, much of which has been discovered since Cauvin wrote.

CHANGING PERSPECTIVES ON THE CHANGES IN SETTLEMENT, ECONOMY, AND CULTURE

From the beginning of field research on the Neolithic, the period was characterized in terms of simple, small villages, utilitarian material culture, and the beginnings of farming. But anomalies were present from the start. For example, in the 1950s the aceramic Neolithic settlement at Tell es-Sultan, ancient Jericho, was as large as the overlying, urban, Early Bronze Age site; it had a circuit wall and rock-cut ditch, a massive circular tower, hundreds of intramural burials, many of which had had the skulls removed, and caches of modeled skulls (Kenyon 1957). And shortly after, Mellaart's excavations on the enormous settlement of Çatalhöyük, in south central Anatolia, produced extraordinary architecture, full of vivid and dramatic wall paintings and sculptures built into the walls and floors (Mellaart 1967). From the mid-1960s, Robert Braidwood and Halet Çambel began work at the site of Çayönü, near Diyarbakır in southeast Turkey, where a long stratigraphic sequence promised valuable data that might chart the process of plant and animal domestication. The series of special buildings that were discovered at the center of the settlement had a low priority in the preliminary publications of the project. Only later did we began to read about the culturally stereotyped, monumental houses (Schirmer 1990) and the series of unique, special-purpose buildings (Erim-Özdogan 2011), especially the 'skull building' (Croucher 2003).

There is now plentiful discussion of the social and symbolic importance of houses in the Neolithic of southwest Asia (for example, Byrd 1994; Byrd and Banning 1988; Kuijt 2000; Watkins 1990; 2004b). The most thoroughgoing exploration of houses as households, embodying the continuity of social lives memorialized within the house through time, is to be found in the idea of 'history houses' at Çatalhöyük (Hodder 2006; Hodder and Cessford 2004; Hodder and Pels 2010). My own contribution to this subject was to apply the term 'theatres of memory' to both the domestic and the communal structures, by which I sought to emphasize the importance of performance, whether in the acts of construction and making, or in the repeated ceremonies, such as those that accompanied the burial of the dead within or close to the house, and the subsequent retrieval, curation, and caching of skulls (Watkins 2004a). As excavators, we had direct experience of the continual and active attention to the houses at Qermez Dere, in north Iraq (Watkins 1990). The interior surfaces of the subterranean houses were repeatedly replastered; the pairs of pillars were replastered, reshaped, and even moved within the house; and a house could be abandoned, dismantled, and filled in, only for a new house to be excavated within the fill of the previous house. At Çayönü, the foundations of a monumental house were built on the foundations of the previous house, when those already existing foundations might have been reused or the materials recycled. In the southern Levant, the pier-houses of the later aceramic Neolithic were repeatedly replastered with thick coats of white lime plaster that was costly in terms of time and labor to produce (Byrd and Banning 1988). In addition to the structuring of whole settlements and the sometimes elaborate formulation of conventions in domestic architecture, from at least the latter part of the Epi-palaeolithic period, and particularly in the early aceramic Neolithic (often labeled the PPNA [Pre-Pottery Neolithic A] in the Levant), there is a remarkable series of public, or communal, buildings in a number of settlements that have been excavated in recent years.

Following on from the discovery of a series of public buildings at Çayönü (Croucher 2006; Özdoğan 1999; 2007), a single, small, rectangular, semi-subterranean building was found during salvage excavations at the site of Nevalı Çori, on a terrace above a small, left-bank tributary of the Euphrates in southeast Turkey (Hauptmann 1993; 1999; 2011). In almost every detail of its design, it was different from the long, rectangular, multichambered domestic buildings of the settlement. The building had been substantially rebuilt in exactly the same spot. Both excavated phases of the building (there was probably at least one earlier phase of construction that was not investigated for lack of time) were almost square in plan, their subterranean terrazzo floors reaching down stone stairs; the later building was slightly smaller, as it was constructed within the walls of its predecessor. Both buildings had a stone-slabbed 'bench' around the base of the walls. A pair of stone monoliths stood in the center of the floor, and there were another dozen monoliths set into the walls, projecting up through the bench. At least one of the central pairs of T-shaped monoliths, about 2.5 m tall, survived in situ in the later building; the peripheral set of monoliths had been deliberately broken, and only stumps survived below the surface of the bench. The surviving monolith has bent arms and hands, which tell us that the stones were seen as (faceless) anthropomorphs. The two central monoliths in one of the mid-tenth millennium BC circular enclosures at Göbekli Tepe (Figure 6.1, a and b) are likewise shown with arms and hands; they also wear neckbands, belts and loincloths.

One of the communal structures at Çayönü was somewhat similar, in that it was semi-subterranean, had a terrazzo floor, and was furnished with a pair of monoliths. The most recent structure so far dated at Göbekli Tepe, near Urfa in southeast Turkey, was also a subterranean, rectangular building with stone walls. It dates to about the same time, around or soon after 8500 BC. The Göbekli Tepe structure, however, had two pairs of monoliths, each about 2.5 m tall, set in its floor. The monolithic pillars, though their tops were damaged, were T-shaped, with flat surfaces and sharp facets, like those at Nevalı Çori. Characteristic of all the many monolithic pillars uncovered

Figure 6.1, a & b The pair of limestone pillars at the center of Structure D, Göbekli Tepe, southeast Turkey (Pillars 18 and 31). Each is 5.5 m tall and is set into a pedestal formed from the living rock. The T-shaped pillars have bent arms in low relief; the fingers of their hands meet above an elaborate belt with a clasp at the front. From the belt a fox skin is suspended. At its 'throat' each monolith wears a collar, and each has a distinctive pendant. Photographs by Nico Becker © Deutsches Archäologisches Institut, by kind permission of Prof. Klaus Schmidt.

Figure 6.1, a & b (continued)

at Göbekli Tepe are the low-relief animals and other creatures with which they were furnished. In the latest, rectangular structure, one of the monoliths has a fierce male lion, whereas another has a bull, above a fox, above a crane (Schmidt 2011: Figure 10 and Figure 14 respectively).

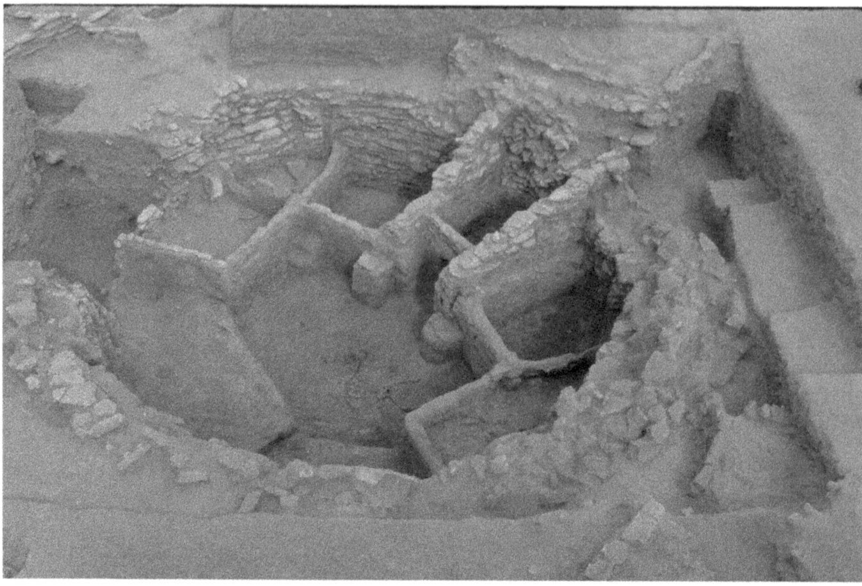

Figure 6.2 EA30 is a communal structure, more than 7 m in diameter and 2 m deep, at the center of the village of Jerf el Ahmar. The cell-like chambers were storage bins for cereals and lentils. At the end of its life, a decapitated body was placed in the middle of the floor, the roof was set on fire, and the cavity was refilled with soil. By kind permission of Danielle Stordeur.

Göbekli Tepe is unlike other sites of the early aceramic Neolithic period, in that the others were settlements, some of which were inhabited over many centuries, whereas Göbekli Tepe was some sort of 'central place', set apart on a bare limestone ridge. Although it is a mound of anthropogenic material, full of structures, those structures (at least, those excavated to date) are not houses in the normal sense (contra Banning 2011), and the debris of which most of the mound is composed is not normal occupation debris. The main focus of the ongoing excavations has been on an area at one side of the mound, where four or five massive, circular stone-built enclosures were constructed (Schmidt 2000; 2006; 2007; 2011). Each enclosure has a pair of T-shaped monoliths set in the middle of the floor, and between ten and thirteen more pillars set into the stone walls and stone-slabbed benches around the foot of the walls. The variety of imagery on the monoliths continues to expand with each season of excavation, and their superhuman size (up to 5.5 m for the largest pair in Enclosure D) is awesome. The first direct radiocarbon date on the construction of one of these circular enclosures supports the earlier *terminus ante quem* series of dates, placing the enclosures at the very beginning of the early aceramic Neolithic period, in the middle and second half of the tenth millennium BC (Dietrich and Schmidt 2011).

Figure 6.3 EA53, a second communal structure, was constructed after EA30 had been destroyed and the center of gravity of the village had drifted southward. A kerb of decorated limestone slabs formed the front face of a bench at the base of the wall. Wooden posts were set between the facing slabs to support the roof, and modeled plaster was formed around the base of each post. By kind permission of Danielle Stordeur.

A series of subterranean, circular, communal buildings contemporary with the Göbekli Tepe circular enclosures can be cited at a number of settlement sites, such as Hallan Çemi (Rosenberg 1999; Rosenberg and Redding 2000), also in southeast Turkey, and Tell Mureybet (Stordeur and Ibañez 2008: 62–69, especially Figure 25), Jerf el Ahmar (Stordeur et al. 2000, and Figure 6.1 and Figure 6.2 here), and Dja'dé (Figure 6.4) and Tell 'Abr (Yartah 2004; 2005) in the Euphrates valley in north Syria. In fact, communal buildings seem to have become almost an epidemic: a large, circular, semi-subterranean structure with extraordinary internal features has been found at the center of the early aceramic Neolithic settlement of WF16 in southern Jordan (Finlayson et al. 2011; Mithen et al. 2011), and another circular, semi-subterranean communal structure is reported from the slightly later Neolithic settlement of Ayios Tychonas–Klimonas in Cyprus (Vigne et al. 2012). For reasons of space, it is impossible to begin discussing the numerous stone sculptures and the smaller stone plaques and shaft-straighteners that bear what a number of archaeologists are now calling 'signs', what are possibly elements in a nontextual writing system (Morenz and Schmidt 2009).

Figure 6.4 Dja'de-el-Mughara, on the Euphrates in north Syria. One of the three internal buttresses, or stub walls, of a circular, subterranean communal building at the center of the early aceramic Neolithic settlement. A similar geometric abstract pattern is painted on the other faces of the buttress, and on a second buttress. The third buttress awaits excavation but is assumed from the symmetry of the structure. By kind permission of Eric Coqueugniot.

THE ECOLOGICAL-ERGONOMIC PARADOX

There is a paradox behind the simple story of increasing group size, increasing sedentism, and more and more elaborate symbolic practices, architecture, and sculpture. On the one hand, the trend toward life in larger and more permanent communities required increasing investment in hunting smaller prey, trapping and netting, and investment of time and labor in the harvesting and processing of nutritious, hard-seeded plants. On the other hand, the monumental public buildings, sculpture, and the elaborate rituals of maintenance and renewal of the house also imply great investment of labor, time, materials, and imagination. Why were these two trends pursued when each cost so much? Why were both trends occurring at the same time, doubling (squaring?) the demands on time, labor, resources, and skills?

Before the discovery of sites like Jerf el Ahmar, Nevalı Çori, and Göbekli Tepe, the standard view was that the changes in subsistence strategy (and therefore of settlement strategy) were adaptive responses to external factors. It was argued first by Binford (1968) and then refined by Flannery (1969) that increasing pressure on available resources was caused by population.

After some time, the story was rewritten in terms of climate change and environmental reversal, particularly the effects of the Younger Dryas phase at the very end of the Epi-palaeolithic period. Those accounts have recently been reviewed and comprehensively demolished by Zeder (2012). She showed that the changes in subsistence strategy do not make sense as adaptations to a changing ecological balance. They do not make sense either in terms of the long-term human evolutionary dynamic, which was always in the direction of larger and more cohesive social groups (Dunbar 2004; Tomasello 2009). The proper context for the increasing population density and increasing size of social group that is charted in the Epi-palaeolithic and early Neolithic of the Levant is the extraordinary evolutionary success of *Homo sapiens* within the long-term history of the genus Homo (spelled out in some detail in Watkins 2008).

THE DEMOGRAPHIC DYNAMIC

From the beginning of the Upper Palaeolithic onward, the evidence indicates that the *Homo sapiens* hunter-gatherer population of the Levant lived at densities that caused the progressive reduction of the prime sources of meat, the intensification of reliance on the smaller gazelle, and the growing reliance on small mammals and birds (Davis 1983; 2005; Davis, Lernau and Pichon 1994; Munro 2004; Stiner 2001; Stiner and Kuhn 2002; Stiner and Munro 2002; Stiner, Munro and Surovell 2000; Stutz, Munro and Bar-Oz 2009). The implication is that the benefits of living at higher densities in larger groups outweighed the costs of hunting and trapping smaller prey and harvesting and processing plant foods such as cereals and pulses.

In spite of the costs, the evidence is that Epi-palaeolithic (between approximately 23,000 and 12,000 years ago) hunter-gatherers in the southern Levant lived together in larger and larger numbers for longer and longer periods. The sites increase in average size, and also increase in average thickness of accumulated occupational debris (Henry 2002: 23). In parallel with that indicator of generally increasing population density, there is a trend toward more sedentary occupation of a number of sites (Maher 2010). By the late Epi-palaeolithic period, permanently resident communities occupied a number of settlements. In the early aceramic Neolithic period (between approximately 12,000 and 10,500 years ago), permanent settlements increase in number and also increase in size by a factor of eight or ten within three millennia (Kuijt 2000). The density of human settlement is actually even greater than that, since, as Ian Kuijt has shown, the ratio of roofed space (buildings) to open space also increases by a similar factor. So, as the aceramic Neolithic progresses, we have a growing number of settlements, which become on average larger and more densely occupied by domestic buildings. Even before they began to cultivate crops or herd animals, people's use of space, both in terms of the intensity with which the immediate territory around the settlement was exploited and in terms of the space occupied by their permanent settlements, changed completely from what it had been for tens and hundreds of thousands of years (Watkins 2009). Estimates of population for the new, permanent settlement sites are tentative, of course, but it is simple to observe that the mobile hunter-gatherer groups of the Upper Palaeolithic would be counted in tens, whereas the settlements of the early aceramic Neolithic numbered in the hundreds. Some archaeologists have suggested populations of between five and ten thousand for the larger settlements in the time when farming began to replace hunting and gathering in the Neolithic.

Although it had its costs, there are good evolutionary reasons that explain the adaptive advantages of a strategy concentrating larger numbers of people living permanently together in strongly cooperative communities. Recent studies have shown that human demography was a critical factor in facilitating intergenerational cultural transmission, especially as human culture became more complex and involved symbolic components (e.g., Kuhn and Stiner 2007; Powell, Shennan and Thomas 2009; Shennan 2001).

THE ESSENTIAL ROLE OF SYMBOLIC MATERIAL CULTURE

So, was it simply a question of time that the first, large, permanently coresident communities emerged only at the end of the Pleistocene and the beginning of the Neolithic? The answer, I believe, is a partial 'yes': what we see developing through the Epi-palaeolithic and early Neolithic indeed follows from what had gone before, and it could not have happened earlier. The up-scaling of population units could not have happened without the development of intensive harvesting, storage, and investment in food processing and could not have continued unless supported by cultivation and investment in animal management. However, the richly symbolic archaeology of these first settlements, so radically different from what had gone before in the Palaeolithic, tells us that something more was involved. Research on the evolution of the hominin brain has placed its growth, and especially the growth in the relative volume (and therefore presumably the cognitive power) of the cortical areas of the brain, in the context of the growth in the scale and cohesiveness of human social groups. It has become accepted that *Homo sapiens* has a brain size and complexity that makes it capable of managing and manipulating social relations with up to about 150 others ('Dunbar's number') (Aiello and Dunbar 1993; Aiello and Wheeler 1995; Dunbar 1997). This has been developed by Dunbar as the 'social brain' theory (Dunbar 1998; Dunbar, Gamble and Gowlett 2010). But what we can observe in the Epi-palaeolithic and early Neolithic of southwest Asia is, in evolutionary terms, an extraordinarily rapid rise in the size of coresident groups: within a few thousand years, far too rapidly for biological evolution to operate, human social group size expanded massively beyond Dunbar's theoretical capacity of the biological brain. Enhancing the cognitive and cultural capacity of people's minds, deployed in symbolic behaviors and practices and systems of symbolic material culture, facilitated the expansion of social group size. As the populations of those communities exceeded Dunbar's number, at first by a little, later by a factor of five or ten, individuals would not know personally many of the people among whom they lived; yet it was of critical importance that the members of a community could trust that everyone would observe the conventions of living together, and that everyone could cooperate, and that they could identify any cheats or freeloaders. The means to build and sustain such community identity were forms of symbolic exchange and interaction.

Gamble (1998: 429–431; 1999) has shown us the beginnings of the symbolic construction of community as early as the Upper Palaeolithic of Europe, emphasizing the active role that the individual plays in performing society and its structures into existence. Gamble used social network theory to show how *Homo sapiens* in the Upper Palaeolithic in Europe had begun to use materials such as marine shells and high-quality flint in long-distance exchanges in the maintenance of extended networks. But those networks still numbered around only one or two hundred individuals. But in the Epi-palaeolithic and early Neolithic of southwest Asia, we see the construction of whole systems of symbolic representation, in the form of settlement layout and monumental and elaborate forms of domestic architecture, marked with images and accommodating ceremonies. The symbolic construction of large, permanently coresident communities had begun. Anthropologists working among contemporary small-scale societies have remarked the general phenomenon. The most often quoted is Geertz, who wrote that 'man is an animal suspended in webs of significance he himself has spun' (Geertz 1973: 5). In his slim but powerful book on the symbolic construction of community, Cohen shows how communities are symbolically constructed and maintained through the continual manipulation of their symbols in the minds of their members (Cohen 1985: 15). Cohen was concerned to show that 'community' was a valid concept, although anthropologists and sociologists found it hard to categorize within their schemes; they were attempting to fit societies within the categories that they had defined, whereas, Cohen explained, members of a community were quite clear about their belonging to a community, and who did and did not belong. In other words, community is defined from the bottom up—by its members.

Social network theory likewise gives a bottom-up perspective on the individual's world, analyzing the individual's relationships with others in terms of their nature, intensity, and robustness. Gamble (1998; 1999, Chapter 8) used it in his social archaeology of the European Palaeolithic, and I have borrowed from him for a discussion of the changes in social networks between the Upper Palaeolithic and the complex, nested networks of the early Neolithic (Watkins 2008). Turner and Maryanski (1991), in their key text on social network theory, define three levels of social network—the intimate, the effective (those with whom the individual regularly interacts, such as work colleagues or church members), and the extended (acquaintances and friends-of-friends, people with whom we have only occasional dealings). Among contemporary small-scale societies, the individual's extended network ranges between 100 and 400, but there are examples where it may be as high as 1,000. Beyond that is the 'global network', people who may be thought of as 'them', and not 'us'; even so, there will be occasions when the individual may engage with individuals or groups beyond her extended network. Whereas the individual invests emotional capital in the intimate network of close relatives, social network theory indicates that engagement with the extended or global network involves symbolic investment. And that is what we see in the archaeological record left by prehistoric long-distance exchange networks.

The small Upper Palaeolithic communities, spread at low densities across the landscape, used mainly symbolic resources, that is, material things with symbolic values, to obtain what Gamble (1998) calls 'release from proximity', the ability to network with other groups and form larger, extended communities across considerable distances. When large, permanently coresident communities came into existence and grew in size in southwest Asia during the Epi-palaeolithic and early Neolithic periods, the individuals found themselves in day-to-day contact and in permanent dependence on those extended networks, and, as settlements grew in scale, on numbers of people equivalent to a global network. The spatial parameters of social networks had changed utterly: what Gamble had labeled 'the release from proximity' became the anonymity of propinquity, whereby the individual would know little or nothing about many of those who inhabited the same settlement and lived within a hundred meters or so. It was of paramount importance that an individual could convince others that she was a good member of the community, and that she could recognize others as reliable, trustworthy, and unlikely to cheat. Social relations in networks of that scale were necessarily sustained by means of symbolic rather than emotional or material resources, and at intensive levels. Symbolic cultural resources were therefore of critical importance, and these were deployed in ways that showed how people belonged together, in that place, and over time.

If the widely distributed, small-scale, extended networks of the Upper Palaeolithic period were drawn by centripetal force into extended networks made up of many people all living together, and those coresident communities found the means to grow and grow in numbers, it would be remarkable enough. However, the emergent coresident communities focused close attention and great effort in the formation and maintenance of regional networks of sharing and exchange. That process was already under way in the Levant during the Epi-palaeolithic period (Richter et al. 2011). In the early aceramic Neolithic period, we can begin to see signs of a regional network in southeast Turkey and north Syria, centered on Göbekli Tepe, that shared signs and symbols, as well as many cultural traits. And during the early Neolithic, the networking among sedentary communities grew spectacularly in both extent and intensity (Watkins 2008).

The place where people lived was a shared space, a richly symbolic built environment; the land around the settlement, on which the community was totally reliant, was equally important. Although we live our lives differently (for many of us, in a new kind of fission-fusion mode of commuting out to a place of work), it is not difficult to imagine why the individual and the community necessarily identified with that place. For the individual and for the community, life depended on their ability to live and work together, observing the norms of behavior, cooperating with and trusting one another. But how does memory, whether collective, social, or cultural, relate to community identity? Assmann (1995) has shown how collective memory is the essential underpinning of cultural identity, just as personal memory underpins the sense of individual identity. He tells us that cultural memory 'preserves the store of knowledge from which a group derives an awareness of its unity and peculiarity', defining who 'we' are, and how 'they' are different. The knowledge preserved in cultural memory is both formative (educative, civilizing, and humanizing) and normative (in that it provides rules of conduct) (Assmann 1995: 130–132).

COLLECTIVE MEMORY AND COMMUNITY IDENTITY

'Collective memory' is perhaps an unhelpful term, since the memories that members of a community share and that are fundamental to community identity are distributed among them. If collective memory exists distributed in and through the memories of individuals, then it is important to know something of how individuals remember, and how memory contributes to personal identity. A less abbreviated discussion of this subject forms part of a recent publication (Watkins 2012), and more authoritative discussions can be recommended (Assmann 2006; Boyer 2009; Wertsch 2009). Let us be clear that memory exists in the act of remembering; one of the foundation texts on memory has 'Remembering' as its title (Bartlett 1932), and the senior figure in the field of collective memory studies today likewise titles his book 'How Societies Remember' (Connerton 1989). Our sense of identity is built from episodic memory, reexperiencing images from our past; it is about feeling and emotion, rather than factual recall from semantic memory, although we may use factual recall in support. Boyer (2009: 3) notes how remembering an episode from the past uses similar cognitive resources to imagining a potential future event. The psychologists Suddendorf and Corballis (1997; 2007) put these two mental activities together under the general capacity for 'mental time travel'.

If individuals need autobiographical memory in order to have a sense of self, it follows that collective memory, distributed among the individuals who form the collective, is essential to community identity. And if autobiographical memory is based on episodic memory, with its images and emotional affects, it follows that the collective memory of the community is based on the foundation of shared acts of remembering that are imagistic and affecting. The shared practices, mostly consisting, according to Connerton, of 'commemorative ceremonies' and 'bodily acts', were rituals that were repeated with varying degrees of frequency. Some of the rituals may have been religious, in that they assumed the existence of participation of superhuman agents; others, such as the ritual of eating together, might be described as familial; some, such as feasting on the meat obtained in occasional and dangerous corporate hunting of wild cattle at Çatalhöyük, are difficult to define as religious but were clearly of great symbolic significance (Whitehouse and Hodder 2010). In general, Whitehouse (2004) contrasts an 'imagistic' with a 'doctrinal' mode of religiosity; the imagistic mode of religiosity, he says, depends on relatively infrequent and sometimes traumatic ritual events that can trigger enduring and vivid episodic memories, memories that are shared among all those who were present. Perhaps imagistic religious episodes are more dramatic and impressive because they involve more memorable, dramatic, and impressive supernatural beings or events.

Connerton (1989) particularly emphasizes the role of 'commemorative ceremonies' and 'bodily acts' in the formation and maintenance of collective memory, and Whitehouse focuses on the nature of ritual performance. But we should also recognize the importance of a proper place for ceremonies, the physical context that frames and reinforces the 'bodily acts', the gestures and rituals, raising them to a level that resonates. We archaeologists uncover the monuments and other cultural artifacts, the places where the ceremonies took place, and the things that were essential parts of the gestures and rituals. And it is worth noting that, just like the recursive structuring of personal memory (Corballis 2011), these things may also be hierarchically, or recursively, structured (e.g., at Göbekli Tepe (Morenz and Schmidt 2009; Schmidt 2000; 2006; 2007; 2011), the site in its setting in the landscape, the great circular structures set within the historical stratigraphy of the site, the monoliths set within the structure in a very definite order, and the sculpted reliefs of animals, snakes, birds, or spiders on the monoliths).

There were two main concerns in the maintenance of stability in the new, large, permanently coresident communities. As well as celebrating an ethos of cooperation and altruism, communities and their members needed a strong sense of collective or cultural identity; these concerns with community and identity were closely interlinked, of course, in the ways that cultural memory (time) was embodied in the physical structures and structuring of the settlement and its architecture (space).

We have seen that, just as autobiographical episodic memory is essential to the individual's sense of identity, collective (shared) cultural memory is essential to sustaining community identity. It is obvious that an individual's autobiographical memory, his or her sense of self, is generally constructed within cultural conventions; and the individual's sense of identity and the collective (distributed) sense of community identity are closely interwoven. The new, permanently coresident communities built their sense of identity (literally) in the architecture of their settlements (Watkins 1990; 2004a; 2004b). And the essential component of memory—remembering—was built into the settlement in many different ways, whether in the house, the cluster of houses, the settlement, or the settlement at the center of its territory. Memory was evoked continually, whether in the observance of the niceties and conventions of living together in the house, sharing the settlement, or living in the surrounding landscape (Ingold 1993). The echoing of memory is seen clearly, and has been carefully explored, in the houses of Çatalhöyük (Düring and Marciniak 2006; Hodder and Cessford 2004; Hodder and Pels 2010; Russell, Martin and Twiss 2009), the ceremonies when an old house was abandoned and closed and a new house was started, or the large-scale and exciting festivals when hundreds of kilograms of wild beef was consumed (Whitehouse and Hodder 2010). The house, the settlement, and the community's land were essential to life, and by the same token, borrowing from Claude Lévi-Strauss (1963), they were 'good to think'. Following another structuralist insight that we owe to Lévi-Strauss, these were 'house societies', for whom the language of architecture embodied their ideas of who they were (Carstens and Hugh-Jones 1995; Russell, Martin and Twiss 2009).

MEMORIALS, MEMORY, AND THE SENSE OF COMMUNITY IN TIME AND SPACE

As a final thought, it is worth noting that, for many people, in many languages and cultures, time and space are closely related. Time is often envisaged in thought, and expressed, both verbally and in body language, in spatial metaphors (Boroditsky 2000; 2011; Casasanto and Boroditsky 2008; Lakoff and Johnson 1999). When we encounter a new author, we sometimes ask ourselves *where he or she comes from,* by which we mean what can his or her own history tell us about how he or she came to write this book, rather than where exactly he or she was born. Or you, the reader, may be feeling that the end of this chapter is a *long* time coming. If we prefer spatial terms, which are readily imagined, when it comes to thinking about that immaterial dimension of time, it is perhaps easy to appreciate how closely place and memory can be interwoven in the cause of sustaining the sense of identity.

REFERENCES

Aiello, L. and R. I. M. Dunbar. 1993. Neocortex size, group size and the evolution of language. *Current Anthropology* 36: 184–193.

Aiello, L. C. and P. Wheeler. 1995. The expensive tissue hypothesis: the brain and the digestive system in human and primate evolution. *Current Anthropology* 36(2): 199–221.

Assmann, J. 1995. Collective memory and cultural identity. *New German Critique* 65: 125–133.

Assmann, J. 2006. Introduction: what is "cultural memory"? In: J. Assmann (ed.), *Religion and Cultural Memory: Ten Studies,* 1–30. Stanford, CA: Stanford University Press.

Banning, E. B. 2011. So fair a house: Göbekli Tepe and the identification of temples in the Pre-pottery Neolithic of the Near East. *Current Anthropology* 52(5): 619–660.

Bartlett, F. C. 1932. *Remembering: A Study in Experimental and Social Psychology.* Cambridge: Cambridge University Press.

Binford, L. R. 1968. Post-Pleistocene adaptations. In: S. R. Binford and L. R. Binford (eds.), *New Perspectives in Archaeology,* 313–341. Chicago: University of Chicago Press.

Boroditsky, L. 2000. Metaphoric structuring: understanding time through spatial metaphors. *Cognition* 75(1): 1–28.

Boroditsky, L. 2011. How language shapes thought. *Scientific American* (February): 63–65.

Boyer, P. 2009. What are memories for? Functions of recall in cognition and culture. In: P. Boyer and J. W. Wertsch (eds.), *Memory in Mind and Culture,* 3–28. Cambridge: Cambridge University Press.

Byrd, B. F. 1994. Public and private, domestic and corporate: the emergence of the southwest Asian village. *American Antiquity*, 59(4): 639–66.

Byrd, B. F. and E. B. Banning. 1988. Southern Levant pier-houses: intersite architectural patterning during the Pre-Pottery Neolithic B. *Paléorient* 14(1): 65–72.

Carstens, J. and S. Hugh-Jones, eds. 1995. *About the House: Levi-Strauss and Beyond.* Cambridge: Cambridge University Press.

Casasanto, D. and L. Boroditsky. 2008. Time in the mind: using space to think about time. *Cognition* 106(2): 579–593.

Cohen, A. P. 1985. *The Symbolic Construction of Community.* Chichester and London: Ellis Horwood, Tavistock.

Connerton, P. 1989. *How Societies Remember.* Cambridge: Cambridge University Press.

Corballis, M. C. 2011. *The Recursive Mind: The Origins of Human Language, Thought, and Civilization.* Princeton: Princeton University Press.

Croucher, K. 2006. Death, display and performance: a discussion of the mortuary remains at Çayönü Tepesi. In: M. Georgiadis and C. Gallou (eds.), *The Archaeology of Cult and Death*, 11–44. Budapest: Archaeolingua.

Davis, S. J. M. 1983. The age profiles of gazelles predated by ancient man in Israel: possible evidence for a shift from seasonality to sedentism in the Natufian. *Paléorient* 9: 55–62.

Davis, S. J. M. 2005. Why domesticate food animals? Some zoo-archaeological evidence from the Levant. *Journal of Archaeological Science* 32(9): 1408–1416.

Davis, S. J. M., O. Lernau and J. Pichon. (1994). The animal remains: new light on the origin of animal husbandry. In M. Lechevallier and A. Rosen (eds.), *Le Gisement de Hatoula en Judée Occidentale, Israël,* 83–100. Paris: Mémoires et Travaux du Centre de Recherche Français de Jerusalem, No. 8. Association Paléorient.

Dietrich, O. and K. Schmidt. 2011. A radiocarbon date from the wall plaster of enclosure D of Göbekli Tepe. *Neo-Lithics: A Newsletter of South-west Asian Lithics Research* 2(10): 82–83.

Dunbar, R. I. M. 1997. *Grooming, Gossip and the Evolution of Language.* London: Faber.

Dunbar, R. I. M. 1998. The social brain hypothesis. *Evolutionary Anthropology* 6(3): 178–190.

Dunbar, R. I. M. 2004. *The Human Story: A New History of Mankind's Evolution.* London: Faber.

Dunbar, R. I. M., C. Gamble and J. A. J. Gowlett. 2010. *Social Brain, Distributed Mind*. Oxford: Oxford University Press and British Academy.

Düring, B. and A. Marciniak. 2006. Households and communities in the central Anatolian Neolithic. *Archaeological Dialogues* 12(2): 165–187.

Erim-Özdogan, A. 2011. Çayönü, in *The Neolithic in Turkey. New Excavations and New Research—The Euphrates Basin*, eds. M. Özdoğan, N. Başgelen and P. Kuniholm. Istanbul: Arkeoloji ve Sanat Yayinlari, 185–269.

Finlayson, B., S. J. Mithen, M. Najjar, S. Smith, D. Maričević, N. Pankhurst and L. Yeomans. 2011. Architecture, sedentism, and social complexity at Pre-Pottery Neolithic A WF16, Southern Jordan. *Proceedings of the National Academy of Sciences* 108(20): 8183–8188.

Flannery, K. V. (1969). The origins and ecological effects of early domestication in Iran and the Near East. In: P. J. Ucko and G. W. Dimbleby (eds.), *The Domestication and Exploitation of Plants and Animals,* 73–100. London: Duckworth.

Gamble, C. 1998. Palaeolithic society and the release from proximity: a network approach to intimate relations. *World Archaeology* 29(3): 426–449.

Gamble, C. 1999. *The Palaeolithic Societies of Europe.* Cambridge: Cambridge University Press.

Geertz, C. 1973. *The Interpretation of Cultures: Selected Essays.* New York: Basic Books.

Hauptmann, H. 1993. Ein Kult Gebaüde in Nevali Çori. In: M. Frangipane, H. Hauptmann, M. Liverani, P. Matthiae and M. Mellink (eds.), *Between the Rivers and over the Mountains,* 37–69. Rome: Università di Roma La Sapienza.

Hauptmann, H. 1999. The Urfa region. In: M. Özdoğan and N. Başgelen (eds.), *Neolithic in Turkey: The Cradle of Civilization, New Discoveries,* 65–86. Istanbul: Arkeoloji ve Sanat Yayinlari.

Hauptmann, H. 2011. The Urfa region. In: M. Özdoğan, N. Başgelen and P. Kuniholm (eds.), *The Neolithic in Turkey: New Excavations and New Research—The Euphrates Basin,* 85–138. Istanbul: Arkeoloji ve Sanat Yayinlari.

Henry, D. O. 2002. Models of agricultural origins and proxy measures of prehistoric demographics. In: R. T. Cappers and S. Bottema (eds.), *The Dawn of Farming in the Near East,* 15–26. Berlin: Ex Oriente.

Hodder, I. 2006. *The Leopard's Tale: Revealing the Mysteries of Turkey's Ancient 'Town'*. London: Thames & Hudson.

Hodder, I. and C. Cessford. 2004. Daily practice and social memory at Catalhoyuk. *American Antiquity* 69(1): 17–40.

Hodder, I. and P. Pels. 2010. History houses: a new interpretation of architectural elaboration at Çatalhöyük. In: I. Hodder (ed.), *Religion in the Emergence of Civilization: Çatalhöyük as a Case Study,* 163–186. Cambridge: Cambridge University Press.

Ingold, T. 1993. The temporality of the landscape. *World Archaeology* 25(2): 152–174.

Kuhn, S. and M. Stiner. 2007. Palaeolithic ornaments: implications for cognition, demography and identity. *Diogenes* 54: 40–48.

Kuijt, I. 2000. People and space in early agricultural villages: exploring daily lives, community size and architecture in the late Pre-Pottery Neolithic. *Journal of Anthropological Archaeology* 19(1): 75–102.

Lakoff, G. and M. Johnson. 1999. *Philosophy in the Flesh: The Embodied Mind and Its Challenge to Western Thought.* New York: Basic Books.

Lévi-Strauss, C. 1963. *Totemism.* Boston: Beacon Press.

Maher, L. 2010. People and their places at the end of the Pleistocene: evaluating perspectives on physical and cultural landscape change. In: B. Finlayson and G. Warren (eds.), *Landscapes in Transition: Understanding Hunter-Gatherer and*

Farming Landscapes in the Early Holocene of Europe and the Levant, 34–45. London: Levant Supplementary Series and CBRL.

Mellaart, J. 1967. *Çatal Hüyük: a neolithic town in Anatolia*. London: Thames and Hudson.

Mithen, S., B. Finlayson, S. Smith, E. Jenkins, M. Najjar and D. Maričević. 2011. An 11,600 year-old communal structure from the Neolithic of southern Jordan. *Antiquity* 85(328): 350–364.

Morenz, L. D. and K. Schmidt. 2009. Grosse Releifpfeiler und kleine Zeichentäfelchen. In: P. Andrássy, J. Budka and F. Kammerzell (eds.), *Non-textual Marking Systems, Writing and Pseudo Script from Prehistory to Present Times*, 13–31. Göttingen: Lingua Aegyptia—Studia Monographica.

Munro, N. D. 2004. Zooarchaeological measures of hunting pressure and occupation intensity in the Natufian: implications for agricultural origins. *Current Anthropology* 45: S5–S33.

Powell, A., S. Shennan and M. G. Thomas. 2009. Late Pleistocene demography and the appearance of modern human behavior. *Science* 324(5932): 1298–1301.

Richter, T., A. N. Garrard, S. Allock and L. A. Maher. 2011. Interaction before agriculture: exchanging material and sharing knowledge in the Final Pleistocene Levant. *Cambridge Archaeological Journal* 21(1): 95–114.

Rosenberg, M. 1999. Hallan Çemi. In: M. Özdoğan and N. Başgelen (eds.), *Neolithic in Turkey, the Cradle of Civilization: New Discoveries*, 25–33. Istanbul: Arkeoloji ve Sanat Yayinlari.

Rosenberg, M. and R. Redding. 2000. Hallan Çemi and early village organization in Eastern Anatolia. In: I. Kuijt (ed.), *Life in Neolithic Farming Communities: Social Organization, Identity, and Differentiation*, 39–62. New York: Kluwer Academic.

Russell, N., L. Martin and K. C. Twiss. 2009. Building memories: commemorative deposits at Çatalhöyük. *Anthropozoologica* 44(1): 103–125.

Schirmer, W. 1990. Some aspects of the building in the "aceramic neolithic" settlement at Çayönü Tepesi. *World Archaeology*, 21(3): 363–87.

Schmidt, K. 2000. Göbekli Tepe, southeastern Turkey: a preliminary report on the 1995–1999 excavations. *Paléorient* 26(1): 45–54.

Schmidt, K. 2006. *Sie Bauten die Ersten Tempel: Das Rätselhafte Heiligtum der Steinzeitjäger*. Munich: Beck.

Schmidt, K. 2007. Göbekli Tepe. In: C. Lichter (ed.), *Vor 12,000 Jahren in Anatolien: Die Ältesten Monumente der Menschheit*, 74–75. Stuttgart: Badisches Landesmuseum, Karlsruhe.

Schmidt, K. 2011. Göbekli Tepe. In: M. Özdoğan, N. Başgelen and P. Kuniholm (eds.), *The Neolithic in Turkey: New Excavations and New Research—The Euphrates Basin*, 41–83. Istanbul: Arkeoloji ve Sanat Yayinlari.

Shennan, S. 2001. Demography and cultural innovation: a model and its implications for the emergence of modern human culture. *Cambridge Archaeological Journal* 11(1): 5–16.

Stiner, M. C. 2001. Thirty years on the "Broad Spectrum Revolution" and paleolithic demography. *Proceedings of the National Academy of Sciences of the United States of America* 98(13): 6993–6996.

Stiner, M. and S. Kuhn. 2002. Paleolithic demography, small animal exploitation, and the proliferation of personal ornaments. *AAAS Annual Meeting and Science Innovation Exposition* 168: A20.

Stiner, M. and N. D. Munro. 2002. Approaches to prehistoric diet breadth, demography, and prey ranking systems in time and space. *Journal of Archaeological Method and Theory* 9(2): 175–208.

Stiner, M. C., N. D. Munro and T. A. Surovell. 2000. The tortoise and the hare: small-game use, the broad-spectrum revolution, and paleolithic demography. *Current Anthropology* 41(1): 39–73.

Stordeur, D., M. Brenet, G. Der Aprahamian and J.-C. Roux. 2000. Les bâtiments communautaires de Jerf el Ahmar et Mureybet, horizon PPNA (Syrie). *Paléorient* 26(1): 29–44.

Stordeur, D. and J. J. Ibañez. 2008. Stratigraphie et répartition des architectures à Mureybet. In: J. J. Ibañez (ed.), *Le site néolithique de Tell Mureybet (Syrie du Nord),* 33–94. Oxford: Archaeopress.

Stutz, A. J., N. D. Munro and G. Bar-Oz. 2009. Increasing the resolution of the broad spectrum revolution in the Southern Levantine Epipaleolithic (19–12 ka). *Journal of Human Evolution* 56(3): 294–306.

Suddendorf, T. and M. C. Corballis. 1997. Mental time travel and the evolution of the human mind. *Genetic Social and General Psychology Monographs* 123(2): 133–167.

Suddendorf , T. and M. Corballis. 2007. The evolution of foresight: What is mental time travel, and is it unique to humans? *Behavioral and Brain Sciences* 30(3): 299–313.

Tomasello, M. 2009. *Why We Cooperate.* Cambridge, MA: MIT Press.

Turner, J. H. and A. Maryanski. 1991. Network theory. In: J. H. Turner, A. Maryanski and S. Fuchs (eds.), *The Structure of Sociological Theory,* 540–572. Belmont, CA: Wadsworth.

Vigne, J.-D., F. Briois, A. Zazzo, G. Willcox, T. Cucchi, S. Thiébault, I. Carrèe, Y. Franel, R. Touquet, C. Martin, C. Moreau, C. Comby and J. Guilaine. 2012. First wave of cultivators spread to Cyprus at least 10,600 y ago. *Proceedings of the National Academy of Sciences,* published ahead of print May 7, 2012.

Watkins, T. 1990. The origins of house and home? *World Archaeology* 21(3): 336–347.

Watkins, T. 2004a. Architecture and 'theatres of memory' in the Neolithic South West Asia. In: E. DeMarrais, C. Gosden and C. Renfrew (eds.), *Rethinking Materiality: The Engagement of Mind with the Material World,* 97–106. Cambridge: McDonald Institute of Archaeological Research.

Watkins, T. 2004b. Building houses, framing concepts, constructing worlds. *Paléorient* 30(1): 5–24.

Watkins, T. 2008. Supra-regional networks in the Neolithic of Southwest Asia. *Journal of World Prehistory* 21(1): 139–171.

Watkins, T. 2009. Ordering time and space: creating a cultural world. In: J. Córdoba, M. Molist, C. Pérez, I. Rubio and S. Martínez (eds.), *Procedings of the 5th International Congress on the Archaeology of the Ancient Near East,* 647–659. Madrid: Universidad Autónoma de Madrid.

Watkins, T. 2012. Household, community and social landscape: building and maintaining social memory in the early Neolithic of southwest Asia. In: M. Furholt, M. Hinz and D. Mischka (eds.), *'As Time Goes By' Monuments, Landscapes and the Temporal Perspective: Socio-environmental Dynamics over the Last 12,000 Years,* 23–44. Kiel: Bonn: Rudolf Habelt.

Wertsch, J. W. 2009. Collective memory. In: P. Boyer and J. W. Wertsch (eds.), *Memory in Mind and Culture,* 117–137. Cambridge: Cambridge University Press.

Whitehouse, H. 2004. *Modes of Religiosity: A Cognitive Theory of Religious Transmission.* Walnut Creek, CA: AltaMira Press.

Whitehouse, H. and I. Hodder. 2010. Modes of religiosity at Çatalhöyük. In: I. Hodder (ed.), *Religion in the Emergence of Civilization: Çatalhöyük as a Case Study,* 122–145. Cambridge: Cambridge University Press.

Yartah, T. 2004. Tell 'Abr 3, un village du Néolithique précéramique (PPNA) sur le Moyen-Euphrate: première approche. *Paléorient* 30: 141–158.

Yartah, T. 2005. Les bâtiments communautaires de Tell 'Abr 3 (PPNA, Syrie). *Neolithics: A Newsletter of Southwest Asia Neolithic Research* 1: 3–9.

Zeder, M. A. 2012. The broad spectrum revolution at 40: resource diversity, intensification, and an alternative to optimal foraging explanations. *Journal of Anthropological Archaeology* 31(3): 241–264.

7 Constructing the Present Past in the Anatolian Neolithic

Bleda S. Düring

INTRODUCTION

In some societies of the Anatolian Neolithic, houses were important spatial loci in which people expressed their collective histories. They were of great significance to the identity and sense of belonging of Neolithic people. Houses then were cultural institutions in which particular forms of space and time were interlinked, and they constituted an important factor in social life in much of Neolithic Anatolia.

Across the Anatolian Neolithic, a remarkable degree of effort is put into the exact reproduction of houses, often producing house sequences lasting centuries, and great care is taken to avoid damaging earlier houses. This chapter seeks to explore the social values embedded in these building practices, which, I will argue, are both the medium and the outcome of social practices.

Whereas there are studies dealing with individual Neolithic sites in Anatolia, little comparative analysis, in which data from different sites are taken into account, has been undertaken. Here, drawing on evidence from disparate sites such as Çayönü, Aşıklı Höyük, Çatalhöyük, and Ilıpınar, I will argue that at these Neolithic sites the reproduction of houses served important social purposes, in which the house served as a manifestation of social groups and the reproduction of houses served to presence an idealized past.

Contrary to the common practice in archaeology, I will start with a brief consideration of the data at hand rather than with a consideration of theoretical concerns and will subsequently address a number of interpretative issues that emerge from such a consideration. The main concern is to understand how we can best understand the phenomenon of building continuity that is in evidence at a range of sites across Neolithic Anatolia and beyond. I will argue that functional explanations are not adequate for understanding this building continuity, and that we should think about how these houses were important to society. Drawing on ethnographic models, in particular the House concept formulated by Lévi-Strauss, and archaeological data, in particular the rich data set from Çatalhöyük, I will argue that (the continuous reconstruction of) the house was of great significance for the articulation and reproduction of social groups.

BUILDING CONTINUITY IN THE ANATOLIAN NEOLITHIC

With 'building continuity' I refer to the fact that buildings were often reconstructed on the same spot, with the same dimensions, and with the same orientations over a series of rebuilding episodes, and over a considerable span of time lasting up to several centuries. To illustrate this building continuity, I will briefly discuss four Neolithic sites across Anatolia where building continuity is well documented.

Çayönü

Çayönü is located near Ergani, not far from Diyarbakır in southeastern Anatolia (Figure 7.1). The site measures about four hectares in size and has a depth of about 4.5 m. The site was excavated in sixteen seasons between 1964 and 1991 by teams from Istanbul, Chicago, Karlsruhe, and Rome. The main sequence is to be dated to the Aceramic Neolithic, but there are also levels dating to the Ceramic Neolithic, Early Bronze Age, and Medieval Period. The Aceramic Neolithic occupation can be subdivided into first, the round building phase; second, the grill building phase; third, the channeled building phase; fourth, the cobble-paved building phase; fifth, the cell building phase; and sixth, the large room building phase (Bıçakçı 1998; Özdoğan 1999; Rosenberg and Erim-Özdoğan 2011; Yakar 1991: 42–56; 1994: 7–9).

The phenomenon of building continuity with which this chapter is concerned is especially well documented during the 'grill' and 'channeled'

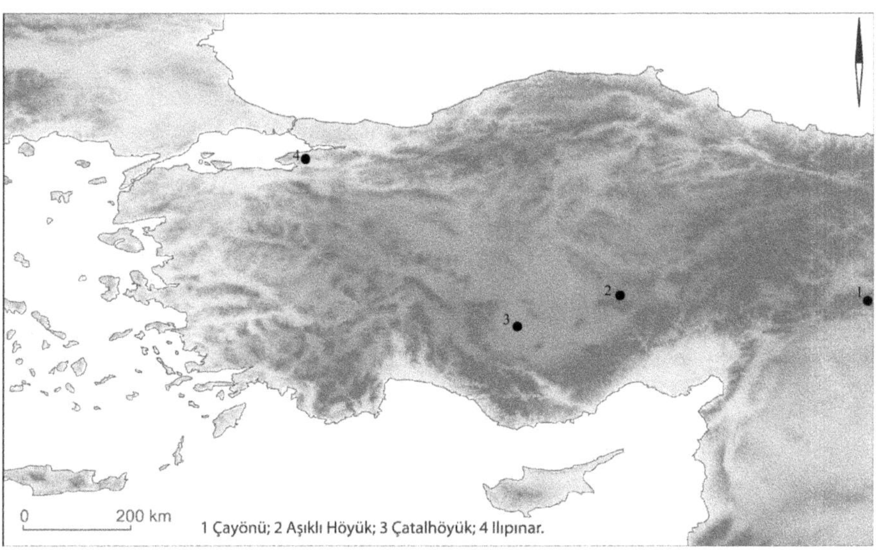

Figure 7.1 Map with sites discussed in text. (Figure prepared by J. Porck.)

building phases at Çayönü, dated to the Early and Middle Pre-Pottery Neolithic B (PPNB), between about 8700 and 7500 Cal BC, but building continuity has also been documented in the later cell building phase, which dates to the Late PPNB, between about 7500 and 7000 Cal BC (Özdoğan and Özdoğan 1998: 590). In the grill building phase, we are dealing with rectangular buildings with foundations built from small stones that served as the basis for a wattle-and-daub superstructure. In the channeled building phase, mud brick superstructures are introduced. The buildings of both phases had a series of parallel walls that served as a platform upon which the superstructure was erected (Bıçakçı 1998).

What is particular about these buildings is that they were rebuilt a number of times with more or less the same dimensions and orientation. Up to five buildings in sequence have been documented stacked each on top of the other, with a slight change in orientation during each rebuilding episode. Thus the foundation walls are frequently found directly overlying those of their predecessors. Whereas it would have been easy to reuse stones from earlier buildings, there are no instances where older walls are disturbed in this process (Özdoğan and Özdoğan 1998: 590).

Aşıklı Höyük

Aşıklı Höyük is located on the eastern fringe of the volcanic plateau of Cappadocia in central Anatolia. It measures about four hectares and is about 16 m high. Aşıklı Höyük was first investigated in the 1960s, and large-scale excavations have been executed by a team from Istanbul University from 1989 onward. The site dates exclusively to the Aceramic Neolithic, with radiocarbon dates between 8200 and 7400 Cal BC (Esin et al. 1991; Esin and Harmankaya 1999; Özbaşaran 2011).

The buildings at Aşıklı Höyük were constructed from mud slabs of irregular sizes that were probably not produced in a mold. The walls were about 30 cm wide, and it has been calculated that they could have supported the ground floor storey only. Buildings encompass between one and three rooms and were normally no more than 12 m^2 in size. They were often constructed directly adjacent to one another and, given the absence of doors, seem to have been accessed from the roofs (Düring 2006; Esin et al. 1991).

The buildings at Aşıklı Höyük were as a rule constructed directly on top of each other, retaining their size and orientation, and using older walls as a foundation. Generally speaking, the excavation strategy employed at Aşıklı Höyük was geared toward opening up large horizontal exposures in order to document the settlement as a whole rather than exploring diachronic developments. Fortunately, a large deep sounding was also undertaken, in trenches 4H/G, providing us with a valuable perspective on the diachronic aspects of buildings at Aşıklı Höyük.

Figure 7.2 represents the building continuity in the deep sounding 4H/G in layers that have been assigned to levels 2I (bottom) up to 2B. Most of the

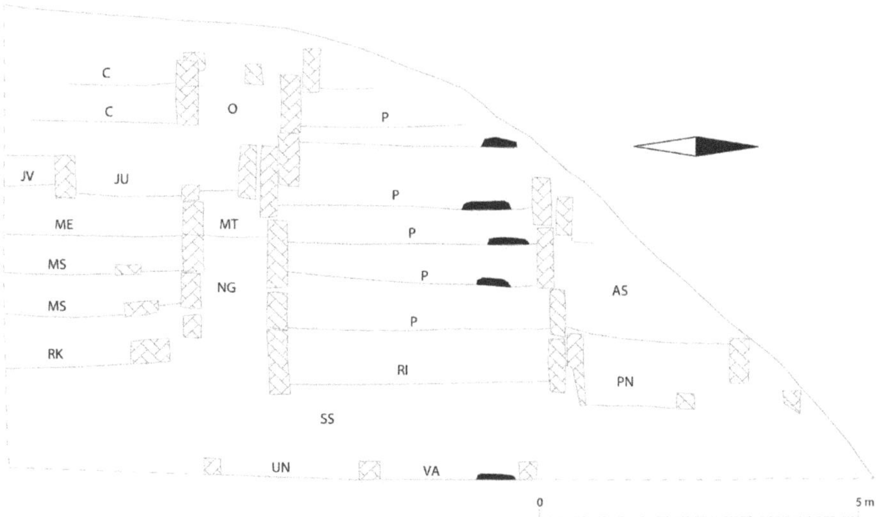

Figure 7.2 Schematic west section of sounding 4H/G at Aşıklı Höyük (redrawn from Esin/Harmankaya 1999, 93, fig. 9). (Figure prepared by M. Oberendorff.)

buildings in sounding 4H/G can be traced through a long development of reconstruction phases, up to eight of which have been documented, amounting to some 5 m of archaeological deposits, before the sequence is cut short by erosion of the uppermost deposits. This is an impressive testimony to the building continuity practiced at Aşıklı Höyük.

Çatalhöyük East

Çatalhöyük East is located on the Konya Plain along the former Çarsamba River. It measures about thirteen hectares and stands some 17 m above the surrounding plain. The site was first excavated between 1961 and 1965 by James Mellaart and his team, and it has been the focus of renewed investigations since 1993 by an international project with teams from Britain, the United States, Poland, and Turkey. The main excavated sequence belongs to the Ceramic Neolithic, with levels XII through VI dating to the Early Ceramic Neolithic, between 7000 and 6400 Cal BC, and levels V through I dating to the Late Ceramic Neolithic, between 6400 and 6000 Cal BC. The building continuity discussed in this chapter only occurs in the Early Ceramic Neolithic at the site (Düring 2006; Hodder 1996; 2011a; Mellaart 1967).

The Çatalhöyük buildings are not dissimilar to those of Aşıklı Höyük. They are also constructed with mud slabs of irregular dimensions. Furthermore, there is a complete lack of exterior doors, at least in the early part of the Çatalhöyük sequence, and buildings were also constructed directly adjacent to one another. This means that here too buildings were accessed from the roof. At Çatalhöyük, negatives of the ladders, where the ladder would have hampered wall plastering, have been preserved in numerous buildings. Within the buildings a more or less standardized set of features was found with a nearly uniform ordering of space within the main rooms (Düring 2006; Mellaart 1967).

In levels XII to VI a pronounced building continuity characterizes Çatalhöyük. The best evidence for this comes from what is now called the 'South Area', where Mellaart excavated in the 1960s, and which has also been a main focus of the new project at the site directed by Ian Hodder. At Çatalhöyük we also see that buildings are in most cases built on top of their predecessors, replicating their dimensions and orientation in the process. In the South Area, a sequence of up to seven buildings in sequence can be documented (Figure 7.3). The stratigraphic deposition over this time period was quite substantial, with a difference of up to 7.5 m between the floors of the earliest and the latest buildings (Düring 2006: 218–229). Common estimates for the use-life of buildings at Çatalhöyük are in the range of 60 years (Cessford 2005; Matthews 2006; Mellaart 1967: 50–51). This would imply that some of the buildings excavated in the South Area have a continuous history of use over a period of about 420 years!

Ilıpınar

The site of Ilıpınar is located in the Marmara region, on the eastern end of the Iznik Lake. The site measures about two and a half hectares and stands some 4 m above the surrounding plain. Excavations at Ilıpınar started in 1987 and lasted until 1997. The site contains layers dating to the Neolithic, Chalcolithic, Early Bronze Age, and Byzantine Period. Of concern to us here are the Neolithic levels X to VII, to be dated between about 6000 and 5700 Cal BC (Roodenberg 1995; 1999; 2011).

The buildings dating to these layers are large rectangular structures of about 30 m^2. They were constructed with either posts in combination with mud slabs or with mud slabs only (less common). These buildings had gabled roofs, a point borne out by the find of a number of post-stands in the center of buildings. Houses are freestanding and separated by narrow alleys. It has been observed that the technology of the Ilıpınar buildings is more akin to that of Neolithic sites in the Balkans than that of Central Anatolian sites (Roodenberg 1995; 1999).

Because we are primarily dealing with wooden architecture at Ilıpınar, building continuity takes on a form that is different from the superimposition seen at Çayönü, Aşıklı Höyük, and Çatalhöyük. In the so-called big square at Ilıpınar, a large number of posthole buildings were found, a few of which had their associated floor surfaces preserved. It is clear that at any one point in time a number of more or less square wooden buildings would have occupied this area. What is of interest here is that buildings were generally rebuilt on the same spot, with similar dimensions and a similar orientation, although they were generally shifted over a short distance in order to avoid

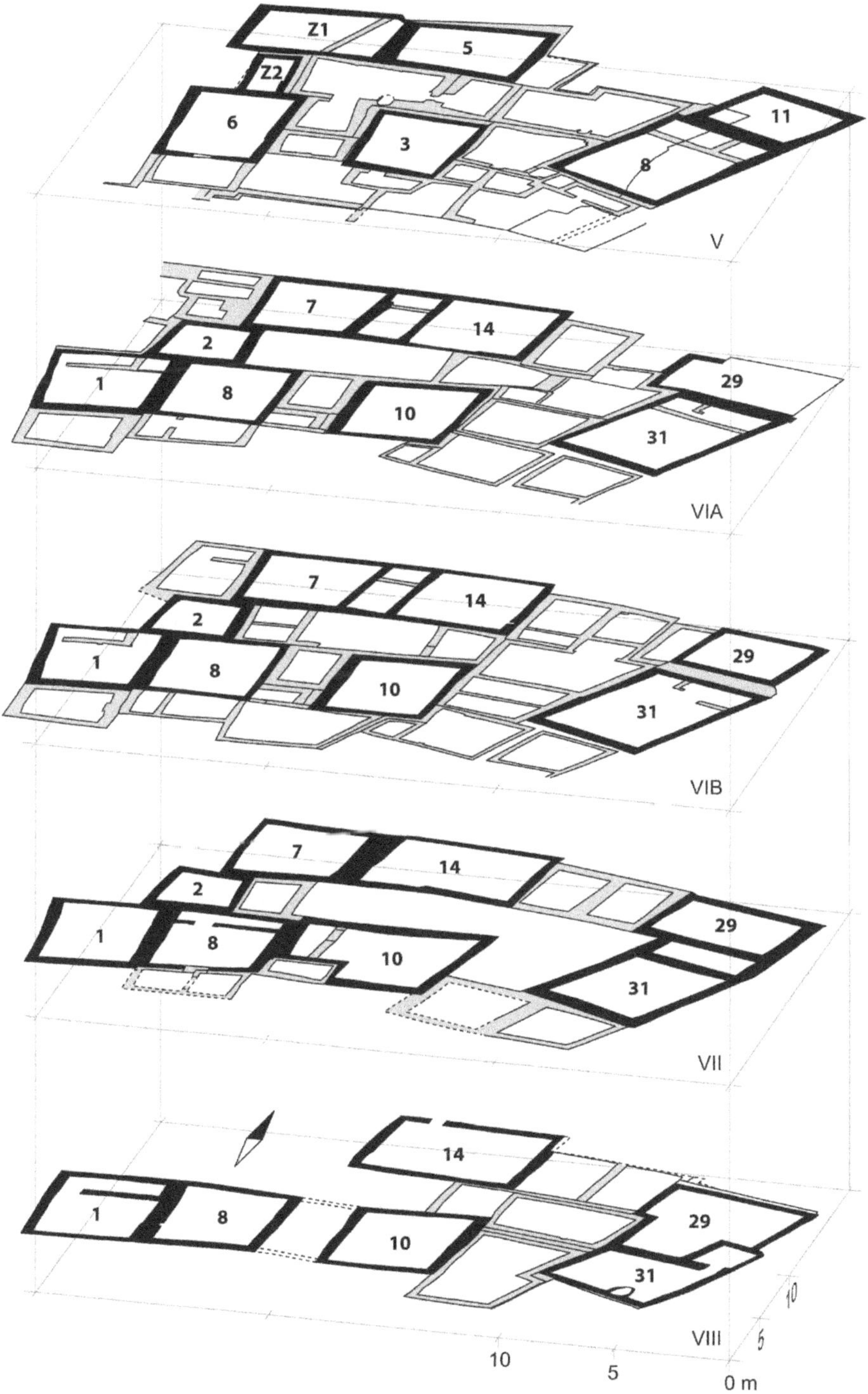

Figure 7.3 Diachronic view of the building sequence in the south area at Çatalhöyük. (Figure prepared by M. Oberendorff.)

the remnants of earlier posts. Up to seven buildings in sequence were documented (Roodenberg 1995: 38). These buildings at Ilıpınar change much more over a series of rebuilding episodes than do the buildings at Çayönü, Aşıklı Höyük, and Çatalhöyük, and in this sense they are the least 'continuous' of the buildings discussed in this chapter.

BUILDING CONTINUITY IN CONTEXT

The fact that there are examples across the Anatolian Neolithic of extraordinary degrees of building continuity is not to be regarded as implying that the Anatolian Neolithic is structured by a pervasive set of values that can be traced across this large territory. If anything, the Neolithic horizon of Anatolia can be characterized as one that is highly diverse and consists of a series of regional cultural traditions (Düring 2011: chapter 5). Moreover, there are many Neolithic sites in Anatolia where building continuity does not seem to have been of great importance. Here mention can be made, for example, of Lake District sites such as Bademağacı;, Kuruçay, and Hacılar; of Ulucak, Ege Gübre, and Hoca Çeşme in Aegean Anatolia; Yumuktepe in Cilicia; and Tepecik-Çiftlik and Köşk Höyük in central Anatolia. On the other hand, there are also sites in Syria with good evidence for building continuity, for example in some layers of Tell Sabi Abyad (Akkermans, Brüning and Kaneda 2011: 2) and at Tell Halula (Guerrero et al. 2009: 386).

Countless vernacular building traditions have been documented both in ethnography and in archaeology, but the degree of building continuity discussed in this chapter seems to be a phenomenon that is particular to the Neolithic in Anatolia and beyond. Rather, in ethnographic research on mud brick buildings and in most archaeological sites, one typically encounters the haphazard use of older walls as a foundation for new buildings (Aurenche 1981: 104; Facey 1997; Horne 1994; Peters 1982). The prevalent practice is one in which buildings are continuously altered to adjust to the changing size, needs, and fortunes of household groups, and building spaces are transferred to other households in exchange for valuables (Peters 1982: 226; Stone 1981). Such dynamic relations between buildings and their inhabitants clearly differ from those documented in Neolithic Anatolia and bring out the exceptionality of the building continuity documented.

UNDERSTANDING BUILDING CONTINUITY

Given the fact that marked building continuity occurs regularly in Neolithic sites in both Anatolia and Syria, one wonders how this phenomenon is best understood. I will start by considering some functionalist explanations and subsequently discuss perspectives that focus more on social aspects. In understanding building continuity, the broad comparative angle taken in this chapter may be helpful. Archaeologists working at particular sites have often come up with functional explanations for building continuities in their data sets, which include first, that building continuity can be related to the need to create secure foundations; second, that building continuity is in part the result of the restrictions of space occurring in densely occupied settlements; and third, that building continuity is related to the emergence of privately owned plots, and thus to rights of inheritance of land and houses on a household level.

First, the idea that building continuity might be related to the need for building on solid foundations appears to make good sense (Esin et al. 1991: 130; Mellaart 1967: 67). Especially in the Central Anatolian Neolithic sites of Aşıklı Höyük and Çatalhöyük, building on top of an older structure would have ensured that buildings were more securely founded, and this would have had beneficial effects on their long-term structural strength. For example, at Çatalhöyük there are a number of cases, such as 'Building 2', in which buildings were constructed on what had previously been empty midden areas, and these buildings often started to sag and became unstable, and consequently had to be abandoned earlier than buildings that were constructed on top of older buildings (Farid 2007). Whereas the need for creating secure foundations is certainly one factor in explaining building continuity in Central Anatolia, it does not sufficiently explain building continuity in the Anatolian Neolithic as a whole. Moreover, the argument works less well for the sites of Çayönü and Ilıpınar. As has already been discussed, the buildings at Çayönü were often slightly shifted when they were rebuilt. If the main concern was to create secure foundations, it would have made more sense to build exactly on top of earlier walls. Further, at Çayönü the buildings are constructed on stone foundations, so there is no real structural need for building on earlier structures. At Ilıpınar building continuity is obviously not related to the need for secure foundation, given that buildings were predominantly built around wooden poles that were sunk into the ground, and that they were not constructed on top of their predecessors.

Second, it has been argued at this point that the buildings under study remained in place and retained their dimensions simply because they were surrounded by other structures on all sides. Again, this argument applies differently to the various sites under discussion. At Ilıpınar buildings were more or less freestanding, and building continuity is less pronounced than at the other sites (buildings move more horizontally and change in dimensions over time). At Aşıklı Höyük and Çatalhöyük the idea that building continuity is determined by the restrictions of space works best. However, at these sites there are also problems with this argument if the data are scrutinized. There are empty spaces in between buildings at both sites, and yet one never sees the expansion of existing buildings onto those spaces. At Çatalhöyük there are a number of buildings that were inserted into a midden area. Typically these buildings often have deviant proportions, few features that relate to the ritual domain, such as burials and moldings, and no successors (Düring 2006: 218–220). These noncontinuous buildings can be contrasted with the continuous buildings that are the topic of this chapter. Furthermore, even in densely built environments one will usually see modifications and reorganizations of buildings, in which rooms of one building may be taken over by the building adjacent to it or an existing building may be split into smaller units (Peters 1982; Wirth 2000: 348). The fact that this does not happen at Aşıklı Höyük and Çatalhöyük over the course of centuries is not explained sufficiently by the fact that buildings are hemmed in by others. At Çayönü all buildings were freestanding on all sides, which means that there was plenty of opportunity to shift the location of buildings. Despite this surrounding space, buildings were constructed directly on top of each other and retained the same dimensions (Özdoğan and Özdoğan 1998). Conditions similar to those in Çayönü occur in almost every village, and it needs to be demonstrated why this leads to building continuity at Çayönü but not in other villages.

Third, it has been argued that building continuity in Anatolia is best understood as the emergence of privately owned family plots. As Roodenberg (2011: 953) put it for the site of Ilıpınar: "houses were rebuilt on the same spot up to five times; from this we can deduce that land parcels remained in the same family for generations." This interpretation is overly simplistic. Ethnographic and archaeological studies have shown that households are very dynamic units, constantly expanding and shrinking in size and adapting their houses accordingly (Goodman 1999; Peters 1982; Souvatzi 2008). Household groups fluctuate considerably in size, and a single household would give rise to multiple households over the course of only a few generations. It is unlikely that households would construct and occupy houses of exactly the same size over hundreds of years. In most societies, the household is a key social institution, and houses are transferred from one generation to the next, but rather than resulting in building continuity, this leads to the constant modification of buildings and settlement space. In addition, the idea that households in Neolithic Anatolia owned plots and houses in the village and transferred these to the next generation is probably a projection of modern ideas concerning ownership onto a period in which such concepts most likely did not exist (Souvatzi 2013). Instead, numerous ethnographic studies of small-scale farming societies show that the allocation of land and houses often took place at a communal level and that the idea of ownership at the household level might have been nonexistent (Lane 1994; Souvatzi 2013).

It would of course be possible to argue at this point that building continuity served different functional purposes at different sites; for example, the need for secure foundations might be argued to explain the practice at Çatalhöyük and Aşıklı Höyük, but not at other sites. Such a position, however, negates the fact that there are important similarities in the building continuities of Neolithic Anatolia. Moreover, this building continuity appears to be particular to the Neolithic and not to later periods in Anatolia. Both considerations underline the point that functional factors can only partly explain the phenomenon of building continuity.

The pertinent question then is what social practices and values might have been connected to these enduring buildings. A second, related issue would be why these social practices and values were no longer relevant after the Neolithic. How can we best understand the extraordinary efforts toward replicating preexisting buildings and keeping older buildings from being damaged? In my opinion, it makes sense to view houses as cultural resources on which people could draw to define individual and group identities (see also Souvatzi 2008). In other words, people used houses as metaphors for social relations, defining themselves and others in terms of membership of one house or another. In this respect, houses may have become the physical and symbolic manifestations of a social group that defined itself in relation to the house. This is not to argue that in places where house reproduction is absent, houses were less meaningful to social groups (Gerritsen 2007; Souvatzi 2008: 199–201), but simply that building continuity was not culturally important.

The idea that building continuity in the Anatolian Neolithic was linked with social practices and values is not original: it has been argued for some sites, such as Çayönü and Çatalhöyük, by various scholars, including Mehmet Özdoğan, Ian Hodder, Peter Pels, Maurice Bloch, and myself, and I will present these ideas shortly. First, however, I would like to note that so far there has been little comparative analysis in which data from different sites are taken into account, which is the main purpose of this chapter.

Pride of place must be given to a paper entitled Building of Cult and the Cult of Building (Özdoğan and Özdoğan 1998), in which it was argued that many buildings of the Anatolian Neolithic were far too well preserved in comparison with building remains from other periods, and that this can only be accounted for by the fact that buildings were carefully 'buried' at the end of their use life. The Özdoğans discuss examples from the sites of Aşıklı Höyük, Canhasan I, and Çatalhöyük in Central Anatolia, where buildings were often preserved up to great heights and were often infilled with clean material and contained no artifacts unless deliberately placed. Further, they discuss the case of Çayönü, where buildings never disturbed the remains of predecessors, despite the fact that it would have been possible to reuse the stones from older walls.

The idea that great care was taken to avoid damaging buildings and that they were carefully 'buried' at many Neolithic sites in Anatolia fits the data remarkably well. At Ilıpınar buildings were generally shifted somewhat when they were reconstructed. Roodenberg (2008) argues that this was done in order to avoid the rotting wooden stakes of older buildings, but it could also be seen as an effort not to disturb them. At Çatalhöyük many examples of behavior related to the cleaning and sealing of buildings have been documented, including the removal of posts, the blocking of passageways and ovens, the removal of all objects, the removal of molded figures attached to the walls, the application of a final plaster layer, the placing of special deposits, and infilling of the building with clean deposits (Düring 2006: 160–162; Hodder and Cessford 2004; Matthews 2006). Similarly, Özdoğan and Özdoğan describe how doorways were carefully blocked of a building dating to the cell building phase at Çayönü, before its successor was constructed on top (Özdoğan and Özdoğan 1998: 590). One other element of interest at Çatalhöyük is that the walls are constructed with highly distinctive soils, with colors and textures easily distinguishable from those of the surrounding matrix. It is likely that the material for these bricks was obtained from the landscape surrounding the site. This is, of course, a remarkable situation. In most excavations, loam walls have matrices that are very similar to the mound as a whole, and one prime reason for this is that soils from tells were often used for making bricks. The fact that bricks at Çatalhöyük were produced from distinctive off-site soils might be related to the desire not to disturb older buildings.

In short, the idea that buildings were buried in the Anatolian Neolithic, as proposed by the Özdoğans, has much going for it. The model successfully identifies certain practices particular to the Anatolian Neolithic. What they do not explain, however, is why the integrity and reproduction of buildings was important to people in this cultural horizon.

More interpretative work on the social meaning of building continuity has been done mainly for the site of Çatalhöyük. For this site, it is possible to document first, that buildings exhibit an extraordinary degree of building continuity, with up to seven buildings in sequence documented that would have encompassed a time span of more than four centuries and a depositional sequence of up to 7 m (Düring 2006: 218–229); second, that at least some buildings appear to have acquired some form of identity, as manifested for example in the reproduction of similar wall painting motifs or molded features in successive buildings in sequence (Düring 2005: 18–20; 2006: 221–222; Hodder 2006: 141–168); and third, that some buildings became more ritually elaborate in the course of their history, something that is expressed especially clearly in the amount and type of burials below their floors, with a few infant burials at first and in the course of various rebuilding episodes leading up to large numbers of subfloor burials including adults of both sexes as well as children (Düring 2005; 2006: 223–225).

Various scholars have argued that the houses at Çatalhöyük are probably best understood as Lévi-Straussian *houses* (Bloch 2010; Düring 2006: 225–226; 2007; Hodder and Pels 2010). Lévi-Strauss suggested that in some societies houses form the central institution around which group membership is defined. The *house* concept of Lévi-Strauss (1983: 174) was initially defined as

> a corporate body holding an estate made up of both material and immaterial wealth, which perpetuates itself through the transmission of its name, its goods, and its titles down a real or imaginary line, considered

> legitimate as long as this continuity can express itself in the language of kinship or of affinity and, most often, of both.

The concept has subsequently been explored in various edited volumes (Beck 2007; Carsten and Hugh-Jones 1995; Joyce and Gillespie 2000). Three key features define *houses:* first, the ideal of continuity; second, the passing down of some form of valued property; and third, the strategic exploitation of the language of kinship or affinity. Houses form hierarchical systems, with some houses with a long history acting as a central focus for an affiliated group of households.

The concept of house societies—in which the house is of key importance in the constitution and reproduction of social groups—is of great value for understanding why building continuity was of importance in the Anatolian Neolithic, beyond Çatalhöyük. First, the *ideal of continuity* is clearly visible in the building continuity evidenced at sites such as Çayönü, Aşıklı Höyük, Çatalhöyük, and Ilıpınar, as is the *passing down of some form of valued property,* which is done by the reproduction of houses in exactly the same place and with the same dimensions as their predecessors. Second, the importance of houses can help us to understand why houses are treated the way they are: that they are carefully sealed and reproduced. Further, it can help us to understand why the interior arrangements of buildings are often reproduced. For example, in sounding 4H/G at Aşıklı Höyük, various hearth sequences have been documented, in which the hearth is consistently placed in the same corner of a room (Düring 2005; 2006). The individual hearths in these sequences are often separated by deposits of up to 40 cm of soil. There is no functional reason why these hearths should consistently be in the same corner of successive buildings. We have to assume, therefore, that continuity in hearth locations was important to society in one way or another.

Finally, the 'house societies' concept can help us to understand why some houses are more ritually elaborate than others at some sites of the Anatolian Neolithic. In particular, at Çatalhöyük there is good evidence for the ritual differentiation of houses. Some houses have large subfloor burial populations, up to sixty-two in one case. The bodies in these burials cannot derive only from the deaths in the associated household. This conclusion, that people from other households were also buried in these buildings, can be drawn on the basis of the use life of these buildings, which is about sixty years, and a postulated household size of about four to six people (Düring 2003). If these houses are analyzed from a diachronic perspective, it can be demonstrated that the buildings richest in burials are generally those with the longest history of buildings in sequence (Düring 2006: 218–229). The pattern that we see is remarkably consistent, with initially limited numbers of burials in these buildings in their 'younger' stages, often predominantly young juveniles, increasing in burial populations over time and incorporating all age categories. Thus, at Çatalhöyük we can demonstrate *the strategic exploitation of the language of kinship or affinity,* the third characteristic of a house society. Through a series of rebuilding episodes, the status of particular buildings became central to social groups that included a number of households and became the locus for various types of ceremonies. These burials are the most visible to us archaeologically, but on the basis of ethnographic parallels we should also think of weddings, initiations, and the like (Kirch 2000). This last argument cannot be made at present for the other sites discussed in this chapter: Çayönü, Aşıklı Höyük, and Ilıpınar.

For all sites, we can ask, however, whether the people living in these settlements were aware of the marked building continuity of the houses they lived in, which must have encompassed centuries for all sites under consideration. Bradley has argued that specific memories and hence specific histories typically fade in a time span from two generations to 200 years at most (Bradley 2003: 221–222). Historical ties with the past over longer periods require special media that can transfer information. In preliterate societies other types of technologies exist, including the construction of impressive monuments and representations, a practice that has been labelled 'inscribed memory' by Connerton (1989: 72–79; cf. Rowlands 1993: 142–146). However, another means of transferring history beyond the limits of oral traditions is through buildings embodying group histories (Waterson 2003: 46). In this sense, it is appropriate that Hodder has referred to Çatalhöyük houses as memory archives and that specific buildings had specific histories that were important to social groups, as exemplified in continuities in iconographical motifs in buildings, and the targeted retrieval of moldings and skulls from earlier building phases (Düring 2006: 218–229; Hodder 2006: 141–168; 2007; Hodder and Cessford 2004).

At present, evidence for historical knowledge of continuous buildings in the Anatolian Neolithic is best documented at Çatalhöyük, with some circumstantial evidence at Aşıklı Höyük in the form of hearth locations. No doubt, similar data could be retrieved at the sites of Çayönü and Ilıpınar if they were more carefully excavated and analyzed. To me, it seems likely that at least some of the older people at the site were aware of longer sequences of buildings. Thus, it may be that by connecting the past to the present, the existing social order was legitimized (Bradley 2002: 11–12; Lucas 2005: 62–64; Van Dyke 2003: 188). In particular, by rebuilding houses in the same place, with the same dimensions, orientations, and internal furnishing, people were actively presencing the past to serve social agendas in the present.

DISCUSSION AND CONCLUSION

Archaeologists are trained to be wary of claims of cultural continuity over long time periods. Even in cases in which we are dealing with uninterrupted cultural sequences, there are instances of practices that have changed enormously in meaning or context, and assuming continuity is a dangerous trap (Kyriakidis 2007). For such reasons, analogical interpretation using the direct historical approach is no longer popular in archaeology (Verhoeven 2005).

By contrast, the building continuity of the Anatolian Neolithic suggests that in this period people might have been actively constructing the present past. Over hundreds of years, buildings were reconstructed in exactly the same place, with the same dimensions, and very often with the same interior arrangement of features. One can only conclude that the exact reproduction of buildings must have had great social significance to societies in the Anatolian Neolithic. These buildings provided a means for groups in society to fix themselves around the material house, and in this manner to objectify their existence (Hodder 2011b: 159–160).

The question why houses with long pedigrees became manifestations of social groups, their history, and their reproduction in various Neolithic sites in Anatolia is difficult to answer. Without doubt, people were involved in new ways with each other, in larger communities than existed previously, and thus the house may have provided an anchor in a vastly more complex social world. Here, one could take the argument further and suggest that houses become most elaborate at sites where the community has reached an unparalleled size, as at Çatalhöyük, which is by far the largest community under consideration here. However, I would be wary of seeing houses simply as means to counteract social stress that came about from living in large communities. A site such as Ilıpınar, for example, is fairly small in scale and in all likelihood would not have outgrown a face-to-face community of about 300 people (Roodenberg 1999: 197), a size typical for Neolithic sites across much of the Aegean and Near East (Düring 2007b: 174). Further, as has been mentioned before, many Neolithic sites in Anatolia lack the type of building continuity that is the topic of investigation here.

Another suggestion could be that Neolithic societies became increasingly complex, and that houses as manifestations of social groups made it possible to have profound power differences between people cloaked in a vocabulary of kinship and belonging (Gillespie 2007: 41–42; Kuijt 2000). If we follow this line of argumentation, then the emergence of houses in the Anatolian Neolithic could be interpreted mainly as an index of socially complex societies without institutionalized hierarchical structures. The fact that we do not appear to have building continuity in the subsequent Chalcolithic in Anatolia could mean that these societies were less socially complex, that is, with fewer power differences, or that they used other means than houses to negotiate such differences. Here, it is of interest to note that *houses* as manifestations of social groups, in the sense proposed by Lévi-Strauss, have been argued to occur in the Chalcolithic and Early Bronze Age elsewhere in the Near East, for example in the southern Levant (Chesson 2007).

To sum up, I have discussed building continuity in a number of sites spread across the Anatolian Neolithic, which are thousands of kilometers and millennia apart. At the same time, there are many other sites in the Anatolian Neolithic where such building continuity is not in evidence. At sites where we do have building continuity, however, such as Çayönü, Aşıklı Höyük, Çatalhöyük, and Ilıpınar, there are striking parallels in how buildings are sealed and reproduced over very long sequences lasting several hundreds of years. To explain this type of building continuity, various functional considerations have been put forward, but none of these can sufficiently explain the extraordinary degree of building continuity documented. In consequence, there must have been social practices and values tied up with these buildings. I have argued that these buildings were most likely manifestations of social groups—in the sense argued for by Lévi-Strauss in his house society model—and that the reproduction of their house was of key importance to these groups. Finally, I have argued that houses might have been important to facilitate the existence of power differences in societies with an egalitarian ethos by cloaking these differences in the language of kinship and belonging. In the Anatolian Neolithic, I would then argue, people actively constructed the present past to objectify and anchor their social order.

In conclusion, in some societies of the Anatolian Neolithic, houses were the spatiotemporal manifestation of social groups that constructed their shared histories and sense of belonging in relation to particular buildings. Space and time, then, were not backdrops to social life but essential ingredients in the materialized order of society.

REFERENCES

Akkermans, P. M. M. G., M. Brüning and A. Kaneda. 2011. Foundation or *rendez-vous?* Constructing platforms in Late Neolithic Syria. In: B. S. Düring, A. Wossink and P. M. M. G. Akkermans (eds.), *Correlates of Complexity: Essays in Archaeology and Assyriology Dedicated to Diederik J. W. Meijer in Honour of his 65th Birthday,* 1–13. Leiden: Nederlands Instituut voor het Nabije Oosten.

Aurenche, O. 1981. *La Maison Orientale: l'Architecture du Proche Orient Ancien.* Paris: Paul Geuthner.

Beck, R., ed. 2007. *The Durable House: Architecture, Ancestors and Origins.* Carbondale: Center for Archaeological Investigations.

Bıçakçı, E. 1998. An essay on the chronology of the Pre-Pottery Neolithic settlements of the East-Taurus region (Turkey). In: G. Arsebük, M. J. Mellink and W. Schirmer (eds.), *Light on Top of the Black Hill: Studies Presented to Halet Çambel,* 137–150. Istanbul: Ege Yayınları.

Bloch, M. 2010. Is there religion at Çatalhöyük . . . or are there just houses? In: I. Hodder (ed.), *Religion in the Emergence of Civilization: Çatalhöyük as a Case Study,* 146–162. Cambridge: Cambridge University Press.

Bradley, R. 2002. *The Past in the Prehistoric Societies.* London: Routledge.

Bradley, R. 2003. The translation of time. In: R. M. Van Dyke and S. E. Alcock (eds.), *Archaeologies of Memory,* 221–227. Oxford: Blackwell.

Carsten, J. and S. Hugh-Jones, eds. 1995. *About the House: Lévi-Strauss and Beyond.* Cambridge: Cambridge University Press.

Cessford, C. 2005. Estimating the Neolithic population of Çatalhöyük. In: I. Hodder (ed.), *Inhabiting Çatalhöyük: Reports from the 1995–1999 Seasons,* 323–326. Cambridge: McDonald Instute for Archaeological Research.

Chesson, M. S. 2007. House, town, field, and wadi: landscapes of the Early Bronze Age southern Levant. In: R. Beck (ed.), *The Durable House: House Society Models in Archaeology,* 317–343. Carbondale: Center for Archaeological Investigations.

Connerton, P. 1989. *How Societies Remember.* Cambridge: Cambridge University Press.

Düring, B. S. 2003. Burials in context: the 1960s inhumations of Çatalhöyük East. *Anatolian Studies* 53: 1–15.
Düring, B. S. 2005. Building continuity in the Central Anatolian Neolithic: exploring the meaning of buildings at Aşıklı Höyük and Çatalhöyük. *Journal of Mediterranean Archaeology* 18: 3–29.
Düring, B. S. 2006. *Constructing Communities: Clustered Neighborhood Settlements of the Central Anatolian Neolithic*. Leiden: Nederlands Instituut voor het Nabije Oosten.
Düring, B. S. 2007a. The articulation of houses at Neolithic Çatalhöyük, Turkey. In: R. Beck (ed.), *The Durable House: House Society Models in Archaeology*, 130–153. Carbondale: Center for Archaeological Investigations.
Düring, B. S. 2007b. Reconsidering the Çatalhöyük Community: From Households to Settlement Systems. *Journal for Mediterranean Archaeology* 20: 155–182.
Düring, B. S. 2011. *The Prehistory of Asia Minor: From Complex Hunter-Gatherers to Early Urban Societies*. Cambridge: Cambridge University Press.
Duru, R. 1999. The Neolithic of the Lake District. In: M. Özdoğan and N. Başgelen (eds.), *The Neolithic of Turkey*, 147–156. Istanbul: Ege Yayınları.
Esin, U., E. Bıçakçı, M. Özbaşaran, N. Balkan-Atlı;, D. Berker, I. Yağmur and A. Korkut-Ali. 1991. Salvage excavations at the Pre-Pottery Neolithic site of Aşıklı Höyük in Central Anatolia. *Anatolica* 27: 123–174.
Esin, U. and S. Harmankaya. 1999. Aşıklı. In: M. Özdoğan and N. Başgelen (eds.), *The Neolithic of Turkey*, 115–132. Istanbul: Arkeoloji ve Sanat Yayınları.
Facey, W. 1997. *Back to Earth: Adobe Building in Saudi Arabia*. London: Al-Turath.
Farid, S. 2007. Introduction to the South Area excavations. In: I. Hodder (ed.), *Excavating Çatalhöyük: South, North and KOPAL Area Reports from the 1995–1999 Seasons*, 41–58. Cambridge: McDonald Institute for Archaeological Research.
Gerritsen, F. 2007. Relocating the house: social transformations in Late Prehistoric Northern Europe. In: R. Beck (ed.), *The Durable House: House Society Models in Archaeology*, 154–174. Carbondale: Center for Archaeological Investigations.
Gillespie, S. D. 2007. When is a house? In: R. Beck (ed.), *The Durable House: House Society Models in Archaeology*, 25–50. Carbondale: Center for Archaeological Investigations.
Goodman, M. 1999. Temporalities of prehistoric life: household development and community continuity. In: J. Brück and M. Goodman (eds.), *Making Places in the Prehistoric World*, 145–159. London: UCL Press.
Guerrero, E., M. Molist, I. Kuijt and J. Afruns. 2009. Seated memory: new insights into Near Eastern Neolithic mortuary variability from Tell Halula, Syria. *Current Anthropology* 50: 379–392.
Hodder, I., ed. 1996. *On the Surface: Çatalhöyük 1993–5*. Cambridge: McDonald Institute for Archaeological Research.
Hodder, I. 2006. *The Leopard's Tale: Revealing the Mysteries of Çatalhöyük*. London: Thames and Hudson.
Hodder, I. 2007. Çatalhöyük in the context of the Middle Eastern Neolithic. *Annual Review of Anthropology* 36: 105–120.
Hodder, I. 2011a. Çatalhöyük: a prehistoric settlement on the Konya Plain. In: S. Steadman and G. McMahon (eds.), *Oxford Handbook of Anatolian Studies, 8000–323 BC*, 934–949. Oxford: Oxford University Press.
Hodder, I. 2011b. Human-thing entanglement: Towards an integrated perspective. *Journal of the Royal Anthropological Institute* 17: 154–177.
Hodder, I. and C. Cessford. 2004. Daily practice and social memory at Çatalhöyük. *American Antiquity* 69: 17–40.
Hodder, I. and P. Pels. 2010. History houses: a new interpretation of architectural elaboration at Çatalhöyük. In: I. Hodder (ed.), *Religion in the Emergence of*

Civilization: Çatalhöyük as a Case Study, 163–186. Cambridge: Cambridge University Press.

Horne, L. 1994. *Village Spaces: Settlement and Society in Northeastern Iran.* Washington and London: Smithsonian Institute Press.

Joyce, R. A. and J. Gillespie, eds. 2000. *Beyond Kinship: Social and Material Reproduction in House Societies.* Philadelphia: University of Philadelphia Press.

Kirch, P. V. 2000. Temples as 'Holy Houses': the transformation of ritual architecture in traditional Polynesian societies. In: R. A. Joyce and S. D. Gillespie (eds.), *Beyond Kinship: Social and Material Reproduction in House Societies,* 103–114. Philadelphia: University of Philadelphia Press.

Kuijt, I. 2000. Keeping the peace: ritual, skull caching, and community integration in the Levantine Neolithic. In: I. Kuijt (ed.), *Life in Neolithic Farming Communities: Social Organization, Identity, and Differentiation,* 137–164. New York: Kluwer Academic Press.

Kyriakidis, E. 2007. Finding ritual, calibrating the evidence. In: E. Kyriakidis (ed.), *The Archaeology of Ritual,* 9–23. Los Angeles: Cotsen Institute of Archaeology.

Lane, P. J. 1994. The temporal structuring of settlement space among the Dogon of Mali: an ethnoarchaeological study. In: M. Parker Pearson and C. Richards (eds.), *Architecture and Order: Approaches to Social Space,* 196–216. London: Routledge.

Lévi-Strauss, C. 1983. *The Way of the Masks* [translated from the French by Sylvia Modelski]. London: Jonathan Cape.

Lucas, G. M. 2005. *The Archaeology of Time.* London: Routledge.

Matthews, W. 2006. Life-cycle and -course of buildings. In: I. Hodder (ed.), *Çatalhöyük Perspectives: Themes from the 1995–1999 Seasons,* 125–150. Cambridge: McDonald Institute for Archaeological Research.

Mellaart, J. 1967. *Çatal Hüyük: A Neolithic Town in Anatolia.* London: Thames and Hudson.

Özbaşaran, M. 2011. The Neolithic on the plateau. In: S. Steadman and G. McMahon (eds.), *Oxford Handbook of Anatolian Studies, 8000–323 BC,* 99–124. Oxford: Oxford University Press.

Özdoğan, A. 1999. Çayönü. In: M. Özdoğan and N. Başgelen (eds.), *The Neolithic of Turkey,* 35–63. Istanbul: Ege Yayınları.

Özdoğan, M., and A. Özdoğan. 1998. Building of cult and the cult of building. In: G. Arsebük, M. J. Mellink and W. Schirmer (eds.), *Light on Top of the Black Hill: Studies Presented to Halet Çambel,* 581–601. Istanbul: Ege Yayınları.

Peters, E. 1982. Landliche Bauweisen im Keban-Gebiet. *METU—Keban Publications* 7(1975): 217–232.

Roodenberg, J. J. 1995. Stratigraphy and architecture: the 'biq square'. In: J. J. Roodenberg (ed.), *The Ilıpınar Excavations I,* 35–76. Istanbul: Nederlands Instituut voor het Nabije Oosten.

Roodenberg, J. J. 1999. Ilıpınar: an early farming village in the Iznik Lake Basin. In: M. Özdoğan and N. Başgelen (eds.), *The Neolithic of Turkey,* 193–202. Istanbul: Arkeoloji ve Sanat Yayınları.

Roodenberg, J. J. 2008. Stratigraphy and architecture: the basal occupation levels (phases X and IX). In: J. Roodenberg and S. Alpaslan-Roodenberg (eds.), *Life and Death in a Prehistoric Settlement in Northwest Anatolia: The Ilıpınar Excavations III,* 1–34. Leiden: Nederlands Instituut voor het Nabije Oosten.

Roodenberg, J. J. 2011. Ilıpınar: a Neolithic settlement in the Eastern Marmara Region. In: S. Steadman and G. McMahon (eds.), *Oxford Handbook of Anatolian Studies, 8000–323 BC,* 950–967. Oxford: Oxford University Press.

Rosenberg, M., and A. Erim-Özdoğan. 2011. The Neolithic in Southeastern Anatolia. In: S. Steadman and G. McMahon (eds.), *Oxford Handbook of Anatolian Studies, 8000–323 BC,* 125–149. Oxford: Oxford University Press.

Rowlands, M. 1993. The role of memory in the transmission of culture. *World Archaeology* 25: 141–151.

Souvatzi, S. 2008. *A Social Archaeology of Households in Neolithic Greece: An Anthropological Approach*. Cambridge and New York: Cambridge University Press.

Souvatzi, S. 2013. Land tenure, social relations and social landscapes. In: M. Relaki and D. Catapoti (eds.), *An Archaeology of Land Ownership*, 21–45. London: Routledge.

Stone, E. T. 1981. Texts, architecture and ethnographic analogy: patterns of residence in Old Babylonian Nippur. *Iraq* 43: 19–33.

Van Dyke, R. M. 2003. Memory and the construction of Chacoan society. In: R. M. Van Dyke and S. E. Alcock (eds.), *Archaeologies of Memory*, 180–200. Oxford: Blackwell.

Verhoeven, M. 2005. Ethnoarchaeology, analogy, and ancient society. In: S. Pollock and R. Bernbeck (eds.), *Archaeologies of the Middle East: Critical Perspectives*, 251–270. Oxford: Blackwell.

Waterson, R. 2003. The immortality of the house in Tana Toraja. In: S. Sparkes and S. Howel (eds.), *The House in Southeast Asia: A Changing Social, Economic and Political Domain*, 34–52. London: Routledge Curzon.

Wirth, E. 2000. *Die Orientalische Stadt im Islamischen Vorderasien und Nordafrika: Städtische Bausubstanz und Räumliche Ordnung, Wirtschaftleben and Soziale Organisation*. Mainz: Philipp von Zabern.

Yakar, J. 1991. *Prehistoric Anatolia: The Neolithic Transformation and the Early Chalcolithic Period*. Tel Aviv: Tell Aviv University Press.

Yakar, J. 1994. *Prehistoric Anatolia: The Neolithic Transformation and the Early Chalcolithic Period, Supplement*. Tel Aviv: Tell Aviv University Press.

8 Communicating over Space and Time in the World of the Iceman

Robin Skeates

INTRODUCTION

In her celebration of the multiple modes of human communicating, the anthropologist Ruth Finnegan (2002) defines communication as a dynamic interactive process, made up of a variety of actions and experiences, created by active participants as they interconnect with and influence each other. Within this incessant, complex, and multisensory process, people draw on a wide array of bodily and environmental resources to communicate with others over space and time. These include sounds, smells, movements, touches, and sights, as well as the material things with which archaeologists are familiar. This perspective certainly highlights some of the limitations of the surviving archaeological evidence, but it should also encourage archaeologists to further develop theories of communication and to explore appropriate questions about the connections between people, places, and things over multiple scales and conceptions of space and time in the past.

This chapter attempts to rise to this challenge, with reference to the archaeology of the north Italian Early Copper Age, and with particular attention to (1) the human body and its biography and mobility; (2) the ordering of meaningful human places and activities across the historic cultural landscape; and (3) the production, movement, and consumption of commodities of exchange, which have the capacity to extend human communicating beyond the constraints of the body by travelling over space and time. A central case study is provided by the well-preserved body and equipment of the 'Iceman'—extensive scientific research on which is helping to transform our understanding of later prehistoric human mobility and connectivity.

THE ICEMAN: WHERE AND WHEN

Scientific studies of the 'Iceman' have provided an excellent source of detailed data concerning the spatial, temporal, and social dimensions of this prehistoric person's life and death. His remarkably well-preserved body, clothing, and equipment were found in 1991, high up in the Ötztal Alps, at an altitude of 3,210 m, in a gully close to the Tisen Pass, just inside the Italian border (Figure 8.1). (For general summaries, see Spindler 1995; and Fleckinger and Steiner 1998). The remains have been radiocarbon dated to between 3350 and 3120 cal BC (e.g., Bonani et al. 1994; Hedges et al. 1992), which situates them within the time span of the Italian Early Copper Age.

Konrad Spindler, who led the scientific investigation of the Iceman, interpreted this extraordinary archaeological find as an 'isolated individual' (Spindler 1995: 236), whose practical equipment, including materials for repair, enabled him to lead a largely self-sufficient life on the margins of society (for example, as an outsider or outlaw, a shaman or priest, a prospector for metal ores, a traveller or trader, a hunter or transhumant shepherd), and as someone who met a 'lonely death' (Spindler 1996: 250) in a remote wild place, high in the Alps. But this influential interpretation of the man as an individual can be questioned, on both theoretical and evidential grounds. The anthropologist Marilyn Strathern (1988) has persuasively contrasted the Western notion of the individual with the Melanesian understanding of persons or 'dividuals' as made up of social relationships with others, to the extent that they are perceived to be composed of parts of many other persons. It follows, then, that the Iceman might best be understood not as an individual but as a man whose body, clothes, and equipment were also

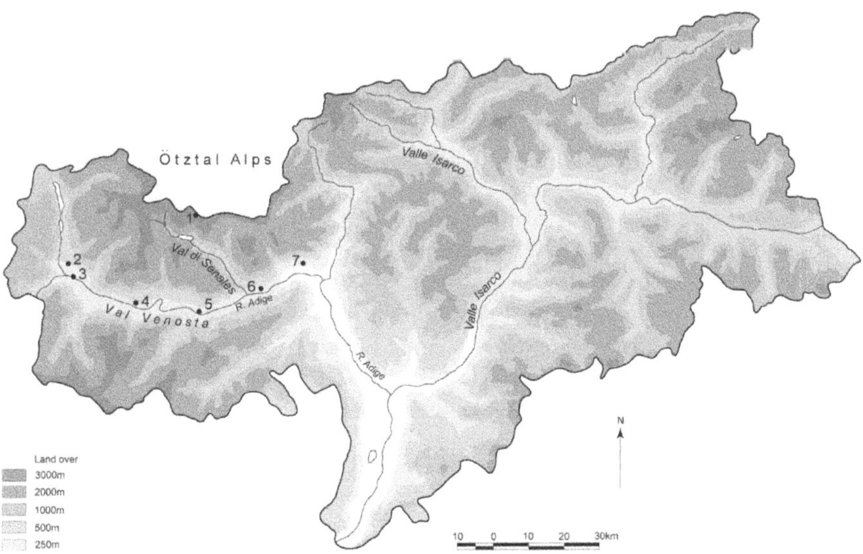

Figure 8.1 Selected Copper Age sites in the Bolzano province mentioned in the text: (1) 'Iceman' (Similaun), (2) Tartscher Bichl (Malles Venosta), (3) Sölesboden (Glorenza), (4) Lasa, (5) Laces, (6) Castel Juvale (Naturno), (7) Lagundo. Drawing by Yvonne Beadnell.

connected in many possible ways to other dividuals. Even Spindler (1995: 236–254), who ultimately believed that the man was an itinerant herder, acknowledged the need to integrate him into the economy and society of contemporary village life. He consequently highlighted four lines of archaeological evidence that indicated the man was, at least temporarily, in close contact with a farming community. First, the fragments of glume and husk, and pollen grains, of einkorn (*Triticum monococcum*) and barley (*Hordeum vulgare var. nudum*) found in his clothes and in one of his birch-bark containers suggest that the man had been at a village at harvest time where grain was threshed, possibly participating in the work at a time in the agricultural calendar when labor was much in demand. Second, the positioning and precise draftsmanship of the tattoos on the man's back show that they were applied by another person, and quite possibly for healing purposes, given their concentration along the lumbar spine where degenerative changes to the bone were also identified. Third, the variable quality of the repairs to the man's clothing, ranging from clumsy to more skilled work, suggest more than one person's work. Fourth, the man's death was immediately preceded by violent interpersonal conflict that left the man wounded. Nevertheless, the connectivity of the man can still be further emphasized (cf. Whittle 1996: 316–317), with particular reference to the ever-growing number and variety of scientific studies relating to the man, his environment, and his communicating over multiple scales of space and time. These studies enable us to work back from the details of his last days and months to more general patterns relating to his relatively long life and ancestry, and to resituate him in a dynamic sociocultural context, characterized by mobility and social interaction—albeit within the constraints of the available data and their sometimes conflicting interpretations.

Precisely when, where, and how the man died remain hotly debated questions. Spindler (1995: 250) originally placed the man's death during late summer or early autumn (i.e., September and October), with reference to the seasonally harvested threshing remains and sloe berry found with him; the presence of autumnal *Artemisia* pollen in an ice sample from the discovery site; and the high-altitude find spot, which would not have been accessible after the onset of the first winter snows. However, an early summer (i.e., July) season of death has since been proposed, with reference to the sloe berry, which can be harvested as early as July; and the chlorophyll content of the maple leaves used to line the man's ember carrier, which suggests they were freshly picked between June and September (Fleckinger and Steiner 1998: 46). A spring (i.e., April or May) season of death has also been argued, on the basis of the flowering season of pollen taxa—notably hop-hornbeam (*Ostrya carpinifolia*)—obtained from food remains in the man's colon and stomach (e.g., European Academy of Bozen/Bolzano 2011; Oeggl 2000). However, a counterargument is that the extremely high percentage of hop-hornbeam pollen found in the man's colon suggests that it was ingested by him—potentially at any time of the year, but arguably during the autumn—by chewing a piece of hop-hornbeam bark or drinking an infusion of such bark for medicinal purposes (Groenman-van Waateringe 2011). The specific details of the pollen data upon which these suggestions are based have, in turn, been challenged (Zink et al. 2011). Nevertheless, I currently favor a September death scenario, because it pulls together the widest range of seasonal indicators proposed so far, although it has to be acknowledged that none of these offer direct and unequivocal evidence of the man's season of death.

As for where and how the man died, Spindler (1994; 1995: 163–165, 250–254) initiated an influential 'disaster' theory, speculating that the man was attacked in his settlement in the Val Venosta (Vinschgau) (pointed to by the threshing remains and pollen characteristic of this valley), suffering personal injuries, but escaping with incomplete equipment to the high alpine pastures that he knew as a transhumant herder, until, in a state of exhaustion, he lay down and froze to death in a gully. Such speculation was added to by Loy's (1998) claim that the blood of up to four other people was found on the man's jacket, knife, axe, and arrows, and that he killed or injured several people before his own death. However, more recent scientific analyses of the man's body have revealed, for example, that his stomach was full of a meal consumed between thirty minutes and two hours before his death (European Academy of Bozen/Bolzano 2011), that he was killed—in less than a few hours—by a flint-tipped arrow shot into his left shoulder that burst a major artery (e.g., Gostner and Egarter Vigl 2002), and that his equipment would have been too bulky to enable him to escape from any opponents. In the light of these findings, the original disaster and flight theory has now been replaced with one that envisages the man's murder at the discovery site, soon after eating what was to be his last meal (e.g., Lippert et al. 2007). It is also worth noting that Loy's (1998) claims about other people's blood have not been substantiated in recent reexaminations (Zink et al. 20010. An alternative to these disaster theories is to envisage the Iceman discovery site as a mortuary site and to place his death in the valley. As a relatively old adult male member of society, who may even have died an esteemed warrior's death, the man might have been accorded a special mortuary rite appropriate to his status. His body may therefore have been carried, ceremonially, to this high place, situated close to an important pass and boundary in the landscape, and then placed on a modified natural stone platform, with equipment (some unfinished, some valuable) arranged around the body, elements of which were later dispersed into the adjacent gully via natural processes (e.g., Vanzetti et al. 2010). Like the autumn death theory, the details of this 'burial' scenario have been challenged, with attention focused on the angular position of the man's arm, in contrast to the more natural placing of arms in contemporary primary burials (Zink et al. 2011). Despite this, I personally find the 'disaster' theory much less persuasive than the 'burial' theory, although certainly a degree of interpretative uncertainly remains.

More can be said about the last days and months of the man's life, particularly with regard to his mobility. Sequential sampling of the food residues found in his colon and small bowel, including their pollen and diatom content, has made it possible to deduce the environments in which his penultimate meals were consumed (e.g., Oeggl et al. 2007; Rott 2000). Apparently, he crossed different habitats, covering considerable distances over the course of his last thirty-three hours or so. The day before he died, he was high up near the tree line (at about 2,500 m) in a subalpine coniferous forest, perhaps along one of the lateral valleys located either east or west of the Senales valley (Schnalstal). Then, about nine to twelve hours before his death, he was in the lower-lying montane zone (at or below about 1,200 m), perhaps in the bottom of the Val di Senales or Val Venosta. Seven to four hours later, he had ascended again to a subalpine coniferous forest (above 1,800 m), where he consumed his penultimate meal. (It is not yet clear where his final meal—represented by his stomach contents—was consumed.)

As he moved during these last days and months, the man would have felt plenty of aches and pains in various parts of his body. He had sustained a deep cut to his right hand about three to eight days before his death (Nerlich, Peschel and Egarter-Vigl 2009). Calcification of his knee joint tendons indicates that he would have suffered from knee pain while walking (Gostner et al. 2011). He also suffered from degenerative osteoarthritis along his spine, on his right hip joint, and on his left little toe (Murphy et al. 2003). It may be in relation to this, and other pains, that tattoos were applied (not by himself) to various parts of his body, and notably to the left and right of his lumbar spine, arguably for a therapeutic purpose (e.g., Pabst et al. 2009). Also, the man's colon contained an infestation of eggs of the intestinal parasite whipworm (*Trichuris trichiura*), which may have resulted in intestinal problems (e.g., Aspöck et al. 2000). In addition, examination of one of the man's fingernails revealed three sets of Beau's lines, indicating that he had been very ill three times in the last six months of his life (Capasso 1995). Yet, despite all this, the man was evidently highly mobile.

Over the rest of his estimated forty-five- to forty-six-year life span (Bernhard 1992), the man's body accumulated further biographical traces, which, together with his artifacts, shed further light on his provenance and mobility. Isotopic data as well as mosses and pollen associated with the man indicate that he spent his entire life to the south of the discovery site, in northern Italy (e.g., Dickson et al. 1996; Kutschera and Müller 2003; Oeggl 2000). His probable birthplace can be traced to a few valleys in the South Tyrol, the Isarco (Eisach) valley being a good candidate. But, during the last two decades of his adult life, he migrated west to the area of the Val Venosta and Val di Senales, at least seasonally if not permanently. The woodlands on the slopes and in the gorges of the latter area also seem to be the most likely source for the wooden equipment that he was found with (Oeggl and

Schoch 2000). He certainly spent some time in this montane zone, including its valley bottoms, to judge by the composition of his grass cape and the cereal remains associated with him; but he also visited the higher-altitude subalpine zone, as indicated by the insulating grasses from his shoes, which would have been collected from subalpine grasslands (Acs, Wilhalm and Oeggl 2005). Indeed, a relatively high degree of mobility over rough terrain is confirmed by the thickening and strengthening of the man's leg bones and muscles (e.g., Ruff et al. 2006).

Broadly contemporary archaeological remains in the area of the Val di Senales and Val Venosta are somewhat limited. They include four poorly understood Copper Age potential settlement sites at Lasa, Castel Juvale, Tartscher Bichl, and Sölesboden; a flint dagger from Laces (Latsch)-Coldrano; and five monumental statue stelae from Laces and Lagundo (Algund) (e.g., De Marinis and Brillante 1998: 152–153; Fantuzzi 2011). This relative paucity of archaeological sites could be taken as an indication that this area was only sparsely populated during the Copper Age, and perhaps especially by mobile groups, who moved in and out of the mountain landscape during the course of activities such as herding, hunting, prospecting, and rites of passage (cf. Whittle 1996: 316). However, it could well reflect the difficulties of identifying prehistoric settlement sites in the uplands (cf. Ayala and Fitzjohn 2007).

The man's clothing and equipment also exhibit stylistic connections to material culture found further afield in the Copper Age of northern Italy. His retouching tool of antler, flint dagger, and marble disc bead have been compared to examples from the rock shelter burial site of Monte Covolo (Barfield 1994), whereas his copper axe, flint dagger, and arrowheads have been compared to examples found in Grave 102 of the Remedello Sotto cemetery (e.g., De Marinis 1997), both sites located in the Brescia province, around 135 to 170 km south of the Iceman find-spot. The flint for the man's dagger may ultimately have been obtained from the Monte Lessini, about 120 km to the south. The man's clothing and weapons can also be compared to depictions of valued weapons and cloth or beads on north Italian statue stelae, erected, sometimes in groups, at ceremonial upland sites, perhaps as potent and durable memorials to prestigious male warrior and female ancestors (cf. Keates 2000). And, over a longer time scale, the man's ancestry can be traced, through the shape of his skull, to people buried at Neolithic and Bronze Age sites located in territories surrounding Lake Garda (Trento, Verona, and Brescia) (Bernhard 1995).

Scientific studies of the Iceman's body and equipment have, then, provided some outstandingly detailed information on the spatial and temporal dimensions of one prehistoric person's life and death. They have, however, been less forthcoming with regard to his social relations, despite Spindler's (1995) initial attempt to integrate the man into the economy and society of a contemporary village.

THE NORTH ITALIAN EARLY COPPER AGE: CROSSING THE CULTURAL LANDSCAPE

Given these limitations to our immediate contextual understanding of the Iceman, it is worth considering the wider contemporary context of the Early Copper Age in northern Italy. For the purposes of this study, I focus below on the excavated remains from a sample of twenty-four relatively well-dated archaeological sites with radiocarbon dates that overlap the estimated time span of the Iceman, that is, 3350 to 3100 cal BC or around 4680 to 4340 BP (see Table 8.1 and Figure 8.2). Ordered according to their positions in the landscape, their forms and their interpreted uses, these sites offer further ideas about space, time, and communicating at the time of the Iceman. To begin with, a basic distinction can be made between those sites located within a core 'domestic domain' and those situated on and beyond its margins in what I term 'the other domain'. From the evidence outlined above, it is clear that the Iceman crossed these two domains throughout the course of his life and death.

The Domestic Domain

A key zone in the north Italian cultural landscape of the Early Copper Age might be described as the domestic domain, comprising a scatter of well-chosen dwelling places and associated land used by small-scale agricultural

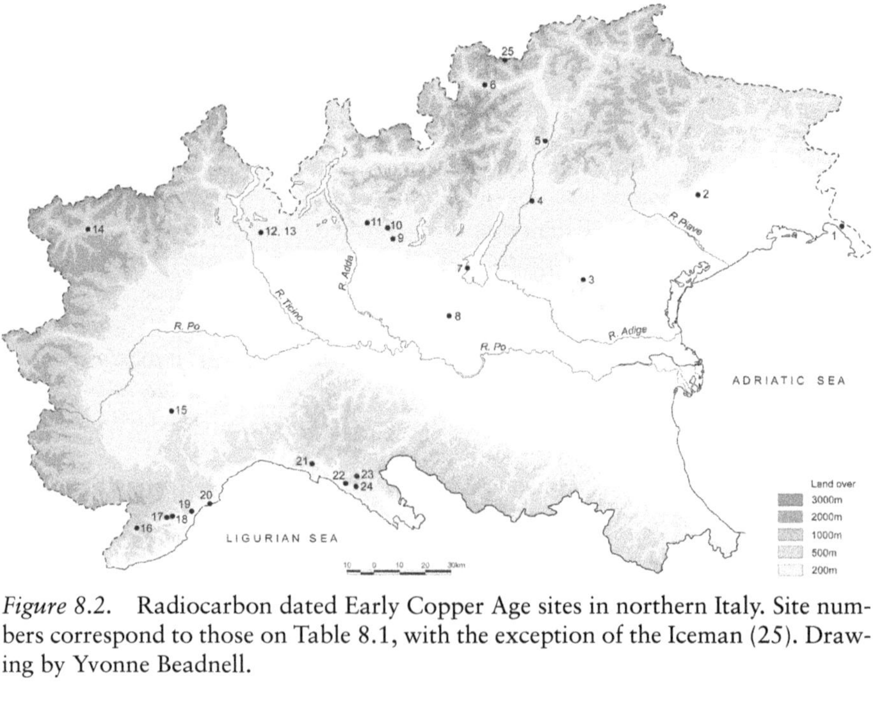

Figure 8.2. Radiocarbon dated Early Copper Age sites in northern Italy. Site numbers correspond to those on Table 8.1, with the exception of the Iceman (25). Drawing by Yvonne Beadnell.

Table 8.1 Early Copper Age sites in northern Italy with radiocarbon dates overlapping those of the 'Iceman'.

Site number	Site name	Location	C14 sample type and context	BP and lab. ref.	Cal BC1σ (CALIB 6.0)	Selected references
1	**Grotta dell'Edera**	Aurisina, Trieste	Charcoal fragments Layer 2, square 4, 230–240 cm depth	4695±165 GX-19021	3656–3120	Spataro 1999–2000: 87
			Layer 2, squares 3–4, 170–180 cm depth	4680±325 GX-19019	3767–2937	Biagi 1996: 14
2	**Palù di Livenza**	Caneva di Sacile, Pordenone	Wood S I anthropic B	4487±115 CRG-930	3355–3024	Corti et al. 1997: 77
			Wood S II post 2	4392±140 CRG-929	3331–3214	Corti et al. 1997: 77
3	**Val di Marca**	Lago di Fimon, Vicenza	Wood from a piragua	4580±50 R-359α	3496–3121	Alessio et al. 1973: 166
4	**Acquaviva di Besenello**	Trento	Human bone Tomb 1	4410±70 ETH-12497	3310–2917	De Marinis and Pedrotti 1997: 255
5	**Tolerait/ Thalerreit (Magrè/Margreid)**	Trentino–Alto Adige	Settlement	4420±65 ETH-13012	3312–2922	Mottes and Nicolis 2002: 245
6	**Lasa/Laas**	Val Venosta, Bolzano	Charcoal 3 m depth in colluvial humus horizon	4350±100 VRI-272	3311–2879	Felber 1974: 280

(*continued*)

Table 8.1 (*continued*)

Site number	Site name	Location	C14 sample type and context	BP and lab. ref.	Cal BC1σ (CALIB 6.0)	Selected references
7	**Riparo Valtenesi**	Rocca di Manerba, Brescia	Oak (*Quercus* sp.) Timber B, Chamber 133	4420±90 Birm-1132	3321–2921	Barfield 1983: 122
			Charred oak Timber F, Chamber 133	4385±33 OxA-17230	3077–2926	Barfield 2007: 427
			Charred oak Timber F, Chamber 133	4363±32 OxA-17229	3012–2919	Barfield 2007: 427
			Charred oak Timber F, Chamber 133	4359±32 OxA-17228	3011–2977	Barfield 2007: 427
			Charred grain Cremation pit	4320±100 OxA-4550	3263–2760	Barfield 2007: 419
8	**Remedello di Sotto**	Brescia	Human bone Tomb 106	4485±65 ETH-12182	3338–3093	De Marinis and Pedrotti 1997: 288
			Human bone Tomb 34	4440±70 ETH-12189	3328–2943	De Marinis and Pedrotti 1997: 288
9	**Canton, Trescore Balneario**	Bergamo	House foundation, US 62	4605±79 GX-19838	3518–3121	Poggiani Keller 2002: 272
			House foundation, US 13C	4515±63 GX-19262	3349–3105	Poggiani Keller 2002: 272

10	**Cap del Pir, Casale, Albino**	Bergamo	Well 1, US 4	4555±65	3482–3105	Ravazzi 2000: 27
				GX-21305		
			Well 2, US 14 (upper part of fill)	4390±90	3312–2903	Ravazzi 2000: 27
				GX-20851		
11	**Castello, Ubiale Clanezzo**	Bergamo	WoodWell beam, US 160	4505±160	3486–2935	Poggiani Keller 2002: 272
				GX-16821		
12	**Lagozza di Besnate**	Varese	Wood (*Conifera*) Pile in lower structure	4580±50	3496–3121	Alessio et al. 1973: 166
				R-78A		
13	**Lagozzetta di Besnate**	Varese	Wooden vessel	4385±50	3087–2918	Alessio et al. 1968: 357
				R-336		
14	**St Martin de Corléans**	Aosta	Charcoal	4450±180	3362–2909	Azzi et al. 1977: 168
			Pile n. 19	F-83		
			Charcoal	4325±170	3331–2697	Azzi et al. 1977: 168
			Pile n. 13	F-79		
15	**Papillon, Alba**	Cuneo	Human tibia	4583±62	3498–3117	Zoppi et al. 2001: 1054
			Collective burial structure	OZE-028		
16	**Grotta del Pertuso**	Goina, Imperia	Charcoal	4410±150	3335–2905	Maggi and Del Lucchese 1988: 338
			Sector 5	HAR-6760		

(*continued*)

Table 8.1 (*continued*)

Site number	Site name	Location	C14 sample type and context	BP and lab. ref.	Cal BC1σ (CALIB 6.0)	Selected references
17	**Grotta del Pertusello**	Imperia	Charcoal	4390±70	3262–2909	Alessio et al. 1967: 348
			Section D–C, Layer III	R-155		
			Charcoal	4650±90	3629–3345	Alessio et al. 1964: 81
			Level IV, 63–70 cm below surface	R-122		
18	**Riparo Arma di Nasino**	Savona	Charcoal	4680±100	3631–3362	Alessio et al. 1968: 353
			Layer VII, 2.3–2.5 m below surface	R-261		
19	**Grotta dell'Olivo**	Toirano, Savona		4470±130	3352–2944	Maggi 1998: 14
				GrN-16143		
20	**Caverna delle Arene Candide**	Finale Ligure, Savona	Charcoal (*Pistacia lentiscus*)	4430±60	3318–2930	Maggi 1997: 36
			Layer 2	Beta-66550/ CAMS-9418		
21	**Castellaro di Uscio**	Genova	Sector H1Q, Stratum 4	4490±70	3339–3095	Maggi 1990: 59
				GrN-14935		
22	**Libiola**	Genova	Oak pick handle	4610±50	3508–3196	Maggi and Del Lucchese 1988: 338
				Bln-3367		
			Oak pick handle	4490±90	3348–3032	Maggi and Del Lucchese 1988: 338
				Gif-7213		

23	**Valle Lagorara**	Maissana, La Spezia	Jasper quarry	4580±50 GrA-5111	3496–3121	Campana et al. 1997–1998: 149
			Jasper quarry	4530±50 Beta-100729	3357–3110	Campana et al. 1997–1998: 149
			Jasper quarry	4520±70 GrA-5151	3353–3104	Campana et al. 1997–1998: 149
			LG2A rock shelter, Layer 9	4510±70 Beta-87979	3348–3102	Campana et al. 1997–1998: 149
			Jasper quarry	4410±60 GrA-5171	3264–2918	Campana et al. 1997–1998: 149
			Jasper quarry	4360±50 GrA-5173	3077–2909	Campana et al. 1997–1998: 149
			Jasper quarry	4360±50 GrA-5179	3077–2909	Campana et al. 1997–1998: 149
24	**Monte Loreto**	Castiglione Chiavarese, Genova	Charcoal (*Prunus* sp.) Layer of debris from trench mine	4600±80 Beta-135160	3517–3119	Maggi and Pearce 2005: 67

communities, extending from the Po plain to the north Italian lakes and river valleys running through the pre-Alps and Alps. Two main types of settlement site can be identified—lake dwellings and river terrace settlements—both of which were established in relatively low-lying and wet environments. At both, powerful, culturally constructed conceptions of social and agricultural time and space are likely to have structured the routines of daily life followed by close-knit communities.

Radiocarbon dated examples of the lake dwelling category of site comprise Lagozza di Besnate (e.g., Cornaggia Castigilioni 1955; Guerreschi 1967), Lagozzetta di Besnate (Cornaggia Castigilioni 1956), and Palù di Livenza (e.g., Corti et al. 1997; Peretto and Taffarelli 1973; Pini 2004; Vitri, Martinelli and Čufar 2002). These agricultural settlements were intentionally established in marshy lake basins, and their remains preserved within peat bogs. Different kinds of durable wooden structure were found at each of them—the timber derived from trees growing in immediately surrounding forests, whose resources appear to have been increasingly managed by the lake-dwelling communities. At Palù di Livenza, for example, three types of structure were identified: an anchorage girder formed by a regular trellis of beams and small horizontal rafters; the remains of flooring made of layers of rafters and branches covered by boarding; and large vertical oak supports for wattle-and-daub walls. The scale of these structures is impressive, and surely implies a significant investment of labor in the physical and social infrastructure of these dwelling places and their close-knit communities. Indeed, a community with a deep and dynamic history is implied by the radiocarbon dates for these wooden structures, which indicate a long-lived settlement and different building phases extending from the Early Neolithic through the Late Neolithic into the Copper Age. Associated cultural materials included pottery vessels, some expressing allegiance to the influential Chassey-Lagozza style, and ceramic pintaderas (or stamps), also exhibiting regionally transmitted decorative motifs. A paddle made of poplar wood was also found at Palù di Livenza, which, combined with the radiocarbon dated evidence of fragments of a dugout canoe (or 'pirogue') found in the Val di Marca (Cornaggia Castiglioni 1967: 43), indicates one of the means of communication used by these lake-dwelling communities. Most of the cultural materials would have been procured and processed by these largely self-sufficient communities, but some (such as flint, rock crystal, and copper) also imply participation in longer distance regional networks of mobility and interaction over land and water.

Radiocarbon dated examples of the river terrace settlements comprise Canton di Trescore Balneare, Castello di Ubiale Clanezzo, and Cap del Pir (e.g., Poggiani Keller 2002), Tolerait (Thalerreit) (e.g., Dal Rì 1973), and Lasa (Laas) (Felber 1974: 280). The first three sites are located in the Bergamo province, on river terraces in pre-alpine valleys. Like the lake dwellings, they were generally long-lived agricultural villages, with deep histories, occupied since the Late Neolithic. Local materials were worked at these places, including pottery that exhibited connections to north Italian (Lagozza and Breno) styles, potters, and communities. The other two river terrace settlements are located further north, in the Bolzano province, in alpine valleys. Both are represented by charcoal-rich soil horizons formed on cones of detritus situated at the foot of steep slopes and on the edge of their respective river valleys, and both contained relatively few artifacts, perhaps suggesting less intensively or lengthily occupied sites than the villages described above. Nevertheless, agriculture was practiced at Tolerait, as indicated by the presence of carbonized cereals and faunal remains. It is likely that the 'Iceman' had direct contact with a small community based at a site such as this.

The Other Domain

Archaeological sites lying on the margins of, or beyond, this domestic domain might be understood as having belonged to conceptually different parts of the cultural landscape, loosely described here as 'the other domain'—a term that plays on the anthropological concept of 'the Other' as a culturally constructed relation of spatiotemporal remoteness and difference (e.g., Hallam and Street 2000: 3). Certainly, the types of sites found here are notably different in character than those of the domestic domain, and it may also be appropriate to think of these places as having been structured and experienced through alternative—slower, more extended, more punctuated, and more dispersed—conceptions of time and space, even though they were ultimately connected to the domestic domain. They can be broadly divided into marginal (or 'liminal') ritual sites and upland subsistence-related sites.

At least four different types of ritual place are represented among the sample of north Italian sites radiocarbon dated to the Early Copper Age: an 'open' cemetery site, rock-shelter burial sites, a ceremonial site with monumental structures and secondary burials, and a burial cairn or tumulus—a list that should be extended to include the naturally monumental discovery site of the 'Iceman' if the 'burial theory' is to be believed (see above). All of them provide information about ritual and social connections and distinctions expressed by the living through the dead over various scales of space and time.

A classic, but also distinctive, example of a north Italian Copper Age cemetery is Remedello di Sotto (e.g., Barfield 1986; Colini 1898; De Marinis 1997). Its location, on the bank of the Chiese river in the central Po plain, perhaps on the margins of a settlement for the living (which may be represented by part of a village ditch recorded nearby), makes it appropriate to be described as a 'settlement for the dead', where ritual notions of time and space replaced daily norms. The cemetery was first established during the Early Copper Age, and later grew to become the largest known Copper Age cemetery in northern Italy, with over 200 burials extending over 4.5 ha. The dominant mortuary rite was the inhumation of crouched bodies accompanied by grave goods in single trench graves. Taken together, the bodies and goods can be seen to have materially symbolized relatively clear-cut social identities and distinctions in terms of gender and age, and—later on in the Copper Age—also wealth. Adult females were generally buried supine, oriented east, and with relatively few grave goods, notably pottery vessels, whereas adult males were generally buried flexed on their left side, oriented north or northwest, and accompanied by more goods, including flint arrowheads and flint or copper daggers, which might have symbolized their status as warriors. Children were buried with only a few flint blades and flakes. Other grave goods included stone and copper axe blades, copper awls, and body ornaments made of a range of sensuous materials (copper, silver, marble, shell, and stone).

By contrast to this relatively fixed ordering of the dead at Remedello di Sotto (spatially, temporally, and socially), a more fluid combination of primary and secondary burial rites was performed at contemporary north Italian rock-shelter burials, by groups who shared some elements of the influential Remedello culture but who also expressed their own 'discrepant' beliefs and customs. These sites might be understood, both as almost timeless natural places and as dynamic cultural places that were gradually modified during the course of previous occupations during the Neolithic, and then ritually transformed into ancestral monuments with another sense of time and space during the Early Copper Age. Acquaviva di Besenello is a relatively simple example, where the body of an adult was placed against the rock-shelter wall, accompanied by two arrowheads and a dagger of flint, and delimited on the three other sides by a cist-like structure, and the skull and long bones later removed, presumably for secondary burial elsewhere (Angelini et al. 1980). A comparable but more elaborate example is Riparo Valtenesi (e.g., Barfield 2007). It is a large natural rock shelter, some 50 to 60 m long, situated on the southwest shore of Lake Garda, at the foot of a cliff below the Rocca di Manerba promontory. This site was previously occupied during the Early and Late phases of the Neolithic, but only began to be used as a burial place during the Copper Age. Radiocarbon dates point to a relatively short period of mortuary activity at the site, spanning just a few centuries (Barfield et al. 2010). A group of rectangular burial chambers was constructed here, perhaps one after the other, with oak plank walls and floors of stone pebbles or slabs. Each measured approximately 2 by 1 m and contained a rich assemblage of disarticulated adult human remains (probably originally deposited sequentially as whole bodies—a minimum number of five in the case of chamber 133—then later reworked into a 'collective' deposit); and scattered artifacts left for the newly dead or ancestors, including necklaces composed of numerous beads of imported marble, steatite, and copper, and ceramics stylistically similar to those found in the Remedello di Sotto cemetery. There was also a cist containing an infant burial positioned in front of one of the adult burial chambers, indicating social differentiation in terms of age.

A ceremonial site with monumental structures and secondary burial deposits was also elaborated at St. Martin de Corléans in the Aosta valley, through a series of laborious construction projects (e.g., Mezzena 1997). A monumental series of large wooden posts was first erected here during the Late Neolithic. After the regular—and probably ritual—plowing of an extensive area at the site, further posts were added during the Early Copper Age, as well as a thought-provoking series of anthropomorphic stone stelae, which might have represented ancestors, and a quadrangular megalithic chamber associated with groups of human remains, which might be compared to the ancestral mortuary chambers at Riparo Valtenesi.

Another kind of collective burial structure was identified at the Papillon site in Alba (Zoppi et al. 2001: 1054). Here, a monumental cairn-like structure of regularly laid stone slabs surrounded and covered a central rectangular burial chamber, which contained the bones of at least ten individuals (four adults and six children—perhaps related to each other) and some minute fragments of copper. The mixing of these bones, and the lack of some long bones, are indicative of extended secondary burial rites.

Upland subsistence-related sites comprise a second set of sites in 'the other domain' of the north Italian Early Copper Age. They include an 'open' settlement, cave shelters, and raw material extraction sites. Castellario di Uscio is a hilltop settlement, located on the summit of Monte Borgo in the Ligurian Apennines, at an altitude of 728 m (e.g., Maggi 1990). The site dominates the meeting point of a ridge descending down to the Mediterranean Sea and another that runs parallel to the coast. The Early Copper Age deposits were disturbed in antiquity. The excavator has, nevertheless, interpreted the site as an agricultural settlement situated on the crossroads of communications routes used historically—and, by extension, in prehistory—by mobile groups practicing some form of pastoralism.

Various upland caves appear to have been used as herders' shelters during the Early Copper Age, occupied temporarily on a daily and seasonal basis by small mobile groups of herder-hunters and their domestic animals. These sites might be thought of as natural monuments within a wider network of culturally significant places scattered across the landscape (including lowland villages, sacred places, upland settlements and shelters, and seasonal pastures), connected by mobile members of society. Radiocarbon dated examples include Grotta del Pertusello and Riparo Arma di Nasino in the Val Pennavaira (Alessio et al. 1967; Leale Anfossi 1974), Grotta del Pertuso (Del Lucchese and Ricci 1987), and Grotta dell'Olivo (Maggi 1998: 14), all in the mountains of western Liguria; Caverna delle Arene Candide, situated on a promontory overlooking the western Ligurian Mediterranean coast (e.g., Bernabò Brea 1946; Maggi 1997); and Grotta dell'Edera in the Trieste karst (e.g., Biagi 1996). Specialist studies have been particularly informative in the case of Grotta dell'Edera. Soil analyses suggest its use as a shelter for humans and as a stable for domestic animals, an interpretation supported by the strong predominance of sheep among the faunal remains, and the relatively small number of pottery sherds found in the Early Copper Age deposits (e.g., Boschian 1997). The presence of carbonized nuts of hazel and Cornelian cherry (Nisbet 2000) might indicate a midsummer to midautumn season of occupation. Connections to another site, perhaps a village, might also be indicated by the presence of one pottery sherd with a distinctively nonlocal mineralogical signature (Spataro 1999–2000).

Contemporary raw material procurement sites have also been identified in the uplands of Liguria. To judge by the large scale of these sites, each represents a significant investment of labor and output of raw materials, probably disproportionate to local needs. In other words, these sites might have been controlled extraction and production centers for valued raw materials that were not only consumed by local communities but also exported over long distances to other people and places. At Monte Loreto and Libiola, miners emptied natural fissures and veins of copper ore, using hammer stones, wooden tools, and fire setting, and also processed the ore on a large scale on adjacent work floors (e.g., Issel 1879: 349; Maggi and Pearce 2005). And in the Valle Lagorara, a large red jasper (radiolarite) quarry was established, where miners extracted rock from two main outcrops and then processed the rock in two rock shelters on the opposite side of the valley to produce bifacially flaked preforms—ultimately intended to be completed elsewhere as arrowheads, dagger blades, and picks (e.g., Maggi, Campana and Negrino 1995). Soapstone beads and pendants were also produced here, probably using a local soapstone outcrop, and copper was smelted in one of the rock shelters—both perhaps opportunistically as sought-after commodities.

CONCLUSION

In this chapter, I have worked between two basic paired categories of time and space: *chronos*—measurable cosmic time, based on physics and mathematical knowledge, combined with the traditional conception of landscape as an objective and measurable physical entity, partly shaped by human action; and *tempus*—culturally determined social time, experienced on different levels and scales by different members of society, paired with an equivalent understanding of landscape as culturally constructed, transformed, and experienced by people.

Chronometric calibrated radiocarbon dates underpin this study, enabling us to compare and contrast broadly contemporary archaeological sites in the Early Copper Age right across northern Italy. Scientific studies have also provided us with some unusually detailed temporal and spatial parameters for the Iceman: from events of his last few days and hours (including estimates of the timing and place of his last meals and death), to his long life (including his re-location as an adult, and the onset of old-age health-related problems—in part stemming from his highly mobile lifestyle in the valleys and mountains of the South Tyrol), to his ancestry in adjacent provinces to the South.

I have also proposed, from a more theoretically informed perspective, that we think about the culturally constructed historic landscape of the north Italian Early Copper Age as structured (both physically and conceptually) by two fundamental time-space zones. There was the domestic domain, characterized by durable settlement structures, sites, and communities, situated in relatively low-lying and wet environments, with long and dynamic histories (of foundation and rebuilding, for example), and by routines of daily life and seasonal work led according to the agricultural calendar (including harvest time and the threshing of grain). There was also 'the other domain': marginal or remote, and characterized by slower, more extended and more punctuated conceptions of time. This was experienced, for example, at mobile herders' shelters in the uplands occupied temporarily on a daily and seasonal basis, and also at cemeteries and monuments established—at historic places in the landscape and then either gradually or more rapidly modified through repeated ritual performances, cumulative construction projects, and spatially ordered symbolic deposits—as memorials to identified members of social groups and to timeless ancestors. Certain mobile members of society and their valued material goods (represented most clearly by the man and his equipment found in the Ötztal Alps, but also by mobile herd animals, portable human bodies and bones, meaningful decorative motifs, and a range of sought-after raw materials), crossed and connected these zones of time and space—sometimes frequently and over long distances—while communicating through a wide array of bodily and material resources with other people and places during the course of their complex lives and deaths.

REFERENCES

Acs, P., T. Wilhalm and K. Oeggl. 2005. Remains of grasses found with the Neolithic Iceman 'Ötzi'. *Vegetation History and Archaeobotany* 14: 198–206.

Alessio, M., F. Bella, F. Bachechi and C. Cortesi. 1967. University of Rome carbon-14 dates V. *Radiocarbon* 9: 346–367.

Alessio, M., F. Bella and C. Cortesi. 1964. University of Rome caron-14 dates II. *Radiocarbon* 6: 77–90.

Alessio, M., F. Bella, C. Cortesi and B. Graziadei. 1968. University of Rome carbon-14 dates VI. *Radiocarbon* 10: 350–364.

Alessio, M., F. Bella, S. Improta, G. Belluomini, G. Calderoni, C. Cortesi and B. Turi. 1973. University of Rome carbon-14 dates X. *Radiocarbon* 15: 165–178.

Angelini, B., B. Bagolini and T. Pasquali. 1980. Acquaviva di Besenello (Trento). *Preistoria Alpina* 16: 67–69.

Aspöck, H., H. Auer, O. Picher and W. Platzer. 2000. Parasitological examination of the Iceman. In: S. Bortenschlager and K. Oeggl (eds.), *The Iceman and His Natural Environment,* 127–136. Wien and New York: Springer Verlag.

Ayala, G. and M. Fitzjohn. 2007. To be seen or not to be. In: M. Fitzjohn (ed.), *Uplands of Ancient Sicily and Calabria,* 99–113. London: Accordia Research Institute.

Azzi, C. M., L. Bigliocca and F. Gulisano. 1977. Florence radiocarbon dates III. *Radiocarbon* 19(2): 165–169.
Barfield, L. H. 1983. The Chalcolithic cemetery at Manerba del Garda. *Antiquity* 57: 116–123.
Barfield, L. H. 1986. Chalcolithic burial in northern Italy. *Dialoghi di Archeologia* 4(2): 241–248.
Barfield, L. 1994. The Iceman reviewed. *Antiquity* 68: 10–26.
Barfield, L. H., ed. 2007. *Excavations in the Riparo Valtenensi, Manerba, 1976–1994*. Firenze: Istituto Italiano di Preistoria e Protostoria.
Barfield, L. H., S. W. Manning, E. Valzolgher and T. F. G. Higham. 2010. A wiggle-matched date for the Copper Age cemetery at Manerba del Garda, northern Italy. In: A. J. T. Jull (ed.), *Proceedings of the 20th International Radiocarbon Conference. Radiocarbon* 52(2–3): 984–1001.
Bernabò Brea, L. 1946. *Gli Scavi nella Caverna delle Arene Candide: Parte I. Gli Strati con Ceramiche*. Bordighera: Istituto di Studi Liguri.
Bernhard, W. 1992. Vergleichende Untersuchungen zur Anthropologie des Mannes vom Hauslabjoch. In: F. Höpfel, W. Platzer and K. Spindler (eds.), *Der Mann im Eis: Band I,* 163–187. Innsbruck: Universität Innsbruck.
Bernhard, W. 1995. Multivariate statistische Untersuchungen zur Anthropologie des Mannes vom Hauslabjoch. In: K. Spindler, E. Rastbichler-Zissernig, H. Wilfing, D. zur Nedden and H. Nothdurfter (eds.), *Der Mann im Eis,* 217–229. Wien and New York: Springer.
Biagi, P. 1996. North eastern Italy in the seventh millennium BP. In: *The Vinca Culture,* 9–22. Timisoara: The Museum of Banat.
Bonani, G., S. D. Ivy, I. Hajdas, T. R. Niklaus and M. Suter. 1994. AMS 14C determinations of tissue, bone and grass samples from the Ötztal Ice Man. *Radiocarbon* 36(2): 247–250.
Boschian, G. 1997. Sedimentology and soil micromorphology of the Late Pleistocene and Early Holocene deposits of Grotta dell'Edrea (Trieste Karst, NE Italy). *Geoarchaeology* 12(3): 227–249.
Campana, N., R. Maggi and F. Negrino. 1997–1998. La cava di Valle Lagorara: estrazione e lavorazione del diaspro dall'Età del Rame all'inizio dell'Età del Bronzo. *Rivista di Studi Liguri* 53–54: 147–160.
Capasso, L. 1995. Ungueal morphology and pathology of the 'Ice Man'. In: K. Spindler, E. Rastbichler-Zissernig, H. Wilfing, D. zur Nedden and H. Nothdurfter (eds.), *Der Mann im Eis,* 231–239. Wien and New York: Springer.
Colini, G. A. 1898. Il sepolcreto di Remedello Sotto nel Bresciano e il periodo eneolitico in Italia. *Bullettino di Paletnologia Italiana* 24: 1–47, 88–110, 206–260, 280–295.
Cornaggia Castiglioni, O. 1955. Nuove ricerche nella stazione palafitticola della Lagozza di Bescnate. *Sibrium* 2: 93–104.
Cornaggia Castiglioni, O. 1956. La Lagozzetta di besnate e gli insediamenti 'palustri' della civiltà di Polada. In: *Studi in Onore di Aristide Calderini e Roberto Paribeni,* vol. 3: 35–58.
Cornaggia Castiglioni, O. 1967. Le piroghe preistoriche italiane: problematica e inventario dei reperti. *Natura* 58: 5–48.
Corti, P., N. Martinelli, M. Rottoli, O. Tinazzi and S. Vitri. 1997. New data on the wooden structures from the pile-dwelling of Palù di Livenza. *Preistoria Alpina* 33: 73–80.
Dal Rì, L. 1973. Tolerait (Magrè all'Adige). *Preistoria Alpina* 9: 252–254.
Del Lucchese, A. and M. Ricci. 1987. Il Pertuso. In: P. Melli and A. Del Lucchese (eds.), *Archeologia in Liguria III.1.* Genova: Soprintendenza Archeologica della Liguria, 169–174.
De Marinis, R. C. 1997. The Eneolithic cemetery of Remedello Sotto (BS) and the relative and absolute chronology of the Copper Age in Northern Italy. *Notizie Archeologiche Bergomensi* 5: 33–51.

De Marinis, R. C. and G. Brillante. 1998. *La Mummia del Similaun*. Venezia: Marsilio Editori.

De Marinis, R. C. and A. L. Pedrotti. 1997. L'età del Rame nel versante italiano delle Alpi Centro-occidentali. In: A. Revedin (ed.), *Atti della XXXI Riunione Scientifica dell'Istituto Italiano di Preistoria e Protostoria,* 247–300. Firenze: Istituto Italiano di Preistoria e Protostoria.

Dickson, J. H., S. Bortenschlager, K. Oeggl, R. Porley and A. McMullen. 1996. Mosses and the Tyrolean Iceman's southern provenance. *Proceedings of the Royal Society B* 263: 567–571.

European Academy of Bozen/Bolzano. 2011. A rest, a meal, then death for 5,000-year-old glacier mummy. *Science Daily,* October 25, 2011. Availablee at: <http://www.sciencedaily.com/releases/2011/10/111025091533.htm> [Accessed January 16, 2012].

Fantuzzi, T. 2011. Continuità e discontinuità nel popolamento della Vinschgau/Val Venosta tra Neolitico ed Età storica. Unpublished dissertation, Università Ca'Foscari, Venezia. Available at: <http://unive.academia.edu/TizianoFantuzzi/Papers/929724/Continuita_e_discontinuita_nel_popolamento_della_Vinschgau_Val_Venosta_tra_Neolitico_ed_Eta_Storica._Unpubl.Diss._Universita_CaFoscari_Venezia_2010_11> [Accessed February 7, 2012].

Felber, H. 1974. Vienna Radium Institute radiocarbon dates V. *Radiocarbon* 16: 277–283.

Finnegan, R. 2002. *Communicating*. London and New York: Routledge.

Fleckinger, A. and H. Steiner. 1998. *The Iceman*. Vienna and Bolzano: Folio Verlag and South Tyrol Museum of Archaeology.

Gostner, P. and E. Egarter Vigl. 2002. Insight: report of radiological-forensic findings on the Iceman. *Journal of Archaeological Science* 29: 323–326.

Gostner, P., P. Pernter, G. Bonatti, A. Graefen and A. R Zink. 2011. New radiological insights into the life and death of the Tyrolean Iceman. *Journal of Archaeological Science* 38(12): 3425–3431.

Groenman-van Waateringe, W. 2011. The Iceman's last days: the testimony of *Ostrya carpinifolia. Antiquity* 85: 434–440.

Guerreschi, G. 1967. *La Lagozza di Besnate e il Neolitico Superiore Padano*. Como: Società Archeologica Comense.

Hallam, E. and B. V. Street. 2000. Introduction: cultural encounters—representing 'otherness'. In: E. Hallam and B. V. Street (eds.), *Cultural Encounters,* 1–10. London and New York: Routledge.

Hedges, R. E. M., R. A. Housley, C. R. Bronk and G. J. van Klinken. 1992. Radiocarbon dates from the Oxford AMS system: *Archaeometry* datelist 15. *Archaeometry* 34(2): 337–357.

Issel, A. 1879. Sulle trace di antichissima lavorazione osservate in alcune miniere della Liguria. *Rassegna Settimanale* 3: 348–349.

Keates, S. 2000. The ancestralization of the landscape. In: G. Nash (ed.), *Signifying Place and Space*. Oxford: BAR International Series 902: 83–102.

Kutschera, W. and W. Müller. 2003. "Isotope language" of the Alpine Iceman investigated with AMS and MS. *Nuclear Instruments and Methods in Physics Research B* 204: 705–719.

Leale Anfossi, M. 1974. L'Arma di Nasino (Savona). *Atti della XVI Riunione Scientifica dell'IIPP in Liguria,* 131–140. Firenze: Istituto Italiano di Preistoria e Protostoria.

Lippert, A., P. Gostner, E. Egarter Vigl and P. Pernter. 2007. Vom Leben und Sterben des Ötztaler Gletschermannes. *Germania* 85: 1–21.

Loy, T. 1998. Blood on the axe. *New Scientist,* 12 September: 40–43.

Maggi, R. 1990. *Archeologia dell'Appennino Ligure*. Bordighera: Istituto Internazionale di Studi Liguri.

Maggi, R., ed. 1997. *Arene Candide*. Roma: Soprintendenza Archeologica della Liguria.

Maggi, R. 1998. Storia della Liguria fra 3.600 e 2.300 anni avanti Cristo (l'Età del Rame). In: A. Del Lucchese and R. Maggi, (eds.), *Dal Diaspro al Bronzo,* 7–28. La Spezia: Luna Editore.

Maggi, R., N. Campana and F. Negrino. 1995. Valle Lagorara (I 28). *Archaeologia Polona* 33: 187–208.

Maggi, R. and A. Del Lucchese. 1988. Aspects of the Copper Age in Liguria. *Rassegna di Archeologia* 7: 331–338.

Maggi, R. and M. Pearce. 2005. Mid fourth-millennium copper mining in Liguria, north-west Italy. *Antiquity* 79: 66–77.

Mezzena, F. 1997. La Valle d'Aosta nel Neolitico e nell'Eneolitico. In: A. Revedin (ed.), *Atti della XXXI Riunione Scientifica dell'Istituto Italiano di Preistoria e Protostoria,* 17–138. Firenze: Istituto Italiano d Preistoria e Protostoria.

Mottes, E. and F. Nicolis. 2002. Il territorio del Trentino tra Neolitico recente ed età del Rame. In: A. Ferrari and P. Visentini (eds.), *Il Declino del Mondo Neolitico,* 237–256. Pordenone: Museo delle Scienze, Comune di Pordenone.

Murphy, W. A., D. zur Nedden, P. Gostner, R. Knapp, W. Rechels and H. Sedler. 2003. The Iceman: discovery and imaging. *Radiology* 226: 614–629.

Nerlich, A. G., O. Peschel and E. Egarter-Vigl. 2009. New evidence for Ötzi's final trauma. *Intensive Care Medicine* 35: 1138–1139.

Nisbet, R. 2000. Nota preliminare sull'antracologia dei depositi olocenici della Grotta dell'Edera, Carso Triestino (scavi 1990–1999). *Quaderno: Società Preistoria Protostoria Friuli-Venezia Giulia* 8: 161–170.

Oeggl, K. 2000. The diet of the Iceman. In: S. Bortenschlager and K. Oeggl (eds.), *The Iceman and His Natural Environment,* 89–115. Wien and New York: Springer Verlag.

Oeggl, K., W. Kofler, A. Schmidl, J. H. Dickson, E. Egarter-Vigl and O. Gaber. 2007. The reconstruction of the last itinerary of "Ötzi", the Neolithic Iceman, by pollen analyses from sequentially sampled gut extracts. *Quaternary Science Reviews* 26: 853–861.

Oeggl, K., A. Schmidl and W. Kofler. 2009. Origin and seasonality of subfossil caprine dung from the discovery site of the Iceman (Eastern Alps). *Vegetation History and Archaeobotany* 18: 37–46.

Oeggl, K. and W. Schoch. 2000. Dendrological analyses of artefacts and other remains. In: S. Bortenschlager and K. Oeggl (eds.), *The Iceman and His Natural Environment,* 29–61. Wien and New York: Springer Verlag.

Pabst, M. A., I. Letofsky-Papst, E. Bock, M. Moser, L. Dirfer, E. Egarter-Vigl and F. Hofer. 2009. The tattoos of the Tyrolean Iceman. *Journal of Archaeological Science* 36: 2335–2341.

Peretto, C. and C. Taffarelli. 1973. Un insediamento del Neolitico recente al Palù di Livenza (Pordenone). *Rivista di Scienze Preistoriche* 28(1): 235–260.

Pini, R. 2004. Late Neolithic vegetation history at the pile-dwelling site of Palù di Livenza (northeastern Italy). *Journal of Quaternary Science* 19(8): 769–781.

Poggiani Keller, R. 2002. Contesti di recente indagine nella Lombardia prealpina, tra tardo Neolitico e prima età del Rame. In A. Ferrari and P. Visentini (eds.), *Il Declino del Mondo Neolitico,* 271–289. Pordenone: Museo delle Scienze, Comune di Pordenone.

Ravazzi, C. 2000. Studio pollinico di un focolare neolitico a Casale (Prealpi Lombarde, bergamo): tafonomia e interpretazione paleoambientale. *Società Preistoria Protostoria Friuli-Venezia Giulia, Trieste, Quaderno* 8: 23–33.

Rott, E. 2000. Diatoms from the colon of the Iceman. In: S. Bortenschlager and K. Oeggl (eds.), *The Iceman and His Natural Environment,* 117–125. Wien and New York: Springer Verlag.

Ruff, C. B., B. M. Holt, V. Sládek, M. Berner, W. A. Murphy Jr., D. zur Nedden, H. Seidler and W. Recheis. 2006. Body size, body proportions, and mobility in the Tyrolean "Iceman". *Journal of Human Evolution* 51: 91–101.

Spataro, M. 1999–2000. An interpretative approach to the prehistory of the Edera Cave in the Trieste Karst (northeastern Italy). *Accordia Research Papers* 8: 83–99.
Spindler, K. 1994. The Iceman's last weeks. *Nuclear Instruments and Methods in Physics Research B* 92: 274–281.
Spindler, K. 1995. *The Man in the Ice*. London: Weidenfeld and Nicolson.
Spindler, K. 1996. Iceman's last weeks. In: K. Spindler, H. Wilfing, E. Rastbichler-Zissernig, D. zur Nedden and H. Nothdurfter (eds.), *Human Mummies*, 249–263. Wien and New York: Springer.
Strathern, M. 1988. *The Gender of the Gift*. Berkeley: University of California Press.
Vanzetti, A., M. Vidale, M. Gallinaro, D. W. Frayer and L. Bondioli. 2010. The ice-man as burial. *Antiquity* 84: 681–692.
Vitri, S., N. Martinelli and K. Čufar. 2002. Dati cronologici dal sito di Palù di Livenza. In: A. Ferrari and P. Visentini (eds.), *Il Declino del Mondo Neolitico*, 187–198. Pordenone: Museo delle Scienze, Comune di Pordenone.
Whittle, A. 1996. *Europe in the Neolithic*. Cambridge: Cambridge University Press.
Zink, A., A. Graefen, K. Oeggl, J. Dickson, W. Leitner, G. Kaufmann, A. Fleckinger, P. Gostner and E. Egarter-Vigl. 2011. The Iceman is not a burial: reply to Vanzetti *et al.* (2010). *Antiquity* 85. Available at: <http//www.antiquity.ac.uk/projgall/zink328/> [Accessed January 13, 2012].
Zoppi, U., E. Fulcheri, F. M. Gambari, Q. Hua, E. M. Lawson, M. Micheletti Cremasco and M. Venturino Gambari. 2001. The Copper Age in northern Italy. *Radiocarbon* 43(2B): 1049–1055.

9 Sharing Space and Time with the Ancestors at an Early Bronze Age Tomb in South Central Crete

Emily Miller Bonney

INTRODUCTION

The editors of this volume emphasize that focusing on the interrelationship between space and time can enhance our understanding of the dynamics of social institutions. They underscore that relationships unfold in real time and space. By following their lead, this chapter demonstrates that examination of the nearly millennium-long transformation of the tomb complex at Lebena Yerokambos (henceforth "Yerokambos") reveals that the relationship between the ancestors and the living in that community was more fluid than traditionally has been acknowledged. Yerokambos, on the south coast of Crete, belongs to a class of communal circular stone tombs known as tholoi that served the population of South Central Crete during the long Prepalatial Period (end of the Neolithic into the early Middle Bronze Age, 3100/3000–1900 BC) (Manning, 2010, Table 2.2) (Figure 9.1).

Although each tomb or tholos at first stood alone, over the centuries the users might construct a secondary tholos, gradually add a small suite of rooms, and formalize the open space by adding paving. Whereas many sites were not in use for the entire period, generally the tholoi constituted spaces where people interacted with the dead and with each other for centuries.

In the literature on Bronze Age Crete, the tombs have been central to discussions not only of mortuary practices and ritual generally (e.g., Branigan 1993; 1998; Murphy 2011; Soles 1992) but of the emergence of elites and the rise of hierarchy and social complexity culminating in the monumental constructions at sites such as Knossos and Phaistos that marked the end of the Prepalatial period (e.g., Murphy 1998; Legarra Herrero 2009; Relaki 2004). In these studies, space and time are secondary considerations deployed to map the tombs onto the larger narrative of Cretan prehistory. To the extent that the authors address the relationship between the living and the dead, it is under the essentialist rubric of ancestor veneration. That is, they do not problematize how the people who built and used the tombs may have perceived the relationship of the deceased to the community of survivors and whether the role of the ancestors changed.

Figure 9.1 Tholos tomb at Lebena Yerokambos. (Author's photograph.)

By contrast, this essay focuses on that relationship, foregrounding the flow of space and time within which the people who built and maintained the tholoi experienced these monuments in the process of constructing social relationships. In its theoretical framework as presented in the next section, the discussion responds to Thomas's (2004) call for a counter-modern archaeology that abandons essentialist tendencies predicated on an atomistic view of humans for acknowledgment of the subjective nature of human experience and the existence of "unfamiliar humanities" (Thomas 2004: 235). The phenomenology of Heidegger (1962) and Merleau-Ponty (2002) and recent advances in cognitive science (Clark 1997; 2008; Noe 2009) argue forcefully for the corporeality of the self and the role of the material world as its extensions. Emerging theories of materiality and of the intersection of space and time complement these ideas, emphasizing the experiential reality of human agents. We cannot separate mind and matter. The object-subject opposition dissolves, and the archaeological record no longer presents itself as a passive sign of human action on the environment but, as we shall see, as the residue of an interactive and dynamic process played out over time and in space. These theoretical considerations require that we recognize that the people whose lives were lived out in and around the tholoi understood these monuments not as two-dimensional floor plans or as markers of complexity but as embodiments of social relationships both within the community of the living and between the living and the dead. Thus we turn from theory to a close reading of the tholos complex at Yerokambos—its architectural history and depositional sequence—and then to a consideration of what these data reveal about the experience of the users. This approach reveals a more fluid relationship between the living and the dead than generally is recognized. The tholoi at first were instrumental for the integration of the dead into the community of the living, that is, the ancestors, although clearly dead, were experienced repeatedly and intimately, always present in the lives of the survivors. Modifications to the spatial arrangement of the tomb over time expressed the gradual distancing of the departed and the emergence of more formalized ancestor veneration.

THEORETICAL FRAMEWORK

Phenomenology

The cornerstone for this investigation is the concept of phenomenology as articulated by Heidegger (1962) and Merleau-Ponty (2002) and as opposed to the ontology and epistemology of Cartesian dualism. Descartes famously viewed the self as separate and apart from the materiality of its own body confronting a world in which the defining attributes of substances are real and susceptible to representation in the mind. Humans were never really in the world, which was a passive object to be studied. For Heidegger, by contrast, human beings were distinguished by their self-awareness and reflexivity, by their being always in and a part of the world which in turn already was there. Through their experiences, they interpreted their world and in turn reinterpreted themselves. They acquired meaning through encounters with objects and people outside themselves. Thus the person always is situated in a particular place and time and is aware of that existence. Because of their being-in-the-world, individuals are historical—aware of their past encounters, engaged in their present, and prepared to plan for a future based on their past experiences. At the same time, their understandings of the full range of human action is situational, a product of their temporal and spatial context. Within this framework, Merleau-Ponty emphasized the role of perception and, thus, of the corporeality of the individual. For Merleau-Ponty, the materiality of the body was not in opposition to the self but thoroughly part of it. Indeed, only through the mediation of its own physicality could the individual be aware of or learn anything (Meillassoux 2008), a view at the core of recent studies of cognitive science (Clark 1997; 2008; Noë 2009). This understanding of the self as existing in and interacting with the world means that the self is always becoming and contingent. There is no clear distinction between subject and object, which are, instead, interacting and engaged. This does not, as Tilley (2004: 29) emphasizes, lead us into an obscure subjectivity, as persons have essentially similar bodies encountering the world physically.

The phenomenological view of the individual has been central in the recent development of notions of agency and of human action in the world. Both Giddens's (1984) theory of structuration and Bourdieu's (1977) concept of *habitus* derive from the experiential foundation of the knowledge that lies behind human action. Individuals acquire through their own behaviors and those of the society around them, discursively and nondiscursively, understanding of a whole range of activity. Although there are differences in the ways in which Giddens and Bourdieu account for and explain these processes, the crucial shared element is a fundamental assumption as to the materiality of the individual, the inextricable coincidence of the self and the body. These ideas are fundamental to the concept of agency, which acknowledges the dynamic nature of human experiences in the world. Although many discussions of agency describe a specifically Western being, that is, as Thomas (2004: 122) observes, an "individual [who] is a self-contained entity, is distinguished by its consciousness and is a center from which agency is emitted," more recent discussions of agency (Boivin 2008; Robb 2010) have broadened the scope of the term, no longer privileging the independent atomistic self. Whereas agency certainly implies the ability to shape the external world, that material externality can itself be an agent mediating and expressing social experiences, as Clark (1997; 2008) and Noë (2009), among others, have argued.

Time, Space, and Materiality

These reconstitutions of the individual as a fully embodied self have played an important part in the reformulation of the fundamental subjects of archaeological inquiry: time, space, and material culture. The essentially Cartesian views of these elements prevailed for much of the early history of archaeology, in which time and space were coordinates that allowed for the proper placement of various material objects in the appropriate typology (Hodder 1993; Jones 2002; Thomas 2004). As Paynter (2002) points out, the New Archaeology or processualism of the 1970s produced a shift in emphasis on the subject of the discipline without a concomitant reassessment of the nature of time. Although engaging with cultural and social processes and thereby moving the inquiry away from 'culture history' (Paynter 2002: S86), the proponents of New Archaeology remained resolutely positivist in their search for universal laws and patterns (Rossignol and Wandsnider 1992). Each developmental stage was a closed, static moment in time. The archaeologist confronted a record of past events clearly demarcated from the present, a situation that Lucas (2005: 36) argues can exist only if we presuppose a universal independent time accessible only through chronology and separable from space.

Post-processualists have rejected this notion of time as disembodied and divorced from actual human experience and in the process squarely affirmed that time and space are inextricably connected. Time is situational, comprised of events that are not neatly marked one from the other. Our experience of time depends on what we are doing (Bourdieu 1977: 103): the hour and a half allowed for an examination may pass quickly or slowly depending on the level of our preparation. Different times (Gell 1992) can coexist, one within the other, ritual and cyclical and codified calendrical, and similarly within any given assemblage, time can be multiscalar (Olivier 1999), ranging from the use-life of the objects deposited and the time of the deposition itself to the long time of subsequent activity at the site. Time is in the end subjective, not objective, because, by virtue of the contingent nature of the human self, it is measured only because we do so, and we determine the ways in which we will mark its passage (Bender 2002). Time, then, is situated and determined by particular historical and political circumstances (Bender 2002), created by social interaction with the material world (Gosden 1994). For the archaeologist, the duration of various phases in a society's life is accessible only through the traces of those activities, that is, the people's interactions with the space they inhabited and of which they were a part (see, for example, Bailey 1981; 2006; 2007; and 2008 for time perspectivism and palimpsests as a way to conceptualize these multiscalar relationships). Most important, because time is not unidirectional and segmented but rather a matter of duration and overlapping events (Olivier 2001), there cannot be a clear demarcation, as Karlsson (2001) has emphasized, between past and present, as being in the present necessarily implies having just been in the past and being bound for the future.

Throughout these reflections runs the acknowledgment that time and space are not separable but intertwined. Articulation of these ideas has coincided with the emergence of landscape archaeology and the recognition of how theories of agency and materiality have altered traditional Western views of landscape. In his seminal article on the relationship between time and landscape, Ingold (1993) laid out the parameters of the interaction between people and the landscape. He distinguished between space more generally and the landscape, which he interpreted as relational and socially constructed. Meanings attached to spaces are gathered from the landscape with boundaries drawn by human activities (Ingold, 1993: 156). Time and space are components for action, not simply a setting (Tilley 1994). Thus humans in the landscape become humans as part of the landscape (Hegmon 2003: 224), and intervention in the landscape, remaking and reshaping it, a conscious decision (Bradley 1993). Like time, the landscape acquires meaning because we read it in, but this is not a purely visual process as Bender (2002: S107) appropriately emphasizes. Rather, the landscape is composed of a full range of sensory experiences—smells and sounds and tactile sensations—creating associations (Thomas 1996: 91) that effectively remake the individual with each encounter, layering the meanings. In this new understanding of materiality (Boivin 2008; Jones 2007; Knappett 2005), the landscape is neither symbolic nor representative but operates as an agent mediating social reproduction. Architectural intervention in the landscape is an act of assertion and insertion of the human presence (Bradley 2000; Cooney and Grogan 1998). Any discussion of the past, then, must take into account the dialectical relationship between individuals and the material world across which and within which their social lives unfolded over time. Likewise there must be acknowledgment that social reproduction occurred on multiple scales—daily, locally, yearly, regionally, and over the centuries.

These perspectives then inform the assessment that follows of the mortuary complex at Yerokambos. The tholos and its subsequent modifications appear as embodiments of the construction of social relationships both within the day-to-day framework of lives lived and in the arc of the centuries over which the tomb was in use. The monumental forms both shaped and were shaped by the contingencies of human life. The transformation of the space of the complex and the experiences of those who used it over the nearly 1,000 years of its history were integral to a process in which the ancestors rather than being immanent in the community of the living were separated and became more distant. This dynamic changed not just the connections between the living and the dead but among the living.

THE THOLOS TOMB AT YEROKAMBOS: THE ARCHAEOLOGICAL DATA

The first step is to lay out the data—the architectural history and the depositional sequence—as established by Alexiou (Alexiou and Warren 2004). Doing so discloses four phases in the complex's role as mediator of social relationships.

Architectural History

The original tholos, tomb II (Figures 9.1 and 9.2), was built at the very beginning of the Early Bronze Age, c.3100 BC (Manning 2010). The builders dug down to the bedrock, and then collected, transported, and positioned large rocks and occasionally massive boulders of local stone, to construct the nearly 2 m thick walls. They filled the interstices and the space between the inner and outer courses with smaller stones and clay, lightly working the interior surfaces of some blocks. Preserved to a height of 1.6 m, the wall is slightly offset at the top to mark the springing of the corbelled vault that originally spanned the 5.1 m wide interior and the remains of which lay on the tomb contents in a circular pattern. The east-facing entrance is composed of a constricted trilithon entrance less than 75 cm square with a lintel that, at a meter and a third long and nearly a meter across, may weigh as much as a ton, and a slightly larger north doorjamb that may weigh even more. An already difficult entry was further restricted by a 30 cm thick stone positioned to cover the entire entrance. When the vault collapsed relatively early in the tomb's history, presumably as a consequence of poor construction, the users made repairs.

Roughly 450 to 500 years later, and while tomb II still was in use, people built a second smaller tholos, IIa (the southwest quadrant of which is still

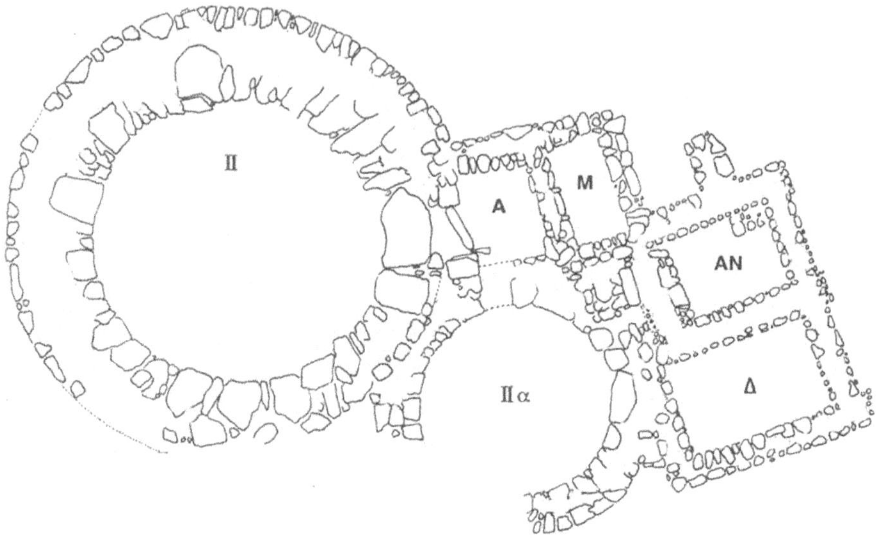

Figure 9.2 Plan of Yerokambos mortuary complex after Alexiou and Warren 2004, figure 12.

under an adjacent building), abutting the main tomb on the southeast. The entrance to IIa, rather than running east and thus parallel to the opening of the main tholos, faces north at a right angle to that original door. Soon after construction the interior of IIa was burned and the floor covered with sand, but people continued to make burials, while at the same time constructing rooms AN and D and perhaps a third room to the east over the next 100 to 150 years. When the people using the complex ceased making burials around 1900 BC, after nearly a millennium of use, they filled the entry to tomb II, that is, the space below the lintel, with bones, sealed the entry to IIa, and built rooms A and M.

Room A appears to have served as an ossuary, not for primary burials, and room M was the site for numerous liquid offerings as evidenced by hundreds of conical cups. Precisely how the users accessed these smaller compartments is unknown, but they may have done so from above through flat ceilings made of perishable materials.

Depositional Sequence

The different stages of architectural intervention are paralleled by shifts in the kinds of objects deposited in each phase of construction. Tomb II contained masses of pottery and badly decayed bones to a height of more than 1.1 m. Nearly 90 percent of the tomb contents consisted of pottery representing nearly 3,000 vases (Alexiou and Warren 2004). Repeated additions to, and reorganization of, the tomb contents meant that Alexiou could not detect distinct strata, but the pottery did appear to form four levels and can be ordered in two chronological groups. The earlier material ranges from Early Minoan (EM) I through EM IIA (3100–2400 BC) (Manning 2010, Table 2.2). The vases include a broad range of shapes and decoration (Miller Bonney 2005–2006) comparable to what one would expect to find in a contemporary settlement. The small finds, except for the 124 pieces of obsidian, are similarly ordinary. By contrast, the later, EM IIB–Middle Minoan (MM) I pottery consists overwhelmingly of cups. The secondary tholos, IIa, primary use of which overlapped with the later levels in the main tomb, also contained almost nothing but EM IIB and later pottery cups. Similarly, rooms AN and D and the poorly preserved room further east contained primarily cups (roughly 70 percent) and almost no small finds. A bench built into the wall in room AN may have been intended for offerings. The last two rooms, A and M, signal another shift in deposition practices. Room A appears to have served as an ossuary filled with bones and a modest amount of pottery after the entrance to the main tomb was sealed. Room M was filled with pottery, primarily the conical cups associated with large-scale drinking and feasting at other sites. By this stage, the diverse pottery of daily life has been replaced by a standardized and shape.

THE LIFE OF THE TOMB

The next step is to consider the archaeological data within the phenomenological framework by reflecting on the potential multisensory experiences of those who built and used the tomb complex. Exploration of the multiple scales of both space and time within which the people engaged with the site discloses ways in which it was part of the mediation and reproduction of social values.

Initial Construction

The decision to construct a monument, as Cooney and Grogan (1998) have noted, involved in the first instance an insertion and assertion of the human collectivity into the landscape. The architecture was a spatial appropriation intended to influence behavior (Thomas 1994) by directing mortuary conduct to a particular location within the landscape. Providing a formalized and monumental structure for the deposition of the dead extended the relationship between the living and the dead (Bradley 1998; Mizoguchi 1993) by regularizing mortuary practices and creating a means (Bradley 2003) for perpetuating the tradition beyond a few generations. The monumentality of the structure ensured that the tomb would continue to convey meaning by impinging on the people's vision at all times (Richards 1994). Whereas the tomb at Yerokambos probably is not the earliest tholos in the Asterousia, nevertheless the decision to situate it in that particular part of the landscape at that time was momentous for the people involved. They situated the tomb in a broad open area so that it always could have been approached from any direction and would have permitted potentially large numbers of people to gather for activities at the complex.

The process of construction maintained the link between the community on the one hand and the landscape on the other. Clearing the site and transporting the large boulders required a collaborative effort and brought those who built the tomb into intimate contact with the landscape. Work penetrated the surface of the earth to the bedrock, a process that, in a similar context, Charles (2005) has likened to exposing the earth and its generative and creative force. Then the team transported the massive local rocks to the site, further grounding the tomb in the landscape. In the process, kinesthetic memories were formed, the physical experiences of clearing the soil and transporting, moving, and raising the boulders were imbedded in the builders. They were working in and on and with the rock of which the landscape was constituted. The erection of the megalithic doorway in particular would have required maximum physical effort. The layering of the corbels gradually hid the bedrock that they had struggled to expose from view. Finally, they placed a massive blocking stone in front of the entrance. What we do not and cannot know is whether the structure immediately went into use as a tomb.

The First 400 Years

For a long time after its construction, the tomb stood as an isolated monument. This tomb, like all the tholoi, which some have analogized to caves (Branigan 1970: 148), seems part of a landscape marked with numerous outcroppings. Only viewed from the east would its character as a built structure have been clear. The massive doorway with its huge blocking stone appeared to give access into the very landscape itself and did not merely stand for or represent the entrance to the tomb, the space on the other side of daily life. Boivin (2008) has described how in a small village in Rajasthan the red soil the villagers applied to their homes was not symbolic of the goddess; it was the goddess. I suggest that we should understand these enormous stones in a similar fashion, not as symbolic of the separation between life and death but as being that boundary. The space inside the tomb thus was substantively different from the external landscape.

During this first phase, the pottery, like the tomb, participated in the complex interplay of landscape and social relations. On the one hand, the simple geometric designs, whether painted or incised, recalled the folded strata of the rock of which both tomb and landscape were composed, as do the black, grey, and reddish brown hues of the unpainted wares. At the same time, these strikingly mundane grave goods also reinforce the impression of the tholos as a collective endeavor in which the corporate entity, the community, included the dead, who occupied a different space but took with them the accoutrements of daily life. Whereas some may have possessed the wealth to acquire the gold beads or Egyptian scarab that accompanied them into the tomb, others took the tools of their trade—fish hooks or nails. Some ceramic shapes appear more often than others, but the diversity suggests there was no standardized mortuary kit. The impression is that the objects deposited with the deceased, even if they may, in some instances, have been manufactured for mortuary purposes only, are the same as those he or she used in daily life. The burials intentionally refer to or quote life outside the tomb, extending these interactions into the tomb.

And that extended interaction was real, repeated and performative as the living entered the tomb over and over for both primary inhumations and the secondary burial within the same tomb until it was full. Initially, the deceased was placed directly on the ground, accompanied by two to three pieces of pottery and perhaps one or two personal belongings. Subsequently, some of the living entered the tombs and rearranged the human and other remains, removing not yet decomposed flesh from the bones. Usually, though not always, they disarticulated the skeleton, occasionally segregating long bones and skulls and apparently disposing of smaller bones, perhaps through simple discard. At this stage, the dead had completed their transition from life and had lost the unique characteristics that distinguished them when alive and fully fleshed. Yet they remained in these rites present in a different sense and required some action to acknowledge their new state. This sequence of procedures was both intensely communal and intimate. Although the state of preservation of the osteological material precludes a statistical analysis by age or sex, it seems clear that both men and women and the whole spectrum of ages, that is, all the members of the community, were buried in the tholoi.

Throughout this phase, the living interacted intimately and intensely with the deceased, eroding the barriers between the two realms in a practice that was profoundly material and sensory. As the living entered the tomb, they left the broader landscape and its monumental reference to the longer time and entered the compressed and immediate space of the interior. At interment, the tight entrance pressed the bodies of the living and the dead close together as the burial party pushed and pulled the body through the opening. Upon reentry, participants had intimate and tactile contact with the bones—scraping off remnants of flesh, rearranging some and removing others—hands touching bones repeatedly. Sound memories were formed as bones rubbed against each other, the hands, the dirt, and the grave goods. The lingering stench of putrid flesh was so intense it produced a taste sensation. In the dark tomb, daylight provided only partial illumination, while a flickering lamp revealed in shifting light the bodies, bones, and grave goods from the past. The viewers saw pottery shapes and styles whose semiotic significance was as hidden from them as the individual identities of the people among whose bones they moved. As the deposit of bodies and grave goods rose toward the ceiling, those participating in the burial would have been unable to stand erect, participants forced to navigate a tangle of bodies, bones, and grave goods. The dead did not occupy a totally separate realm organized according to different rules. Rather, the fundamentally disorderly environment inside the tomb mirrored the untidy daily existence of the living, reasserting the link between the two spaces. Users thus experienced the tholos in the long time of a permanent monument and the greater space of the landscape but also in the brief and intimate spatiotemporal experience of entering the tomb.

Construction of Tholos IIa and Rooms AN and D

The construction of tholos IIa marked an important turning point in the history of the complex. Although tomb II was not yet full, the smaller circular annex was attached to the wall with an entrance that faced north. This addition, by restricting the range within which the megalithic doorway was visible, reduced its impact. Although the treatment of the dead with primary interment followed by secondary manipulation would have continued, the pottery deposited with the body changed. As I have argued elsewhere (Miller Bonney 2005–2006), the diversity of shapes and decoration characteristic of the first 400 years of the tomb's use gives way to a more standardized burial kit of a jug and multiple cups or bowls. Many if not most of the shapes encountered in the earlier phase of the tomb's life are totally lacking. This formalization suggests that a distance was opening between the living and the dead in that the latter no longer took with them the pottery repertoire the survivors enjoyed.

The 'encroachment' of rooms AN and D perhaps 150 years after the construction of IIa further obscured the monumental entrance of the tomb. Although one cannot know the original height of these walls, they certainly impaired the view of someone approaching from the southeast, and the full monumentality of the doorway and of the tholos itself would have been further vitiated by the extension of these rectangular spaces. In addition, their less substantial construction did not engage the significant collaborative effort required by the original construction of the tomb. Of the two rooms, only D contained burials, probably made there after the smaller tholos IIa was full, and there were no burials in room AN, which Alexiou concluded must have been used solely for offerings (Alexiou and Warren: 2004: 19). Thus, whereas during the first phase of the tomb's use offerings had been made in the physical presence of the dead, in this third stage the tendering of a gift occurred in a separate space, underscoring the distance between the living and the dead. So, in the latter part of EM IIB and EM III, some sort of change gradually had taken place at the site in which there were activities associated with the many cups in these smaller rooms and a possible displacement of focus from the tholos. This is a time when, at other tholoi in the region, the users pave the open space around the tomb, perhaps suggesting an intensification of activity outside the tomb. The different behavior associated with rooms AN and

D suggests that feasting and drinking may also have occurred in the broader area of Yerokambos.

Construction of A and M

The construction of Rooms A and M mark the transformation of the tholos from an active mortuary complex to the locus of activities that apparently involved drinking or toasting based on the volume of conical cups. The decision to seal the entrance to the main tholos closed off a space in which people once had interacted intimately with the dead. The deposition of bones in the ossuary that was room A, the lack of primary burials, the total absence of small finds, and the overwhelming number of cups suggest that the patterns of behavior at the tholos had changed. The tomb and the dead now occupied different places in the society. Ritual activity still was directed at the tomb but no longer included the intense interaction with the dead that characterized individuals' presence at the tomb for much of its history.

CONCLUSION

The reading of events sketched out above illustrates the insights that can be gained by the approach that is the core of this volume. By combining the phenomenological perspective with current understandings of time, space, and material culture, we recognize the real, subjective lived experiences of the people who actually were engaged with each stage of the complex's history. Thus, rather than focusing on the addition of the external compartments as strategies for expressing power in the competition that theoretically informed the emergence of hierarchies late in the Prepalatial Period, we see them as shaping and being shaped by changes in the views the survivors had of the departed. In his groundbreaking work on death in the Christian west, one of the themes that Aries (1981) identified was that changes in mortuary activity came about in large part because participants in the rituals had developed new understandings of the condition of the departed and their own relationship with those individuals. Given the enormity of the initial undertaking in building the tholos and the almost certain prolonged impact on the community of the presence of that monument, the decision to alter that structure when the primary tholos was not yet full required the sorts of changes Aries described.

Such a transformation in thinking about the dead seems even more crucial because the evidence from Trypiti (Vasilakis 1989) indicates that the tomb preceded the settlement, perhaps by several centuries, and those who elected to build only 150 m away from the tomb did so because they desired the proximity. Preucel (2006: 259) has argued that architectural agency is intended to promote certain thoughts through materiality, a view also expressed by Jones (2007), for whom monuments are integral to the formation and perpetuation of social memories. As already noted, the decision to build the tomb signaled an intention to promote activities within the space in and around the tomb, which was itself framed so as to assert a connection with the landscape. The Asterousia Massif, which rises immediately north of Yerokambos, is the most arid part of Crete, and the people who built the tomb often would have been forced to move their flocks further north for better grazing. The monumental tomb was a place to which to return. Actions at the tomb were relatively brief—the time it takes for a single burial and the subsequent secondary manipulation—but also were repeated over long stretches of time. There were the actions in the immediate vicinity of the tholos and the longer journeys back. There was a whole array of sensory experiences. For 400 years, there was only the stone tomb and the burials.

Then the users of the tomb began to make changes in ways similar to changes to burial monuments in Neolithic Britain. Bradley (1998) has argued that the process by which Neolithic burial monuments became larger, more elaborate, and composed of stone, not wood, meant that they had become more symbolic. The stone stands for the wood the people once used, and that representational effect applies to the entire tomb. Bradley further contends (2002) that when ultimately those buried in the tombs were not simply deceased members of the community but the ancestors, thereby occupying a different historical position, the monument went out of use. Architectural changes at Yerokambos seem to suggest a similar pattern. The trilithon doorway that, I argue, had served as the actual boundary between this world and the next, lost its visual prominence as IIa and then slightly later the small rooms AN and D further east altered the apparent unity of the original structure. At this same point, there was a certain formalization of the grave goods, as cups and a smaller number of jugs replaced the diverse offerings of the earlier burials, eroding the sense that the pottery in the tomb is essentially the same as that used by the living. The integration of the dead into the community of the living was beginning to weaken even as the collective behavior remained largely intact. The separation of actions in rooms AN and D, such that burials are restricted to D and offerings made only in AN, implies a different relationship than when the survivor defleshes the bones of the ancestors. When rooms A and M were constructed, the occupants in the tomb had become the ancestors, citizens of a very different time and space, and as Bradley suggests, when at this point the complex had become more symbolic and the dead occupied a different historical position, the site was abandoned.

Concentrating on the ways in which the users of the tomb complex increased the number of spaces over time and altered the uses to which they were put reveals a previously unrecognized aspect of Prepalatial social formation. We see for the first time that whatever other forces may have been shaping the community, part of the process that preceded and may have contributed to the rise of monumental architecture and some new social order was a redefinition of the relationship between the living and the dead. The dead with whom the living intimately consorted in the early history of the tomb were close to and part of that society and could not be manipulated to the advantage of one party over another. By contrast, the distant ancestors who are accessed only through libations are more readily available for symbolic wielding as a source of power.

REFERENCES

Alexiou, S. and P. Warren. 2004. *The Early Minoan Tombs of Lebena, Southern Crete*. Sävedalen: Paul Åströms Förlag.

Aries, P. 1981. *The Hour of Our Death*. New York: Alfred A Knopf.

Bailey, G. N. 1981. Concepts, time-scales and explanations in economic prehistory. In: A. Sheridan and G. N. Bailey (eds.), *Economic Archaeology*, 97–117. British Archaeological Reports International Series 96. Oxford: Tempus Reparatum.

Bailey, G. N. 2006. Time's arrow: the measurement and theory of archaeological time. *Antiquity* 80: 717–720.

Bailey, G. N. 2007. Time perspectives, palimpsests and the archaeology of time. *Journal of Anthropological Archaeology* 26: 198–223.

Bailey, G. N. 2008. Time perspectivism: origins and consequences. In: S. Holdaway and L. Wandsnider (eds.), *Time in Archaeology: Time Perspectivism Revisited*, 13–30. Salt Lake City: The University of Utah Press.

Bender, B. 2002. Time and landscape. *Current Anthropology* 43(August–October Supplement): S103–S112.

Boivin, N. 2008. *Material Cultures, Material Minds: The Role of Things in Human Thought, Society and Evolution*. Cambridge: Cambridge University Press.

Bourdieu, P. 1977. *Outline of a Theory of Practice*. Cambridge: Cambridge University Press.

Bradley, R. 1993. *Altering the Earth*. Edinburgh: Society of Antiquaries of Scotland.

Bradley, R. 1998. *The Significance of Monuments*. London and New York: Routledge.

Bradley, R. 2000. *The Archaeology of Natural Places*. London and New York: Routledge.

Bradley, R. 2002. *The Past in Prehistoric Societies*. London and New York: Routledge.

Bradley, R. 2003. The translation of time. In: R. M. Van Dyke and S. E. Alcock (eds.), *Archaeologies of Memory*, 221–227. London: Blackwell Publishing.

Branigan, K. 1970. *The Tombs of Mesara*. London: Duckworth.

Branigan, K. 1993. *Dancing with Death*. Amsterdam: Hakkert.

Branigan, K., ed. 1998. *Cemetery and Society in the Aegean Bronze Age*. Sheffield: Sheffield Academic Press.

Charles, D. K. 2005. The archaeology of death as anthropology. In: G. F. M. Rakita, J. E. Buikstra, L. E. Beck and S. R. Williams (eds.), *Interacting with the Dead: Perspectives on Mortuary Archaeology for the New Millennium*, 15–24. Gainesville: University of Florida Press.

Clark, A. 1997. *Being There: Putting Brain, Body and World Together Again*. Cambridge, MA: MIT Press.

Clark, A. 2008. *Supersizing the Mind: Embodiment, Action and Cognitive Extension*. Oxford: Oxford University Press.

Cooney, G. and E. Grogan. 1998. People and place during the Irish Neolithic: exploring social change in time and space. In: M. E. Richardson and C. Richardson (eds.), *Understanding the Neolithic in North-Western Europe*, 456–481. Glasgow: Cruithne Press.

Gell, A. 1992. *The Anthropology of Time*. Oxford: Berg Publishers.
Giddens, A. 1984. *The Constitution of Society: Outline of a Theory of Structuration*. Cambridge: Polity Press.
Gosden, C. 1994. *Social Being and Time*. Oxford: Blackwell.
Hegmon, M. 2003. Setting theoretical egos aside: issues and theory in North American archaeology. *American Antiquity* 69: 213–243.
Heidegger, M. 1962. *Being and Time*. London: Wiley-Blackwell.
Hodder, I. 1993. The narrative and rhetoric of material culture sequences. *World Archaeology* 25: 268–282.
Ingold, T. 1993. The temporality of the landscape. *World Archaeology* 25: 152–174.
Jones, A. 2002. *Archaeological Theory and Scientific Practice*. Cambridge: Cambridge University Press.
Jones, A. 2007. *Memory and Material Culture*. Cambridge: Cambridge University Press.
Karlsson, H. 2001. Time for an archaeological 'Time-Out'? In: H. Karlsson (ed.), *It's About Time: The Concept of Time in Archaeology*, 45–59. Göteborg: Bricoleur Press.
Knappett, C. 2005. *Thinking Through Material Culture: An Interdisciplinary Perspective*. Philadephia: University of Pennsylvania Press.
Legarra Herrero, B. 2009. The Minoan fallacy: cultural diversity and mortuary behavior on Crete at the beginning of the Bronze Age. *Oxford Journal of Archaeology* 28(1): 29–57.
Lucas, G. 2005. *The Archaeology of Time*. London and New York: Routledge.
Manning, S. 2010. Chronology and terminology. In: E. Cline (ed.), *The Oxford Handbook of the Bronze Age Aegean*, 11–28. New York: Oxford University Press.
Meillassoux, Q. 2008. *After Finitude: An Essay on the Necessity of Contingency*. New York: Continuum.
Merleau-Ponty, M. 2002. *Phenomenology and Perception*. New York: Psychology Press.
Miller Bonney, E. 2005–2006. A reconsideration of depositional practices in Early Bronze Age Crete. *Aegean Archaeology* 8: 31–50.
Mizoguchi, K. 1993. Time in the reproduction of mortuary practices. *World Archaeology* 25: 223–235.
Murphy, J. M. A. 1998. Ideology, rites and rituals: a view of Prepalatial Minoan tholoi. In: K. Branigan (ed.), *Cemetery and Society in the Aegean Bronze Age*, 27–40. Sheffield: Sheffield Academic Press.
Murphy, J. M. A., ed. 2011. *Prehistoric Crete: Regional and Diachronic Studies on Mortuary Systems*. Philadelphia: INSTAP Academic Press.
Noë, A. 2009. *Out of Our Heads: Why You Are Not Your Brain and Other Lessons from the Biology of Consciousness*. New York: Hill & Wang.
Olivier, L. 1999. The Hochdorf 'princely' grave and the question of the nature of archaeological assemblages. In: T. Murray (ed.), *Time and Archaeology*, 13–30. London and New York: Routledge.
Olivier, L. 2001. Duration, memory and the nature of the archaeological record. In: H. Karlsson (ed.), *It's About Time: The Concept of Time in Archaeology*, 61–70. Göteborg: Bricoleur Press.
Paynter, R. 2002. Time in the valley: narratives about rural New England. *Current Anthropology* 43(August–October Supplement): S85–S101.
Preucel, R. W. 2006. *Archaeological Semiotics*. London: Blackwell.
Relaki, M. 2004. Constructing a *Region:* the contested landscapes of Prepalatial Mesara. In: J. C. Barrett and P. Halstead (eds.), *The Emergence of Civilisation Revisited*, 170–188. Oxford: Oxbow Books.

Richards, C. 1994. Monumental choreography: architecture and spatial representation in Late Neolithic Orkney. In: C. Tilley (ed.), *Interpretative Archaeology*, 143–178. Oxford: Berg Publishers.
Robb, J. 2010. Beyond agency. *World Archaeology* 42: 493–520.
Rossignol, J. and L. Wandsnider. 1992. *Space, Time and Archaeological Landscapes*. New York and London: Plenum Press.
Soles, J. S. 1992. *The Prepalatial Cemeteries at Mochlos and Gournia and the House Tombs of Bronze Age Crete*. Hesperia Supplement 34. Princeton, NJ: American Schools of Classical Studies at Athens.
Thomas, J. 1994. The hermeneutics of megalithic space. In: C. Tilley (ed.), *Interpretative Archaeology*, 73–97. Oxford: Berg Publishers.
Thomas, J. 1996. *Time, Culture and Identity*. London and New York: Routledge.
Thomas, J. 2004. *Archaeology and Modernity*. London and New York: Routledge.
Tilley, C. 1994. *A Phenomenology of Landscape*. Oxford: Berg.
Tilley, C. 2004. *The Materiality of Stone: Explorations in Landscape Phenomenology*. Oxford and New York: Berg.
Vasilakis, A. 1989. O protominoikos oikismos Trypitis. *Archaeologia* 30: 52–56.

10 Time and Space in the Middle Bronze Age Aegean World

Ialysos (Rhodes), a Gateway to the Eastern Mediterranean

Toula Marketou

The problem arises of knowing whether the unity of a discourse is based not so much on the permanence and uniqueness of an object as on the space in which various objects emerge and are continuously transformed.

Foucault (2004: 25)

INTRODUCTION: MOVING IN THE LANDSCAPE THROUGH TIME

Wandering in open-air space and careful exploration of the environment gave humans the first glimpse of knowledge and essential learning of the exterior world and nature, forcing them to shelter in the friendly and protected space of caves and landscapes with arable land. By practicing initial body movements from the exterior into the interior, humans managed to conquer, control, and organize space through time by choosing sites for semi-permanent occupation. These complex processes led to the creation of the built environment, the emergence of the first houses, and the genesis of architecture. Moving patterns of several domestic groups in numerous landscapes on the large and fertile island of Rhodes and the long processes that took place before the construction of the major Late Bronze Age IA settlement of Ialysos/Trianda (Furumark 1950; Marketou 1988; 1990b; 1998; 2009; Monaco 1941) provide interesting insights into the continuous communication of prehistoric societies with the environment in intrasite and intersite levels over space and time and offer multidimensional evidence toward explaining the development of Rhodes into one of the most important gateways to the eastern Mediterranean.

Analysis of the results of previous surveys (Blinkenberg 1931: 23–26, 60–67, plates 3.24, 26; Blinkenberg and Johansen 1926: 30, plate 36. 1, 2; Dietz 1974; Hope Simpson and Lazenby 1973: 133–136; Inglieri, 1936; Sampson 1987: 96–105) and most recent large-scale excavations indicate a strong adaptability as well as the ability of people to relocate on the island from the Late Neolithic to the first stages of the Middle Bronze Age. People select their own space related with wealth-producing sources and organize settlement patterns either close to the coastline or on mountains, hills, and fertile plains through time. This long and complex process, closely related with the surrounding environment, led the prehistoric societies of Ialysos to configure and organize their built environment during successive time periods in prehistory.

At the end of the Late Bronze Age IA, after the eruption of the Thera volcano, the prehistoric settlement of Ialysos continued to play an important part until the end of Late Helladic III A:2/beginning Late Helladic IIIB:1. At this time, the Ialysians were forced to abandon their land due to a devastating flood, the outcome of which is attested through various daily-life objects left in situ and buried beneath the alluvial deposits of the present-day flood plain of Trianda. Life continued at other locations of the island, where the Mycenaean civilization spread and flourished until the twelfth century BC. The number of burials increased in the cemeteries at Ialysos in the early twelfth century BC, but the new site of the associated settlement has yet to be discovered (Figure 10.1).

Most recent excavations in the wider area of the territory of Ialysos provided strong evidence for chronological questions concerning the so far obscure transition from the end of the Early Bronze Age to the nearly unknown Middle Bronze Age and the Late Bronze Age IA in the Aegean (Marketou 1990b; 1998; 2007; 2009). A set of radiocarbon dates provided a rather absolute chronological framework for the aforementioned periods and the transition from one period to the next (Marketou, Fakorellis and Maniatis 2001; Manning et al. 2002). In addition, detailed and contextual evidence derived from several rich 'closed' floor deposits found at the Early Bronze Age site of Asomatos (Marketou 1990a; 1997) and at the Middle Bronze Age settlement at Ialysos/Trianda (Marketou 2009: 79–81, Figures 9, 11) has provided indications for the use of space over time in both these cultural/chronological assemblages.

All these complex processes, which forced people to move to different places and to organize their built environment, give archaeology a new interpretative perspective. However, the development of elaborate architecture could be seen as a result of transformations that took place in preexisting affluent societies, before the Late Minoan IA period on the island. From this perspective, the origins of emblematic Late Bronze Age Aegean architecture and great achievements in building techniques took place in the Early and Middle Bronze Age. Consequently, all these successive achievements resound a continuous and harmonic interaction between the local societies and the surrounding landscape.

Architecture and order of the Early Bronze Age II and Early Bronze Age III spacious rectangular buildings at Asomatos, the earliest proto-urban center in the northwest part of the island (Marketou 1997; Marketou 2010: 776–777), could be perceived as a result of continuous movements of people and collectively complex processes, which took place after the abandonment

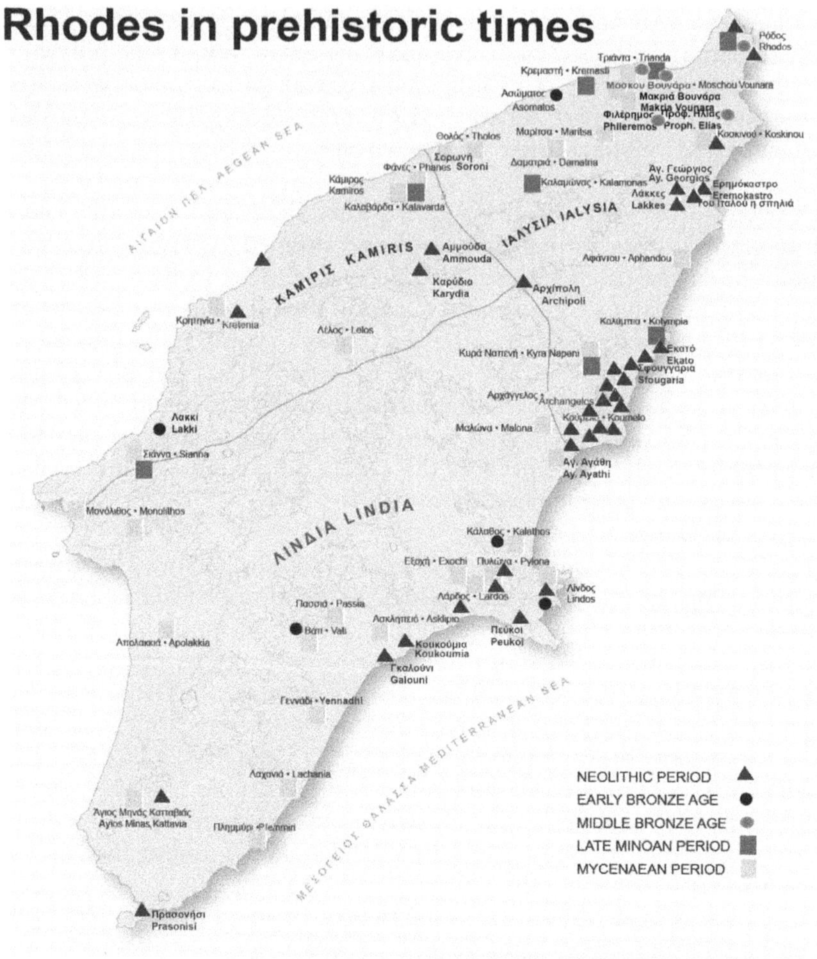

Figure 10.1 Movements of people on the island of Rhodes during prehistory.

of the Late and Final Neolithic caves and some open-air spaces in other locations of the island (Sampson 1987: 96–105). This concept is reinforced from evidence of some sporadic Early Bronze Age sites on the island (Blinkenberg 1931: 23–26, 61–68, plates 3.24, 26; Blinkenberg and Johansen 1926: 30, plate 36. 1, 2; Dietz 1974; Hope Simpson and Lazenby 1973: 133–136). Analytical accounting of material culture found nearly intact on floor deposits maintains that over the course of at least the last phase of the occupation, in the end of the third millennium BC, the freestanding buildings of Asomatos could be seen as visual symbols of society (Hillier and Hanson 1984: 1). In light of the above evidence, it could be suggested that the residents of the settlement shared common characteristics with other societies in the wider area of the northeastern Aegean and Anatolia (Marketou 1997: 407–411).

Figure 10.2 Distribution of material culture in the Early Bronze Age III megaroid building at Asomatos *in situ*.

SETTLEMENT PATTERN AND ARCHITECTURE IN THE MIDDLE BRONZE AGE

The domestic groups of the coastal Early Bronze Age site of Asomatos were forced to flee their homes due to a devastating flood, resulting in a wide shift of population in other locations. Hence, Middle Bronze Age communities appeared in rather safe places, namely on the top of Mount Phileremos (Benzi 1984; Coldstream 1969) and on the northeast peak of the same mountain, near the old chapel of Prophetes Elias (Marketou 1988: 28, Figures 2–3; 2009: 73, Figure 1), whereas evidence for Middle Bronze Age habitation is also documented in the lower levels of the well-known Late Bronze IA settlement (Marketou 1998: 62; Marketou 2009: 77–82). This wide expansion has also been identified in a short distance of about 600 to 620 m east/southeast of the main Late Bronze Age settlement of Trianda, as well as in the area of the port of Acandia, beneath the layers of the ancient city of Rhodes (Dreliosi-Irakleidou 1999: 24–25). However, the nucleus of the above regional Middle Bronze Age settlements occupied a low hill or most probably a slope in the southern part of what is the present-day flood plain of Trianda, south/southwest of the well-known Late Bronze Age I and Mycenaean settlement (Furumark 1950; Marketou 1988; 1990b; 1998; 2010: 779–786; Monaco 1941).

A series of several salvage excavations, within the nucleus of the above described regional pattern of the Middle Bronze Age expansion in the wider area of Ialysos, pointed to several urban characteristics that define a major significant settlement that predates the well-known coastal Late Bronze Age city. Organized with *insulae* delimited by streets, alleys, and open spaces, the site presents a sophisticated and nucleated layout. Streets are paved with pebbles, similarly to open-air areas and streets around the Early Bronze Age III freestanding buildings of Asomatos. However, the idea of the oblong house subdivided along its length into two uneven sectors continued earlier traditions of Asomatos (Marketou 1997: 396, 403–407; 2009, Figure 10; 2010, Figure 58.1), although Middle Bronze Age buildings have lost the neat and elegant appearance of the Early Bronze Age III previous period. In some cases, a main spacious room occupied one part of the building, whereas the other was divided into smaller spaces for storing.

Pebble-paved streets border several *insulae* forming indentations across the buildings. Uncovered in a length of 12 m, its width about 3 m, one such street was most probably one of the main circulation arteries of the settlement leading from the coastline to the slope, near the present-day church of the Virgin Mary (Figure 3, number 8). The buildings revealed so far across this street were associated with workshops occupied by skilled technicians and wealthy storehouses. However, indentations were also revealed across a partially uncovered building in the east sector of the settlement (Figure 3, number 5) across a pebbled open-air area.

Buildings shared more or less similar storage practices. However, spacious rooms have so far suggested a degree of differentiation in terms of architecture, size, or contents within the Middle Bronze Age settlement of Ialysos. In a small storeroom, for example, an assemblage of imported Kamares ware was uncovered (Figure 10.3, number 6). In other cases, groups of jars and jugs and other pottery types found fallen on the floor of storage areas, most probably coming from the cupboards opened into the mudbrick wall, show individual storage practices in the settlement and more or less similar spatial uses within individual households. Smaller storerooms were full of closed vessels, such as jars and pithoid jars with everted rims. Tritons among jugs, knobbed jars, other closed vessels, and carinated cups were found in two other rooms farther north, below the early Late Bronze Age IA walls in the southernmost sector of the major Late Bronze Age settlement. However, contrary to the evidence of the Early Bronze Age building at Asomatos, where the main large room with the hearth, nearly in the middle of it, was used for everyday activities according to the contextual meaning of several daily-life finds in situ (Marketou 1997), some spacious rooms of the nucleated Middle Bronze Age settlement were found empty.

The walls, about 40 to 45 cm thick, were built on low foundations of rough stone. The top of the foundations was leveled off with a layer of mud mixed with pebbles, whereas the upper structure was made of mud brick and irregular lumps of clay mixed together with small stones, pebbles, and organic material. Timbers that run horizontally and transversely through the upper structure and shape the openings of the rooms reinforced the buildings,

Figure 10.3 Layout of the excavated parts of the Middle Bronze Age settlement.

and strong layers of fine and burnished plaster, normally painted dark red, strengthened and protected the walls from moisture. Walls built entirely of mud brick, especially those that do not preserve the plaster on their interior surface, are not easily identified in excavations, as collapsed mud brick tends to form patches of dark brown earth. However, in certain areas, collapsed mud brick walls are easily recognized from their orange-reddish color, a result of regional conflagration. Within these solid layers, the shape of some mud bricks could be still distinguished.

Reddish plastered benches and other features built with mud brick existed within the rooms too. Building materials and consequently sophisticated techniques were totally harmonized with the recourses of the surrounding area, mainly the clayish earth found in abundance in Ialysos, which compensated for the lack of stone in the area. Building techniques as well as the main layout of the buildings continued earlier traditions of Asomatos. A strong layer of mud plaster, overlaid by fine and burnished like plaster, usually painted dark or orange red, helped consolidate the wall and protect it from structural dampness. The technique recalls the Early Minoan III houses at Vasiliki (Zois 1976; 1982) and other sites in Crete, a tradition that also continued into the Middle Minoan phases in Crete with the best examples in the Palace and Quartier Mu at Malia (Poursat 1996; Poursat and Schmid 1992).

On the other hand, the layout in the southernmost part of the east sector of the nucleated Middle Bronze Age settlement shows a remarkable change of the standard settlement pattern. Located nearly in the south limits of the settlement, a partially revealed spacious room (4.50 by 6.15 m) with red polished floor connected with similarly plastered walls and benches appears quite different compared with other buildings of the settlement revealed so far (Figure 10.3, number 3). A pinkish-orange plastered bench is built at the west side of the red room, in front of which are the remnants of a large freestanding plastered elaborate platformlike feature set on the floor. The west side of this room is attached to a wide structure (1.20 m) made of cobbles aligned along two parallel rows. A door with a monolithic, also plastered, threshold at the southwest wall of the room provided access into a second room (2.30 by 2.65 m). Remains of plaster on the walls connected with a plastered floor were also found in the second room. The luxurious appearance of both rooms indicated that they were part of an important building complex located at the south edge of the settlement (Marketou 1999: 946–947, Figure 20; 2009: 79–80, Figure 8).

CHANGE IN ARCHITECTURE AND ORDER: THE EMERGENCE OF DIFFERENTIATION

Similar spaces with mostly red plastered walls and floors appeared also a short distance north of the previously mentioned rooms. The walls were made of mud brick on stone foundations, according to the normal building techniques that used the material of the surrounding environment due to the lack of stones. The decayed upper structure from the walls was found fallen on the plaster floors. Patches of burnt mud bricks fallen in situ give a dramatic picture of the catastrophe of the site that indicates the final abandonment of the building in its heyday (Figure 10.4). However, it is obvious that the

Figure 10.4 Middle Bronze Age building with the fallen upper structure of the walls *in situ*. The room with the pier-and-door partitions in the middle. View from the north.

abandonment was the result of a devastating catastrophe that affected all the peripheral Middle Bronze Age settlements as well, the majority of which were not occupied over the rest of the Bronze Age, apart from the top of Mount Phileremos and the location of the Late Bronze Age settlement farther north. The abandonment is also obvious from the rest of the material culture, pottery, tools, and other objects left forever on floors, and mainly the dense deposits within the storing spaces, some of them collapsed from cupboards and benches (Marketou 2009: 80–82, Figures 2 and 3).

The collapsed building consists of at least nine spaces, some of them rearranged and reconstructed over a short time during the same chronological phase according to the typology of the associated material culture (Figure 10.3, number 7; Figure 10.5). However, the most intriguing architectural innovation is the presence of an impressive and emblematic polythyron, nearly in the middle of the structure, that gave access to a wide rectangular room to the south (37.8 m^2). The pier-and-door partitions are restricted in the west part of the north side of the room, whereas the east part is closed by a wall, most probably following a rearrangement. The doorjambs[1], made of carefully hewn T-shaped porous stone, give a monumental appearance to the polythyron that is quite unusual compared with the known Neopalatial examples in Crete, Akrotiri, and the Late Bronze Age IA settlement on Rhodes (Driessen 1982; Marketou 1998: 48–49; McEnroe 2010: 54, 77, 88- 89, 94, 97).

Apart from the known polythyra of Malia, namely a Middle Minoan I polythyron in Crypte Hypostyle (Allegrette and Schmid 1997; Amouretti 1970) and the Middle Minoan II of Building A in Quartier Mu (Poursat and Schmid 1992; Schmid 1983), this is the earliest polythyron in the Aegean known so far. Moreover, its appearance, with all the carefully carved stone parts, comprising heavy stone, T-type doorjambs and thresholds, decorated with fine painted plaster, constitutes the best known predecessor of the Neo-palatial residential quarter type. The doorjambs, decorated with elegant schematic plants and reeds on the surface of fine and burnished white plaster, stand as emblematic pillars seen from the north for those entering the building.

The elegant decoration on all the sides of the doorjambs most probably continued on their upper part covering the wooden parts and the mud brick of the piers in harmonic coexistence with the decorated plaster on the emblematic wide thresholds associated with the plaster floor. However, similarly remarkable is the decoration of the plaster floor in the spacious room south of the doors. Decorated with imitations of rectangular veined slabs painted either black on violet or violet on white in an antithetical arrangement, the floor possesses a luxurious and impressive appearance of polychrome marble slabs.

Figure 10.5 Middle Bronze Age building with pier-and-door partitions dominating in the middle, after the removal of the fallen upper structure on the plaster floors. Detail of the south part of the painted floor.

Only the south wall of the polythyron space is preserved, whereas both the west and east walls are missing. Adjoining to the south is a benched red room, similar to the spacious one found farther south in another excavation. The monolithic stone base of a pillar found below the level of the reddish plastered floor is located in the same axis with the base of the east doorjamb of the polythyron. Similarly, the same axis follows the location of another circular opening found in the plaster floor of the narrow space north of the polythyron. This is a smaller room, most probably the anteroom of the polythyron, with red plaster floor, its walls decorated with frescoes depicting red linear motifs on fine white plaster. East of the anteroom is a smaller slightly upraised space, divided with a very thin parapet at its eastern end. This appears to be another rearrangement and segmentation of space that divides the previously wider areas of both the polythyron and the anteroom.

The room to the east of the polythyron, although not completely revealed, also gives a striking impression due to the painted plaster on the floor in a style similar to the adjacent main room with the polythyron. However, the rectangular room to the west of the main room, with the pier-and-door partitions, might have been an early type of light room[2] that gave light to the main room. Paved with a heavily burnished red plaster, it creates a low cavity in its northeast part; in the middle of this a clay vessel serving as a collector of liquids was found. Of utmost interest is the west wall of this space, because the interior face of the north part is constructed with porous stone blocks, which are shaped nearly triangular to get attached to each other at their edges, whereas the irregular back side of the wall is filled with smaller stones. Nevertheless, the south part of the wall was made of mud brick covered with the usual red plaster, continuing the red floor adjoined also with the similarly plastered south wall. Similarly, all the other walls of the building complex were covered with plaster joined with the plaster floors. This early appearance of ashlar masonry could be perceived as a forerunner of Neopalatial ashlar masonry that also existed in the Late Bronze Age IA settlement of Ialysos (Marketou 1998: 48). All other rooms of the building have strong monochrome reddish burnished plaster floors in continuity with the walls.

However, most intriguing for the use of space and the short history of the building is the presence of a series of four pier-and-door partitions found below the south wall of the anteroom. Projected in the south side of the mud brick wall, three roughly shaped stone doorjamb bases were found; the west of them is Γ-shaped, whereas the rest were plain rectangular. Apart from the strong evidence for the development of this architectural feature and within the same space and in a short period of time, the row of the stone bases indicated that the main entrance into the main rooms was always from the north, signifying the similar movement of bodies during the short history of the building.

All the rooms of the complex were found nearly empty, apart from two small storing spaces north of the anteroom that provided rich floor deposits of pottery. However, both the northeastern and northwestern parts of the building were badly damaged. Quantities of broken and intact *murex brandaris* shells were concentrated in the southwestern part of the revealed building complex, which was most probably an open-air area, like a narrow part revealed north of the storing rooms.

The emergence of emblematic architecture in the Middle Bronze Age settlement of Ialysos attests to great achievements in building techniques, which most probably derived from an increasing knowledge of technology, as presented in the successive construction of two polythyra. Both the polythyra are connected with a complex wooden framework that gave stability and elasticity in spacious buildings. Thus, the wooden framework in the upper structure of the polythyron provided a stable and elastic wall connected in axis with other features within the room and the rest of the building. However, the above evidence strongly suggests early communication with Middle Minoan I/II Crete and other sites in the Aegean, which were also in contact with Crete, such as Miletus III (Niemeier 2005; Raymond 2007), which is documented as an indigenous community in contact with Protopalatial Crete.

Nevertheless, the processes and dynamics of architecture and order at Ialysos, seen from the perspective of early influence and interactions with Middle Minoan I/II Malia in Crete, are still under investigation. However, this is also indicative of quick developments and tradition within the site and points to a quick development of technology in parallel with the elegant style of architecture and decoration, the origins of which had appeared earlier toward the close of the third millennium at Asomatos.

COMMUNICATING BETWEEN THE EXTERIOR AND INTERIOR BUILT ENVIRONMENT AT IALYSOS IN THE MIDDLE BRONZE AGE: TECHNOLOGY AND COMPLEXITY IN MIDDLE BRONZE AGE IALYSOS

Life during the transitional periods from the Early Bronze Age to the beginning of the Late Bronze Age I developed in a continuous movement of social groups in the landscape and in between the interior of the house and the exterior. Similarly to the Early Bronze Age households, Middle Bronze Age houses or buildings at Ialysos, though partially revealed, provide evidence for their systematic study as basic units of social structure (Lévi-Strauss 1987; Souvatzi 2008) and in a wider consideration of the scheme *Domus and Agrios* (Hodder 1990; 1998).

Memories of this shift in habitation patterns and the effect of population expansion in the wider area in the northwest regions of the island, mainly due to devastating natural catastrophes, had been embodied in the societies through time and still resounded as mythology (Lévi-Strauss 1987) narrated by later historians, such as Diodorus of Sicily (80–20 BC) (Marketou 2007).

The material culture at Ialysos and the wider area in the northeast part of the island of Rhodes during the Middle Bronze Age forms a particularly interesting case study that interrelates architectural and natural space through time. Use of space within individual buildings and households shows an interesting patterning of social relations integral to the material culture found in situ (Giddens 1984) that varies from sector to sector. This development signifies the identity and the increasing complexity of social behavior. On the other hand, segmentation either for organizing storage spaces or for privacy and growth of the domestic group over time shows a degree of differentiation (Kent 1990; 1991).

The long tradition of this opposition between the interior and the exterior led people to the selection of the location of the doors and the existence of courtyards and other open-air areas, such as streets or squares, and first of all it inspired them to create their own symbolic communication code between the inside and the outside world in the small scale with the surrounding landscape and in the large scale with the wider world of the Aegean and the eastern Mediterranean. The number and location of doors and thresholds bear the symbolic meaning of passages of human bodies from the outside world both at the intersite and intrasite levels (Stavrides 1999). This is emphasized by the extra labor required for construction with carefully hewn stones, for making thresholds or isodomic façades, given the lack of stones in the vicinity, as mentioned earlier.

Thus, the emergence of the well-known major Late Bronze Age coastline settlement of Trianda could be seen as the result of the aforementioned complex processes already occurring during the Middle Bronze Age on the island rather than as a sudden result of Minoan Thalassocracy (Branigan 1981; Hägg and Marinatos 1984; Schofield 1982; Wiener 1990), according to previous theories that still remain intact despite accumulating new evidence (Knapp 1993). The Middle Bronze Age expansion in the northwest part of the island, after the sudden abandonment of Asomatos, could be seen as a result of powerful control of space and an increase of social power that led to the emergence of local elites. However, analysis of settlement patterns has shown that there existed a variety of factors and complex interrelations that contributed to the development of a nucleated settlement, such as agriculture and stock raising in the fertile landscape of the north part of the island, its proximity to the opposite Western Anatolia mainland (Asia Minor) already developed in the Early Bronze Age, the development of local production of textiles (Marketou 2009: 78), and mostly the progress in metallurgy. However, peripheral installations participated in the establishment of the nucleated settlement interrelating and sharing common characteristics, for example, similar pottery and building techniques (Trigger 1967: 157).

These processes gave a great stimulus for interrelations with other sites in the Aegean and the opposite mainland as well as for early contacts with the Eastern Mediterranean and Egypt, as suggested by the presence of a fragmentary faience object, found in the building with the polythyra. However, imports of luxury items, such as Kamares pottery and stone vases, either by traveling abroad or by the presence of foreigners on the island, indicated that local communities participated in the complex exchange systems of the Middle Bronze Age period. This evidence opens a new chapter in Aegean prehistory against previous theories of isolation and gaps in the Middle Bronze Age period of the island (Davis 1982) due to lack or scarcity of Middle Bronze Age evidence (Papagiannopoulou 1985). According to the above evidence and despite their bad preservation, the existence of monumental buildings in Middle Bronze Age Ialysos along with the affluence of material culture in the context of their floor deposits emphasizes the cosmopolitan role of the large and fertile island as a gate into the Mediterranean. Undoubtedly, the Late Bronze Age IA settlement played an important part in parallel with the Cretans in the Aegean, as is depicted in the exquisite wall paintings of the settlement, some of them made by itinerant painters travelling in the Mediterranean, before the Thera eruption. It could be assumed that the foundations of this 'international' role of the island were laid in the Middle Bronze Age.

Thus, the emergence of monuments in the wider Mediterranean area (Kolb 1991), such as the early appearance of spaces with pier-and-door partitions in the case of Ialysos, along with the development of a sophisticated architecture and order, had been recorded among the histories of the landscape (Souvatzi 2008: 27; Watkins 2005: 97–105) that affected as well the succeeding periods. In the same sense, several scenes and movements of people between different locations are inscribed on the landscape, narrating aspects of temporality and remembrance (Tilley 1994: 27–34). On the other hand, similar monumental architecture and elaborate early floor paintings that appear simultaneously on Crete and at other sites in the Mediterranean, if we just mention the Middle Bronze Age monumental building at Tell Kabri in western Galilee, Israel (Yasur-Landau et al. 2012), are indicative of large-scale communication and interrelations of the island of Rhodes with the Mediterranean.

In addition, the settlement pattern in the wider area of Ialysos from the coastal Early Bronze Age site of Asomatos to the northern part of the island, shows the use of this distinctive marshy land with small rivers reaching the sea through deltas, and Mount Phileremos and other hills dominating in the landscape. The merchants, sailors, craftsmen, and envoys might have arrived at the Delta harbor of Ialysos (Marketou 2007: 161) before the foundation of the Late Bronze Age I settlement. Riverine plants such as edges, papyrus, and reeds depicted in the subtropical landscape with the river on the east wall of the flotilla fresco from Thera (Morgan 1988: 38–39, 160–161) still grow today in the small rivers of Ialysos. This continuous interrelation with the surrounding magnificent nature could be easily recognized in some fragments of the elegant frescoes found among the ruins of the devastating earthquake at the end of the Late Bronze Age IA, before the tephra eruption, a fact that proves in a dramatic manner that the people of Ialysos were betrayed by the nature they loved so much.

TECHNOLOGY, RITUAL, AND AFFLUENCE IN THE MIDDLE BRONZE AGE

Recent excavations in areas of the Middle Bronze Age Ialysos provided evidence for the accumulation of wealth in several floor deposits that were distributed in the nucleated settlement and the peripheral installations. These data point to a flourishing settlement and a society of affluence. Apart from some imports, such as fragments of Kamares pottery mentioned above, a few pieces of Middle Cycladic pottery, and also some stone vases indicative of external relations, the majority of pottery, stone tools, loomweights, and other small objects were made locally using local resources. The characteristic clay of the area (Ware A, according to Marketou et al. 2006: 9–20) was intensively used for the construction of mud bricks and the majority of ceramics. The rich repertory of pottery types configures one of the basic aspects of the cultural identity of the Middle Bronze Age culture within the long process of some traditional Early Bronze Age techniques and types, whereas some of them recall the traditional Anatolian prototypes, such as those known from Miletus (Niemeier 2005; Raymond 2007: 223–227) and Iasos (Momigliano 2009: 134, Figure 26). Dull brownish or reddish slip is applied smearily on larger closed vessels in continuity of the Early Bronze Age tradition of Asomatos, a technique that survived also during the Late Bronze Age period of Ialysos (Marketou et al. 2006: 20–23). However, the majority of the pottery comprises numerous carinated cups, in several dimensions, and also some plain straight-sided cups; within the repertory of open vessels, there are small and large bowls with spreading, vertical, inverted, or flat rims and deep conical basins made according to the earlier tradition of Asomatos. Among the open vessels, a reddish stemmed carinated goblet resembles Anatolian and Minyan pottery (Marketou 2009: 82, Figure14a). Closed vessels, mainly round-mouthed and beak-spouted jugs with high narrow necks (Marketou 2009: 82, Figure 13), large round-mouthed jugs, wide-mouthed cylindrical jars, and in some cases bridge-spouted with basket handles occur as well. Less frequently, however, appear askoi and scuttles similar to the examples of Miletus (e.g., Niemeier 2005: 3, Figure 5) and an early form of a fire-stand.

Nevertheless, the most intriguing evidence for the development of technology is derived from the southwest sector (Figure 10.3, number 8), as part of a most recent excavation at Apostolidis square.[3] A pebble-paved street, one of the main circulation arteries of the settlement leading from the Bronze Age coastline area of Ialysos to the north, up to a Bronze Age low hill or slope to the south, near the church of the Virgin Mary, borders two distinctive building complexes *(insulae)* of the settlement. The buildings revealed thus far in both *insulae* are associated with workshops and storehouses, obviously used by skilled technicians and perhaps merchants. The east border of the N-S street forms four indentations similar to others revealed at the east sector of the settlement (Figure 10.3, number 5). They define at least four spaces from north to south, which continue beyond the excavated area. Within the corner of the second indentation, a curvilinear working area with a floor made of beaten limestone material seems to belong to a small working area, whereas at the opposite side of the street a probably open-air spacious working area is located too.

However, a fascinating and unexpected deposit was found in the third of the above spaces (an elongated room measuring 10.50 by 3.60 m). Within an area around 3.60 by 3.00 m, a deposition of 195 boar tusks[4] was found, along with thousands of gold foils, gold discs, and miniature rivets, of either gold or silver, and in many cases bronze, wrapped on the top with gold foils. There were also simple spiral-shaped bronze ornaments wrapped with gold foils, as well as bone rosettes with gold-capped rivets in the middle. A bone ornament with voluted end, decorated with a gold band attached in a low channel, was also found within the deposit. At the western part of the same deposition, a group of extremely high-necked ritual spouted jugs, either double- or single-necked along with carinated or rounded cups, was also deposited. Two intact boar tusks wrapped with sheets of bronze were thrown up in between the pots in a distance from the dense mass of tusks. Numerous pins, made of either gold or silver, bronze, and even bone, might have been used for the suspension of the decorative precious ornaments found among the tusks. However, some of the ornaments might have been suspended from twisted gold or bronze wires. Polychrome decoration seems to have been also employed by inlaying jewelry with colored stones, a method that has a long history in Mesopotamia and in Syria, whereas in the Middle Bronze Age Aegean it is attested in the *Aegina Treasure*.

Nevertheless, the only find so far that could be associated with boar tusks and gold ornaments comes from a Middle Bronze Age tomb at Kolonna on Aegina, although the boar tusks found there were cut to make boar-tusk helmets (Kilian-Dirlmeier 1997). It is obvious that the Aegina find is a forerunner of the Late Minoan IA boar-tusk helmet illustrated in the Theran fresco and the later well-known Mycenaean ones. Moreover, the find sounds examples of gold hilts found either later in the Royal Tombs of Mycenae or the ritual Middle Minoan gold hilt with the acrobat of the Malia sword.

Whereas until now this hoard, unique in the literature, remains a mystery in terms of the use of space, it apparently displays power and wealth. But above all, it points to the presence of extremely skilled technicians. Thus, the presence of a lead weight found in one of the storing spaces at the opposite side of the street, along with a balance weight yielded among jars, jugs, and cups in another storage space, are indicative of deposits associated with the rapid development of technology and long-distance exchange patterns.

On the other hand, the final abandonment of the site, which forced people to run off leaving behind their valuable material culture with symbolic meanings that indicate spatial practices and their dimensions in space and time, give important insight into issues of identity, memory, and in general the development of social networks on intra- and intersite levels (Flannery and Winter 1976; Souvatzi 2012).

CONCLUSIONS

Monumental architecture and display of wealth in Middle Bronze Age Ialysos attest the presence of greatly skilled technicians and artisans. The study of the significant Middle Bronze Age expansion of people from the nucleated Early Bronze Age settlement of Asomatos to many dispersed settlements and clusters of installations in the northeast part of the island has produced a wealth of new data that show the emergence of local elites whose desire was to acquire raw materials both from the island and through long-distance trade. Thus, they developed new technologies and were responsible for a massive production of pottery, textiles, and precious metals distributed in space and storage areas. Over time, they also achieved a high level of technological expertise that enabled them to construct sophisticated timber frames for pier-and-door partitions for monumental polythyra, which allowed a direct connection with the surrounding natural environment and the landscape. Monumental buildings were then decorated with elegant red stylized plants on white fine plasters joined with painted plaster floors that imitate veined marble slabs. These elegant internal architectural details show many aspects of the social structure and the ability of people to obtain a peaceful life in harmony with nature, which was unfortunately interrupted by natural disasters. Since life continued over time in the same place, it seems obvious that within the collective memory of next generations, no matter how strongly or not they were influenced by the Minoans or the Mycenaeans, narration about old monuments, art, and mainly natural catastrophes remained.

The early development of metallurgy and jewelry making, as shown by the eccentric and even exotic artifacts such as the elaborately decorated boar tusks derived from a hecatomb of animals, demonstrate the great achievements of the local communities during a time that had been previously conceptualized as Dark Ages in the Aegean (Hadji 2004: 45–56).

Hence, the story of the experienced Ialysian craftsmen are reminiscent of the mythical *Telchines,* the extremely experienced technicians "who were the discoverers of certain arts and that they introduced other things which are useful for the life of mankind. Men say that they were also wizards and could summon clouds and rain and hail at their will and likewise could even bring snow". According to mythology, the island of Rhodes was first inhabited by them "and at a later time, perceiving in advance the flood that was going to come, forshook the island and were scattered" (Diodorus of Sicily, History, Bibliotheca Historica lib. Book 5.55–56.1).

I would argue that what happened in this space over the span of the first two and nearly a half centuries of the second millennium BC had been depicted in the collective memory *via* successive narrations from generation to generation, according to Diodorus of Sicily, which almost always related to the surrounding landscape and the physical catastrophes. It remains to the contextual interpretation of the archaeological material recovered from current and future excavations to indicate whether myths represent the truth or not.

NOTES

1. Three out of four doorjambs' bases are found, whereas the west adjacent to the northwest corner of the space is missing already from antiquity.
2. Personal communication with Prof. Joseph W. Shaw.
3. I am particularly grateful to the Institute of Aegean Prehistory for generously funding the excavation conducted in 2012.
4. The number could obviously be associated with nearly a hecatomb of animals, since some of the tusks were found in decayed pieces.

REFERENCES

Allegrette, A. and M. Schmid. 1997. Malia, Crypte hypostyle (bâtiment KB). *Bulletin de Correspondance Hellenique* 121: 790–792.

Amouretti, M.-C. 1970. *Fouilles exécutées à Mallia: Le centre politique II. La crypte Hypostyle (1957–1962), Études crétoises* 18. Paris/Limoges: École française d'Athènes-Bontemps.

Benzi, M. 1984. Evidence for a Middle Minoan Settlement on the Acropolis at Ialysos (Mt. Philerimos). In: R. Hägg and N. Marinatos (eds.), *The Minoan Thalassocracy: Myth and Reality,* 93–105. Proceedings of the Third International Symposium at the Swedish Institute in Athens, May 31 – June 5, 1982. Stockholm: Swedish Institute at Athens.

Blinkenberg, C. 1931. *Lindos I: Fouilles de l'acropole 1902–1914: Les petit objects.* Berlin: Walter de Gruyter and Cie. Libraires-Éditeurs.

Blinkenberg, C. and K. F. Johansen. 1926. *Corpus Vasorum Antiquorum.* Danemark, fasc. 2. Copenhague: Musée National. Fasc. 1. Paris: Librairie ancienne Éduard Champion.

Bourdieu, P. 1977. *Outline of a Theory of Practice.* Cambridge: Cambridge University Press.

Branigan, K. 1981. Minoan colonialism. *The Annual of the British School at Athens* 76: 23–33.

Coldstream, N. J. 1969. The Phoenicians at Ialysos. *Bulletin of the Institute of Classical Studies* 16: 1–8.

Davis, J. 1982. The earliest Minoans in the South-East Aegean. A reconsideration of the evidence. *Anatolian Studies* 32: 33–41.

Dietz, S. 1974. Two painted duck-vases from Rhodes. *Acta Archaeologica* 45: 133–143.

Dreliosi-Irakleidou, A. 1999. Palaia kai nea ebremata pro tou synoikismou apo ten pole tes Rodou. In: *Rhodos 2.400 chronia: E pole tes Rhodou apo ten idryse tes mechri ten katalepse tes apo tous Tourkous (1523)*. Rhodos, October 24–29, 1993, Vol. A, 21–28. Athens: Archaeological Receipts and Expropriations Fund.

Driessen, J. 1982. The Minoan hall in domestic architecture on Crete: to be in vogue in Late Minoan IA? *Acta Archaeologica Lovaniensia* 21: 27–92.

Flannery, K. V. and M. Winter. 1976. Analysing household activities. In: K. V. Flannery (ed.), *The Early Mesoamerican Village*, 34–47. New York: Academic Press.

Foucault, M. 2004. *The Archaeology of Knowledge*. London: Routledge.

Furumark, A. 1950. The settlement at Ialysos and Aegean history c. 1550–1400 B.C. *Opuscula Archaeologica* 6: 150–271.

Hadji, A. 2004. The construction of time in Aegean archaeology: archaeological time and third millennium BC Aegean chronology. Ph.D. dissertation, UC Berkeley. UMI publ. no. 3146870 (www.umi.com).

Hägg, R. and N. Marinatos, eds. 1984. The Function of the Minoan palaces. *Proceedings of the Fourth International Symposium at the Swedish Institute at Athens*, 245–249. Göteborg: Paul Åström's Förlag.

Giddens, A. 1984. *The Constitution of Society: Outline of the Theory of Structuration*. Cambridge: Polity Press.

Hillier, B. and J. Hanson. 1984. *The Social Logic of Space*. Cambridge: Cambridge University Press.

Hodder, I. 1990. *The Domestication of Europe: Structure and Contingency in Neolithic Societies*. Oxford: Blackwell.

Hodder, I. 1998. The *domus:* some problems reconsidered. In: M. Edmonds and C. Richards (eds.), *Understanding the Neolithic of North Western Europe*, 84–101. Glasgow: Cruithne Press.

Hope Simpson, R. and F. J. F. Lazenby. 1973. Notes from the Dodecanese III. *The Annual of the British School at Athens* 68: 127–179.

Inglieri, R. U. 1936. Carta Archeologica dell' Isola di Rodi: Istituto Storico-archeologico 'Fert'. Florence: R. Istituto Geografico Militare.

Kent, S. 1990. A cross-cultural study of segmentation, architecture, and the use of space. In: S. Kent (ed.), *Domestic Architecture and the Use of Space: An Intedisciplinary Cross-Cultural Study*, 127–152. Cambridge: Cambridge University Press.

Kent, S. 1991. Partitioning space: cross cultural factors influencing domestic spatial segmentation. *Environment and Behavior* 23: 438–473.

Kilian-Dirlmeier, I. 1997. *Das mittelbronzezeitliche Schachtgrab von Ägina: Römisch-Germanisches Zentralmuseum, Forschungsisntitut für Vor-und Frühgeschichte*. Mainz: von Zabern.

Kolb, M. 2011. The genesis of monuments among the Mediterrannean islands. In: M. E. Smith (ed.), *The Comparative Archaeology of Complex Societies*, 138–164. Cambridge: Cambridge University Press.

Knapp, A. B. 1993. Thalassocracies in Bronze Age eastern Mediterranean trade: making and breaking a myth. *World Archaeology* 24(3): 332–347.

Lévi-Strauss, C. 1987. *Anthropology and Myth: Lectures 1951–1982*. Oxford: Blackwell.

Manning, S. W., C. Bronk Ramsey, C. Doumas, T. Marketou, G. Cadogan and C. L. Pearson. 2002. New evidence for an early date for the Aegean Late Bronze Age and Thera eruption. *Antiquity* 76: 733–744.

Marketou, T. 1988. New evidence on the topography and site history of prehistoric Ialysos. In: S. Dietz and I. Papachristodoulou (eds.), *Archaeology in the Dodecanese*, 27–33. Copenhagen: Department of the Near Eastern and Classical Antiquities and The National Museum of Denmark.

Marketou, T. 1990a. Asomatos and Seraglio: EBA production and interconnections. *Hydra* 7: 40–48.

Marketou, T. 1990b. Santorini tephra from Rhodes and Kos: some chronological remarks based on the stratigraphy. In: D. A. Hardy and C. Renfrew (eds.), *Thera and the Aegean World III,* 100–119. London: The Theran Foundation.

Marketou, T. 1997. Asomatos Rodou: Ta megaroschema kteria kai oi scheseis tous me to Boreioanatoliko Aegaeo. In: C. Doumas and V. La Rosa (eds.), *Poliochni e L' Antica Età del Bronzo nell' Egeo Settentrionale,* 395–413. Athens: Scuola Archeologica di Atene and University of Athens, Department of Archaeology and History of Art.

Marketou, T. 1998. Excavations at Trianda (Ialysos) on Rhodes: new evidence for the Late Bronze Age I Period. *Rend. Mor. Acc. Lincei* s. 9 9: 39–82.

Marketou, T. 1999. Ialysia. Trianda. *Archaeologikon Deltion. Chronica* 54 (1999): 946–947.

Marketou, T. 2007. Ephaesteiake tephra kai plemmyres ston oikismo tes Ysteres Chalkokratias tes Ialysou. In: Archaeologikes ereunes kai euremata sta Dodecanesa: Rhodos, Ialysos, Kos, Nisyros kai Gyali. Archäoligische Forschungen und Funde in der Dodekanes Rhodos, Ialyssos, Kos Nisyros und Giali, 157–180. Weilheim/Ob: Verein zur Förderung der Aufarbeitung der Hellenischen Geschichte e. V.

Marketou, T. 2009. Ialysos and its neighbouring areas in the MBA and LB I periods: A chance for peace. In: E. Hallager, C. F. Macdonald and W. D. Niemeier (eds.), *The Minoans in the Central, Eastern and Northern Aegean: New Evidence,* 73–96. Acts of a Minoan Seminar January 22–23, 2005, in collaboration with the Danish Institute at Athens and the German Archaeological Institute at Athens. Monographs of the Danish Institute at Athens, Volume 8. Århus: Aarhus University Press.

Marketou, T. 2010. Rhodes. In: E. Cline (ed.), *Oxford Handbook of the Bronze Age Aegean,* 1178–1204. Oxford: Oxford University Press.

Marketou, T., G. Fakorellis and Y. Maniatis. 2001. New Late Bronze Age chronology from the Ialysos region, Rhodes. *Mediterranean Archaeology and Archaeometry* 1: 19–29.

Marketou,T., E. Karantzali, H. Mommsen, N. Zacharias, V. Kilikoglou and A. Schwedt. 2006. Pottery wares from the prehistoric settlement at Ialysos (Trianda), Rhodes. *The Annual of the British School at Athens* 101: 1–55.

McEnroe, J. C. 2010. *Architecture of Minoan Crete: Constructing Identity in the Aegean Bronze Age.* Austin: University of Texas.

Momigliano, N. 2009. Minoans at Iasos? In: E. Hallager, C. F. Macdonald and W. D. Niemeier (eds.), *The Minoans in the Central, Eastern and Northern Aegean: New Evidence,* 121–140. Acts of a Minoan Seminar, January 22–23, 2005, in collaboration with the Danish Institute at Athens and the German Archaeological Institute at Athens, Monographs of the Danish Institute at Athens, Volume 8. Århus: Aarhus University Press.

Monaco, G. 1941. Scavi nella zona Micenea di Jaliso (1935–36). *Clara Rhodos* 10: 41–183. Rhodes: Istituto Storico Archeologico.

Morgan, L. 1988. *The miniature wall paintings of Thera: A study in Aegean culture and iconography.* Cambridge: Cambridge University Press.

Niemeier, W. D. 2005. Minoans, Mycenaeans, Hittites and Ionians in western Asia Minor. In: A. Villing (ed.), *The Greeks in the East,* 1–36. The British Museum Research Publication Number 157. London: The British Museum Press.

Papagiannopoulou, A. G. 1985. Were the S.E. Aegean islands deserted in the MBA? *Anatolian Studies* 35: 85–92.

Poursat, J. C. 1996. *Artisans Minoens: Les Maisons-Ateliers du Quartier Mu. Fouilles Exécutées à Malia: Le Qurtier Mu III.* Études Crètoises 32. Paris: Ecole Française d' Athènes.

Poursat, J. C. and M. Schmid. 1992. Guide de Malia au temps des premiers palais: Le Quartier Mu. *Sites et Monuments VIII,* École Française d'Athènes. Paris: Diffusion de Boccard.

Raymond, A. 2007. Minoanization at Miletus: The Middle Bronze Age ceramics. In: P. Betancourt, M. C. Nelson and H. Williams (eds.), *Krinoi and Limenes: Studies in Honor of Joseph and Maria Shaw,* 221–229. Prehistory Monographs 22. Philadelphia, PA: Institute of Aegean Prehistory Academic Press.

Sampson, A. 1987. *E Neolithike katoikese sta Dodecanesa.* Athens: Archaeological Receipts and Expropriations Fund.

Schmid, M. 1983. Les portes multiples au 'Megaron' du Palais de Malia. *Bulletin de Correspondance Hellenique* 107: 705–716.

Schofield, E. 1982. The Western Cyclades and Crete: a special relationship. *Oxford Journal of Archaeology* 1: 9–25.

Souvatzi, S. 2008. *A Social Archaeology of Households in Neolithic Greece: An Anthropological Approach.* Cambridge and New York: Cambridge University Press.

Souvatzi, S. 2012. Space, place and architecture: a major meeting point between social archaeology and anthropology? In: D. Shankland (ed.), *Archaeology and Anthropology Past, Present, and Future,* 173–196. Association of Social Anthropologists Monographs number 48. London: Berg.

Stavrides, S. 1999. Pros mia anthropologia tou katophliou. *Outopia: Epitheorese Theorias kai Politismou* 33: 107–121.

Tilley, C. 1994. *Place, Paths, and Monuments: A Phenomenology of Landscape.* Oxford: Berg.

Trigger, B. 1967. Settlement archaeology: its goal and promise. *American Antiquity* 32(2): 149–160.

Watkins, T. 2005. Architecture and 'theatres of memory' in the Neolithic of southwest Asia. In: E. De Marrais, C. Gosden and C. Renfrew (eds.), *Rethinking Materiality: The Engagement of Mind with the Material World,* 97–106. Cambridge: McDonald Institute for Archaeological Research.

Wiener, M. 1990. The isles of Crete? The Minoan thalassocracy revisited. In: J. A. Hardy, C. Doumas, J. Sakellarakis and P. M. Warren (eds.), *Thera and the Aegean World* vol. 3.1, 128–161. London: The Thera Foundation.

Yasur-Landau, A., E. H. Cline, N. Goshen, N. Marom and I. Samet. 2012. A MB II orthostat building at Tel Kabri, Israel. *Bulletin of the American Schools of Oriental Research* 367: 1–29.

Zois. A. 1976. Basiliki I. *Nea archaeologiki ereuna eis to Kephali plesion tou choriou Vasiliki, Ierapetras.* Vol. I. Athens: E en Athenais Archaeologike Etaireia.

Zois, A., 1982. 'Gibt es Vorlaüfer der minoischen Paläste auf Kreta?' Ergebnisse neuer Untersuchungen. In D. Papenfuss, H. Prückner and V.M. Strocka (eds), *Palast und Hütte. Beiträge zum Bauen und Wohnen im Altertum:* 207–15. Mainz am Rhein: von Zabern. 16.

11 Space and Temporality in Herding Societies

Exploring the Dynamics of Movement during the Iberian Late Prehistory

Patricia Murrieta-Flores

INTRODUCTION

Movement is fundamental for all human activities, and as such, it comprises the essential act of weaving places into the web of landscape. Places are not so much defined by their location, their boundaries or shape, but through the flows and convergences experienced within them. Whereas places are gatherings continuously on the move, herding is embodied movement. Pastoralist societies guide animals between pastures, settlements, water sources, and salt licks. Through mobility, they weave relations between distant places, things, and people. For herders, to travel through the landscape is also to travel through time, as movement resonates with the seasonal changes in the landscape.

Approaches that treat the landscape dimension as more than joined 'dots', but as a continuum where it is acknowledged that places are related through entangled experiences over space and time, are especially useful in the study of herding societies, as the journeys these societies embark upon are the core of their social and economic strategies. In this chapter, I present a theoretical-methodological approach for the study of herding societies during prehistory, using the theoretical concepts proposed by time-geography[1], in combination with geographic information systems[2] (GIS). The chapter uses the Copper Age (c.3100–2100 BC), the Early Bronze Age (c.2100–1550 BC), and the Late Bronze Age (c.1550–850 BC) societies of western Sierra Morena in Andalucía in Spain as case studies. The archaeological evidence available is analyzed in an attempt to understand the relationship between the spatial dimension of mundane and sacred places and the role of temporality and memory in herding cycles and prehistoric pastoral orbits.

PREHISTORIC HERDING SOCIETIES OF IBERIA: THE CASE OF WESTERN SIERRA MORENA

The Iberian Peninsula is a highly fragmented territory. It is characterized by the harsh configuration of its topography, where a contrast of extended valleys and high mountains dominate the landscape. These characteristics made mobile pastoralism the main economic strategy and mode of subsistence during historical times in this area. Transhumant practices in Iberia can be traced at least to the Iron Age through historical sources, such as the Roman historian Livy, who described how Phoenicians used local herders as guides through the Silver Route and the mountain range of Sierra Morena during their incursions (Alfaro Giner 2001). The herding practices in Iberia are also evidenced by the many important pastoral legends and traditions that seem to have originated mainly in the mountains. This is the case of the several mythological allusions related to these activities, such as the herds of *Gerión* (Almagro-Gorbea 2005; Estévez Escalera 1983; Torres Ortíz, 2002).

The origins of the adoption of a pastoral mode of subsistence in Iberia have been highly debated. However, during the Iron Age there was an extended path network, and a number of studies present evidence for the practice of pastoralism and of short- and long-distance trade (Criado Boado and Vaquero Lastres1993; Cruz Berrocal 2005; Fairén Jiménez 2004; Fairén Jiménez et al. 2006; Galán Domingo and Ruíz-Gálvez 2001; Galán Domingo and Martín Bravo 1991–1992; Ruíz-Galvez Priego and Galán Domingo 1991). This suggests that Late Prehistory is a credible starting point for these activities.

Since the nineteenth century, the impressive prehistoric landscape of Iberia, particularly the funerary architecture of southwestern Andalucía, has become a center of attention and a focal point of research. Iconic megalithic sites[3] in the Guadalquivir Valley such as Valencina de la Concepcion, Seville, have captured the public imagination. In addition, research in general has focused mainly on the archaeological evidence of the region's valley environment and its funerary evidence. It was only in recent decades that research concerning aspects of daily life in the more marginal neighboring areas, such as Sierra Morena, has been carried out. Constituting the contact zone between the Central Plateau and the Betic Depression, Sierra Morena is a mountain range of 400 km that crosses the northern extreme of Andalucía from east to west (Figure 11.1).

Recently, diverse projects have sought to understand the settlement patterns and megalithic phenomena of certain regions in the western area of this mountain range, such as Sierra de Huelva and Sierra Norte de Sevilla (western Sierra Morena), Andalucia (e.g., Costa Caramé 2011; García Sanjuán 1999, 2011; Linares Catela and Odriozola Lloret 2011; Murrieta-Flores 2007; 2012; Murrieta-Flores, Wheatley and García Sanjuán 2011; Pellicer Catalán and Hurtado Pérez 1980; Pérez Macías 1983; 1999; Piñón Varela 1987; Wheatley et al. 2010).

Although very rich in terms of mineral resources, western Sierra Morena has been characterized by the poverty of its soils, which, in addition, are made very unsuitable for agricultural exploitation by a highly developed superficial drainage (Moreno Rey 1998: 19). This may be why this mountain has been historically a well-established herding region. Pastoral practices in western Sierra Morena can be traced through historical documents during

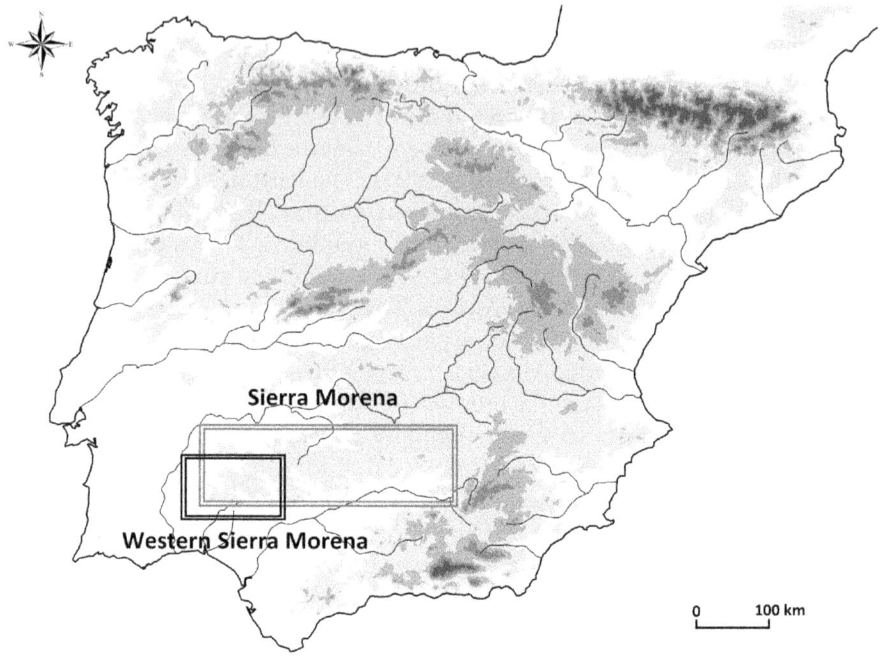

Figure 11.1 Situation of Sierra Morena and the study area.

the Middle Ages (Carmona Ruiz 1993), and possibly back to the Iron Age (Alfaro Giner 2001: 222). Furthermore, ethnographical and archaeological evidence in these regions has suggested that groups living in this mountain range probably practiced a combined agropastoral strategy as their mode of subsistence, adopting it as early as the Copper Age and developing it further throughout the Early and Late Bronze Ages (Pérez Macías 1983; García Sanjuán 1999; Murrieta-Flores 2011). Nevertheless, the harsh conditions in this region, particularly its acidic soils, have resulted in poorly preserved archaeological evidence. In addition, from the few sites excavated, only some have been extensively published (i.e., Hurtado Pérez, 2011). Therefore, the data from the few excavated sites, combined with spatial analysis, and edaphological and palynological studies, constitute the only means of understanding the modes of subsistence of these communities.

In this particular region and the southwest, the available palynological studies suggest that possibly around 4000 BCE, prehistoric communities in these mountains may have begun the development of the *dehesa* systems (Stevenson and Harrison 1992). The dehesa system can be described as an ecosystem of human creation for the management and effective exploitation of agricultural, livestock, and forestry resources in hostile environmental conditions (Martín Galindo 1966). It constitutes an agrosilvopastoral system still used in much of Iberia, where dense forests are cleared and woody vegetation is controlled in order to create and maintain grassland spaces for herding purposes, but at the same time exploiting forest resources and wild animals as game (San Miguel Ayanz 1994).

In the case of Iberia, dehesa systems can be seen as a natural solution invented, adopted, and developed gradually since prehistoric times. The earliest evidence for this system, in its incipient form during the Copper Age (3100–2100 BC), was identified through paleoenvironmental research. The system was more fully developed during the Iron Age (850 ACE) in some parts of southwestern Iberia, such as El Acebrón and Laguna de las Madres in Huelva (Stevenson and Harrison 1992: 243). In other sites, such as Las Viñas, Papa Uvas, and Cantarras, the exploitation of the forest was already in place, and human-created grassland was also present toward the Copper Age (Llergo López and Ubera Jiménez 2008: 2371; López Sáez, López García and Martín Sánchez 2001: 47). Further paleoenvironmental research has also identified that in other southwestern sites such as Cueva de los Covachos (Almadén de la Plata), Coudelaria Alto do Chão (Portugal), and Cueva de la Sima (Constantina), the use of vegetal species and grassland for herding was in place during this period (Duque Espino 2005: 28; Rodríguez Vidal et al. 2001; Rodríguez Vidal et al. 2003). This suggests not only an extended practice of agropastoralism as a mode of subsistence throughout this region, but also an increasing understanding of the exploitation of herding resources and the creation of dehesa environments during this time.

In western Sierra Morena, studies of settlement patterns have suggested that Copper Age societies in this region preferred to establish their settlements in the few areas where there was some agricultural potential (García Sanjuán 1999: 148). Nevertheless, judging from the extreme poverty of the soils and from paleoenvironmental studies, it is highly unlikely that they sustained an economy exclusively based on agriculture (Murrieta-Flores 2011). Toward the Early Bronze Age, local communities occasionally settled where soils were capable of supporting some agricultural activity. However, a clear shift of preference for the establishment of their settlements in what has been regarded as traditional dehesa landscapes can also be observed (García Sanjuán 1999: 146). This is the case for important sites such as La Papúa, El Trastejón, and La Traviesa (García Sanjuán 1999: 146). However, during the Early Bronze Age, there were still some settlements located in areas with soils of higher quality, though this changed toward the Late Bronze Age, when settlements started to be located exclusively in areas that traditionally had been used as grasslands and dehesa environments. Conventionally, this shift in settlement patterns is seen as being primarily caused by an increasing social differentiation that evolved into a tense dynamic between different groups, leading to the need to settle in more secure positions (García Sanjuán 1999: 120). This change may also be due to shifts in the mobility patterns of these herding societies, where the adoption of different strategies of movement could lead them to establish their settlements in different places. In fact, lack of association of many of these sites with other factors, such as the possibility of visual control of the surrounding area or economic resources, has led to them being understood as herding stations (Murrieta-Flores 2011). Paleoenvironmental analyses carried out in sites of long occupation, such as El Trastejón and La Papúa, provided evidence of plants strongly associated with herding and dehesa systems. In my opinion, this mode of subsistence probably gave these societies some independence, allowing the establishment of their habitats in any kind of location, and not necessarily only those with agricultural potential. This change appears to be gradual, and it could be deeply connected to the ability of the Bronze Age societies living in the sierras to manage their environment in a more convenient way, leading finally to an improved understanding of the dehesa systems.

These facts, as well as the ethnohistorical evidence showing a predominant transhumant tradition present at least since the Iron Age, and the results of analyses presented elsewhere (Murrieta-Flores 2011), make it reasonable to suggest that Late Prehistoric communities of western Sierra Morena probably practiced a mode of subsistence based predominantly in pastoralism. Since this region does not produce enough fodder, the historical solution adopted by societies in this region was the practice of transhumance.

It has been vigorously debated whether transhumance was carried out during prehistory in the same fashion as the one observed during historical times (Chapman 1979; Davidson 1980; Walker 1983). One of the main reasons conditioning the historical practice of transhumance was the presence of large herds, requiring long trips to access a vast amount of resources. Nevertheless, during Late Prehistory, it is more likely that herds were smaller, leading to the practice of short-distance transhumance, or *trasterminancia*. Short-distance transhumance implied seasonal movements between diverse altitudes. According to Braudel (1966), the *normal* type can be described as when the main residency of the herder was on the plains or valleys, traveling during summer to higher altitudes. The *inverse* type was when the main residency was on the mountains and the herds were driven to lower altitudes during winter. The *mixed* type was when the settlements were located in between the two altitudes. These types define *trasterminancia* as observed in historical times. The practice of short-distance transhumance assumes a certain degree of sedentism at least for part of the population.

Usually, these types of movement are defined based on the place of family residence, where the settlement to be considered as 'permanent' (in the mountains, in the valleys, or in between) was where the family of the herders or part of the group lived. During their trips, pastoral societies do not wander around aimlessly. They visit specific sites and carry out journeys that are related to the seasonal availability of resources. Throughout these movements, they look for pastures in the valleys during winter, returning to higher altitudes and fresher climates during summer. The habitats or settlements normally also have a seasonal character. The majority of them are revisited periodically, and the paths followed act as a thread connecting the sites that are used by these communities in ecological or social terms. The extension of these journeys and the concept of territory in these societies "takes on a unique definition, in that their movements are almost always confined to an 'orbit' that is regular and repetitive from year to year. This orbit exists as a conceptual space, which may define a large-scale territory, but is experienced incrementally over time as groups pass over it" (Frachetti 2006: 133; 2008). These cycles are spatially known as the *pastoral orbits,* and they can be seen as the *spatial boundaries* established by these groups, which contain many important social and economic locations (Figure 11.2).

Herding stations can consist of places of good pasturage, water, and salt for the animals. However, mobile herders also visit familiar sites, places of symbolic significance such as locations where their ancestors are buried, and others, where they can have social interaction. The decisions to go to the places they visit and to follow the paths they take are deeply influenced by social and environmental factors (Murrieta-Flores 2010), where, for instance, social relationships with other groups and the presence of areas of natural transit can have a great impact on the location of herding stations[4].

The practice of such patterns of movement leaves traces in the archaeological record, not only through portable material such as traded objects and goods but also through fixed features such as landscape markers and paths. In the case of the Iberian landscape, this seems to be reflected in a spatial association between transhumance routes and megalithic monuments, which can be observed in diverse regions (Cara Barrionuevo and Rodríguez

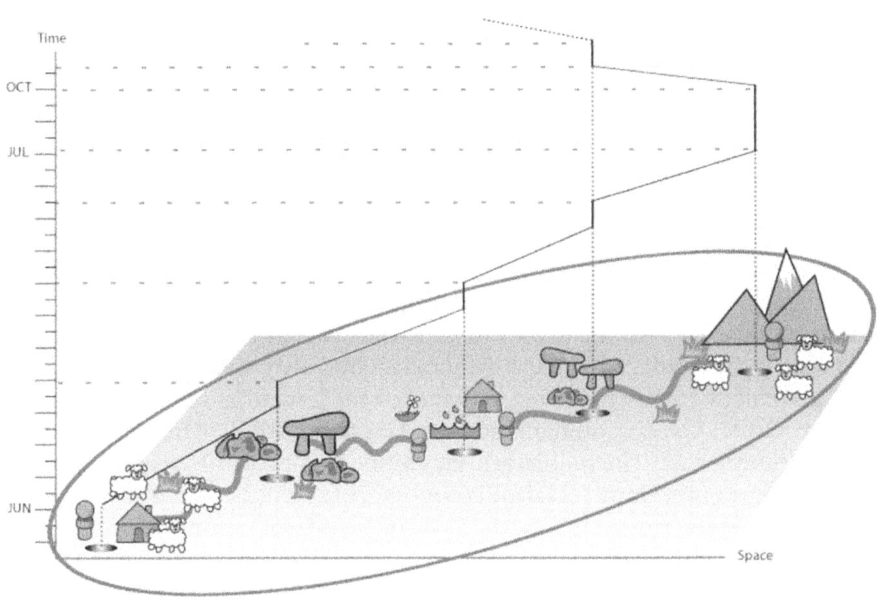

Figure 11.2 Space-time path depicting the pastoral orbit.

López 1987; Fairén Jiménez et al. 2006; Galán Domingo and Martín Bravo 1992; Galán Domingo and Ruiz-Gálvez 2001; Murrieta-Flores 2007; 2010; Ruiz-Gálvez Priego and Galán Domingo 1991; Wheatley et al. 2010). In the specific case of the western Sierra Morena, this relationship has actually been confirmed through statistical tests and two different GIS-based spatial methods. The first looked to identify natural areas of transit and their relationship with prehistoric monuments (Murrieta-Flores 2012), whereas the second was a GIS model based on the idea of "topography of accessibility" (Mlekuž in press). This model defined the most accessible areas, showing the structuring potential of landscape itself and identifying where interaction and activities can be facilitated by the landscape.

These models interpret the connection observed between these elements as being in part a possible fossilization of prehistoric herding pathways. However, these results have also prompted new research questions, suggesting that megalithic monuments may have acted as landscape markers within herding orbits, and particularly as landmarks for territorial definition (Murrieta-Flores 2011). Aiming at exploring this and looking to understand the possible role that megalithic monuments might have played during these cycles, we developed two specific methodologies of spatial analysis with GIS software.

EXPLORING SPACE AND TIME THROUGH GIS-BASED MODELS

Megalithic monuments in Europe have been normally associated with the emergence of agriculture and therefore with a sedentary way of life. However, in the case of Sierra Morena, it is unlikely that this association is applicable. In fact, although it is probable that the spread of the tradition of megalithic constructions might be due to contacts among groups with different traditions or modes of subsistence, their uses and meanings were probably tailored to the ways of life practiced in different regions. In western Sierra Morena, the main way of life was pastoral; therefore, these monuments may have been constructed mainly by herders and formed part of their daily landscape. As perdurable human-made structures, megalithic monuments were possibly a unique and constant reminder of the cycles of life and ancestry. As such, they functioned as markers in the landscape, imbuing it not only with their symbolism but also with a real (and spatial) statement commemorating the time of the ancestors on earth (Cámara Serrano 2001; García Sanjúan and Ruíz González 2009). The role of megalithic monuments in the prehistoric landscape was probably very important, forming part of the daily social and economic life. This becomes evident from the results of the previous models mentioned, which confirmed that megalithic monuments in Sierra Morena were associated with natural corridors and places where movement is funneled by topography (Murrieta-Flores 2011; Mlekuž in press).

Although for now it might not be possible to identify the individual itineraries of single groups, the general patterns observed at a landscape scale might help us establish whether our hypotheses regarding a possible connection between mobility practices, funerary places, and herding itineraries are valid. Archaeologically, the patterns of land use by mobile and semi-sedentary groups can be difficult to grasp, as there is the need not only to combine different data series but also to think about them in terms of diverse scales and temporalities. In fact, the evidence left by mobile groups is actually built on what could be regarded as a continuum, in which the setting is the landscape in its entirety. The study of any space-time path of a pastoral orbit clearly suggests that archaeological research on mobile societies needs to consider both space and time. Seasonal shelters, settlements, paths, social and economic stations, and funerary and ritual monuments form part of the multiscalar phenomena that can be understood as evidence of the activities that took place within the unique social landscape experienced by a mobile society (Frachetti 2006).

Geographic information systems (GIS) can help us understand and identify spatial patterns through the evaluation of models generated from known data sets. We created computational models in order to analyze possible mobility patterns of these groups during the Copper Age, the specific part that funerary monuments might have played within these patterns, and the changes that occurred during the transition to the Bronze Age. Making use of well-established statistical analysis and spatial methodologies, such as cost surface analyses, least cost path and flow accumulation analysis, two different GIS models were created for each period. The first one constituted a travel time model, based on calculating the time expended to traverse the western Sierra Morena from each of the settlements recorded.

The second type is an energy-based model that seeks to understand possible costs of movement, using the most important resources for herding societies. From the time cost model it is possible to establish specific times expended during the journeys and possible walking threshold areas. The energy cost model and flow accumulation analyses can identify (1) the pathways that were most likely used in different seasons, considering the availability of the best mountain or valley pastures and the energy cost of movement; and (2) any possible pattern in the location of the monuments, with respect to those pathways that might indicate their use as landmarks in territorial definition. The technical development and parameters of these models are discussed elsewhere (Murrieta-Flores 2011). However, it is important to note here that to generate the *cost time model* the main variable taken into account was the topography of the terrain. The *potential herding pathways* were calculated taking into account the different types of pasture available during different seasons, the availability of water across western Sierra Morena, the topography of the terrain, and the impossibility of crossing large rivers.

HERDING DYNAMICS AND MOVEMENT DURING LATE PREHISTORY

During the Copper Age, settlement patterns in western Sierra Morena seem to be fairly consistent with seasonal herding patterns. Whereas some of the settlements identified are located in lowlands, where pastures can potentially be used in winter, the majority of the settlements within the inner Sierra (Cortegana, Almonaster la Real, Aracena, and Puerto Moral) are located on higher ground, possibly indicating their use as stations during summer.

This was also suggested by the temporary character that many of the settlements seem to have (Pérez Macías 2010), which shows preference for sites located in the mountains within woodlands and dehesas. In addition, it has been observed that some of the settlements in western Sierra Morena seem to have been inhabited throughout long time spans, like El Trastejón (Hurtado Pérez 2011), whereas others were apparently abandoned after relatively short periods of occupation (Pérez Macías 2010: 276). Iberian ethnographical observations suggest that it is common to observe households or families that establish their residencies as a group in the same places in consecutive years (González Álvarez 2007). However, they can also move to nearby locations, abandoning the site used in previous years as winter or summer stations for a variety of reasons, for example to allow nearby patches of vegetation to recover, because they consider a new patch to be better pasture, or simply to be nearer to other members of the family or neighbors. In this manner, what is observed in the archaeological record might be the product of a pastoral pattern depicting the use of seasonal stations within the orbits.

Moreover, it was also observed that when settlements are clustered, there seems to be a pattern of regular distance between them in terms of travel time. Assuming that many of them were contemporary, they seem to be located at intervals from each other of at least thirty to forty-five minutes' walking time (Figure 11.3).

Although we can only speculate about the possible reasons for this, it has been ethnographically observed that there can exist competition for the best locations among Iberian herders (González Álvarez 2007). In this sense, locating the settlements at particular distances may have been intended to allow an adequate field of action for the animals of each household or settlement and the exploitation of the immediate resources for each family. As demonstrated by cases from historical times, this rivalry could have triggered the desire for space delimitation. The results from the flow path analysis seem to support this, as they established an interesting spatial relationship between the calculated herding pathways and the megalithic monuments. As mentioned earlier, the result from the flow accumulation analysis depicts the pathways of best availability of pastures and water, in combination with the optimal energetic expenditure in terms of the topography of the terrain departing from the prehistoric settlements. Accordingly, the paths simulated

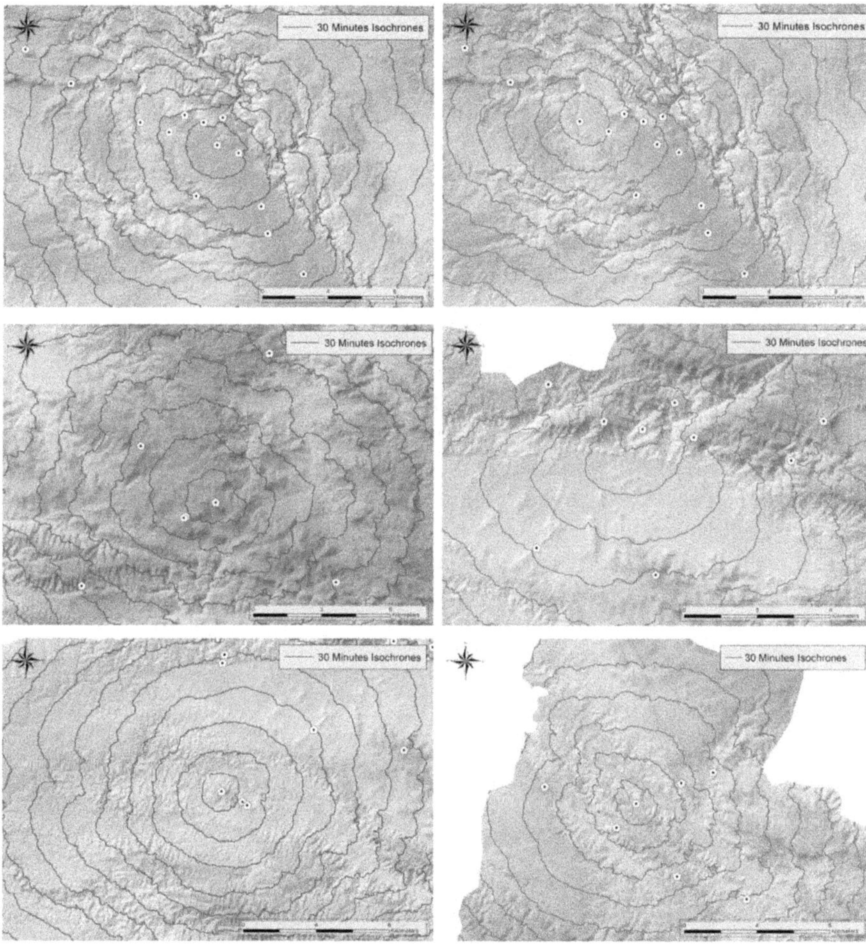

Figure 11.3 Isochrones of thirty minutes from diverse settlements. Notice how the surrounding settlements tend to fall into regular intervals: (a) Dehesa dek Viar, (b) Cerro Ruinas, (c) Bejarano, (d) Santa Bárbara, (e) Peñas de Aroche, and (f) Huerta el Picón.

in this model represent the most likely routes followed by the herders during their journeys, due to the availability of resources useful for their animals.

As seen in historical times, the routes followed by transhumant pastoralists are usually the same through generations, and during the journeys specific places with known resources are used. Over time, one of the most compelling ways to create rights over the use of land has been the establishment of ties linking a place with family and ancestors (Chapman 1995 Saxe 1970) In this sense, it is possible to suggest that the location of the megaliths might have been marking specific areas in the surroundings of the households, as the time model seems to indicate. They might also have acted as landmarks along the routes: a constant reminder of heritage and memory of the groups or families on those lands. If this were the case, the location of the megaliths near herding pathways would serve as a material reminder of specific claims regarding each space and place. In this way, the megaliths seem to have established a relation between community and place. Even when the community was not present, the presence of the monuments would last for generations.

Therefore, if monuments were used as landmarks, it is reasonable to expect them to be related to the routes that were most likely followed by pastoral groups to reach their resources, as seems to be the case for western Sierra Morena. As shown in Figure 11.4, the megaliths tend to be located along the calculated herding pathways. However, in order to clarify whether this is only a matter of chance, a further statistical test of significance (Kolmogorov-Smirnov) was carried out investigating whether the megalithic monuments were randomly distributed with respect to the paths calculated.

The most remarkable result from this test was not only that a statistically significant relationship between the monuments and the paths was indeed established, but that the megaliths are in closer connection to these paths than to simple natural corridors. In the previous experiments mentioned earlier (Mlekuž in press), it was demonstrated on the grounds of

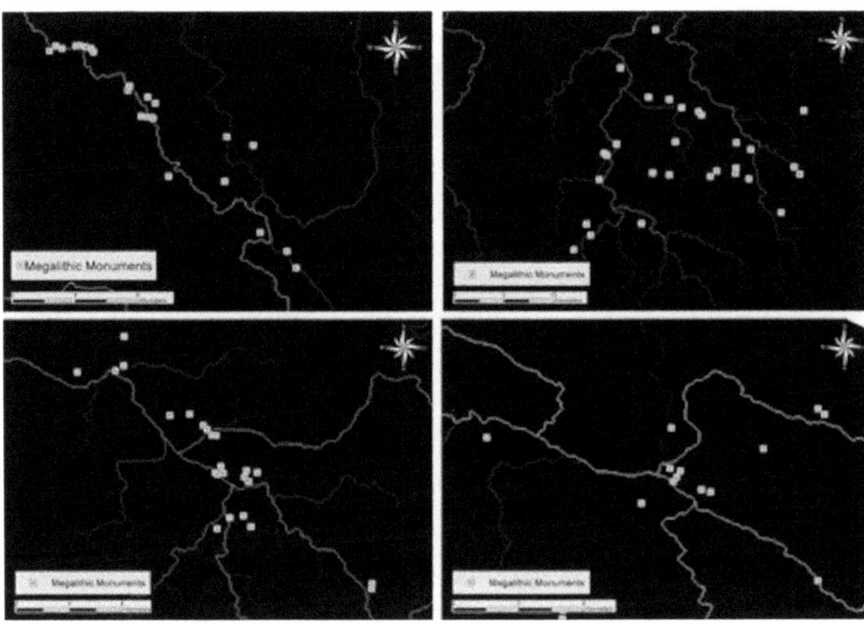

Figure 11.4 Examples of the relationship between the paths calculated and the megalithic monuments of (a) Almadén de la Plata, (b) Zalamea la Real, (c) Castillo de las Guardas, and (d) Aracena.

the topography of accessibility that in some regions, such as Almadén de la Plata, megalithic monuments tend to concentrate in the areas where movement is funneled through generally inaccessible terrain, such as passes, narrow valleys, and difficult topography. In addition, a comparison between monuments and natural corridors was carried out (Murrieta-Flores 2012). Although the monuments were located up to a distance of 6 km from natural corridors, from the paths calculated taking into account herding resources, only one was farther than 2.5 km. In fact, whereas in terms of the comparison with natural corridors, a close relationship with 36 percent of the monuments located in a range of 500 m around them was observed, in this case 62 percent of them fall closer than 500 m from the paths calculated. This is extraordinary, considering that this was the closest range of action to any event or activity related to the monuments.

Megaliths are located along the routes that could be regarded as the most likely herding pathways in this landscape. This suggests that they might have constituted stations along the herding orbits that may have marked the seasonal journeys of the herders. However, the extent of territorial practices within these groups cannot be defined on the basis of the available evidence, and the results from this research suggest that the location of megalithic monuments is related to the demarcation of space. As landmarks, the position of the monuments near to or in the paths frequented was of importance, probably because they acted as reminders during the journeys of the cycle of life, and membership in one group or another might have been sufficient to establish the traditional use of that space by those groups. In addition, the association of megalithic monuments with liminal spaces such as crossroads and passageways would only reinforce their mnemonic properties. As they are associated with pastoral movement, visits to those places would correspond to specific seasons or times of the year. In this manner, the seasonal temporality of the pastoral cycle would attain, through movement, another level of deeper temporality in human life and the time of ancestors.

As prehistoric societies of western Sierra Morena progressively developed the dehesa system, and as exploitation of resources increased, permanent settlements might have become more common. The establishment of settlements in higher altitudes became obvious toward the Early Bronze Age, possibly as the beginning of the inverse type of transhumance that would be carried out during historical times.

Although the reasons for an apparent preference of locations at higher altitudes are debatable, the ethnographic evidence suggests that Iberian transhumant herders prefer settlements on the mountains over those in the valleys for a variety of reasons ranging from the beauty of the landscape and the availability of resources, to the protection that these environments provide, for both animals and members of the group (González Álvarez 2007). In this case, an increasing development of the dehesa system may explain why more permanent settlements in higher locations, where the population seems to concentrate, are observed. This would allow the inhabitants to participate in different tasks, such as small-scale agriculture, hunting, and exploitation of a diversity of other resources.

With a more permanent presence of people, settlements probably started to play a more important role in the definition of space. The construction of megalithic structures as statements of social presence might have been progressively perceived as less necessary, and increased individuality may have been related to the development of new dynamics of movement. Exchange and procurement of goods, in conjunction with traditional herding resources, the acquisition of exotic goods, and the establishment of agreements on, for instance, exploitation of resources and rights of way with other factions and communities, may have been performed from the very beginning by the herders who established contacts and extended social networks. Through the acquisition of knowledge and resources, herders might have been the first in this region to experience an individual accumulation of wealth. Whereas a trend of individual funerary patterns seems to emerge as a reflection of this process, the importance of certain individuals can also be inferred from the work invested in the construction of their burials, and through differentiated burial goods. Although this differentiation in western Sierra Morena seems to be less acute than in other parts of Iberia, there is still a general trend toward an increasingly stratified society. The symbolic significance of movement, however, is never lost. In fact, it seems to be emphasized through the location of the burials in close association with herding routes. This is also demonstrated in our analysis, which shows that when comparing Early Bronze Age monuments with natural corridors, some of the burials were located up to 5 km away from them (Murrieta-Flores 2012), whereas in the case of the herding pathways, the majority are located less than 500 m away, with the farthest being within 1 km.

Toward the Late Bronze Age, the trend observed before in settlement patterns seems to be even more heavily emphasized. Only a few settlements were located in lowlands. An apparent concentration of population in places related to metallurgical activities seems to suggest their increasing importance, whereas the increment in the distances observed between settlements might have had the purpose of avoiding conflicts. With the exception of the metal deposit of Ría de Huelva, and although there is a growth in the extraction of mineral resources, the evidence available now is scarcer than that for previous periods. However, it is very possible that they equally influenced exchange networks, and the perception of wealth in these communities.

A further understanding of the dehesas emerges during this period, and some of the settlements, such as the ones in Aroche, seem to adopt or at least to combine their strategies with agriculture in the Chanza valley. Continuity in the use of the herding paths is highly likely, as this mode of subsistence was maintained. Nevertheless, it is also probable that with the presence of more permanent settlements, new places of exchange, and the exploitation of different resources, these paths started to assume different configurations, as observed during historical times. The use of monuments and graphical markers such as rock art continued, and many of the symbolic depictions seen in Sierra Morena in this period probably derive from earlier traditions (Murrieta-Flores 2011). It would be of great interest to study the spatial distribution of rock art and graphical markers in the area in the future. Such a study would facilitate further exploration of the relationship between herding paths and markers, as the symbols depicted in the rock art, carvings, and possibly also warrior stelae may be related to the markings of the animals as symbols of identity and membership. This hypothesis can only be confirmed through the study of the distribution of the symbols and their possible spatial relationship with different territories. In any case, the research and approach discussed here demonstrate that Iberian societies were fundamentally herding ones during Late Prehistory, at least in western Sierra Morena, and that the evidence we have today makes more sense if seen in this light.

CONCLUSION

Consideration of space and time in prehistoric herding societies, as evidenced in this chapter, should assume a fundamental part in the study of these societies. For herders, to travel through the landscape is also to travel through seasons. These journeys connect different environments and places, upland pastures, sources of water and salt for the animals, but also familiar sites, places of symbolic importance, and gatherings with other groups into a single pastoral orbit. Pastoral movements have always been a means of social interaction between different groups, and a vehicle for material and ideological exchange. Pastoral journeys establish new relations, weaving together places, things, and people. They constitute a fluid phenomenon, a continuous process of becoming.

The analysis presented here confirmed that megaliths in southwestern Iberia are located along the routes that could be regarded as the most likely herding pathways in the landscape. This connection with herding routes suggests that they probably constituted stations along the herding orbits and marked the herders' journeys. Through time, the monuments as works of the ancestors might have served as material reminders of a deep past, of a temporality that extended beyond the seasonal cycle, where every movement acquired time depth, becoming the reiteration of the actual movements of the ancestors. In this way, we might see the emergence of megaliths in association with transhumance paths as a way to stabilize the fluid, generative nature of pastoral movements. Through the use of fixed materiality, such as monuments and symbols, individuals and communities may have influenced and extended their presence, even when they were not physically there. This was probably intensified through the association of these monuments with specific places and seasons. In this manner, funerary monuments probably acted to anyone moving by them as reminders of the cycle of life, as signs of membership and identity, and even as landmarks pointing out certain delimitations of space. Pastoral societies are not conflict free, and disagreements about resources are common, but they can be resolved through a common mechanism that allows the establishment of certain rights. This is ethnographically known as the pastoral pact, and funerary monuments, megaliths, warrior stelae, and rock art in southwestern Iberia could be seen as its material manifestation (Murrieta-Flores 2011).

ACKNOWLEDGMENTS

I would like to thank Dimitrij Mlekuž for his comments, which have greatly improved this chapter, and for sharing with me many thoughts that have become central ideas in my research. My gratitude goes also to David Wheatley and Leonardo García Sanjuán for their help and remarks on the early development of these experiments. This research was funded by the National Council of Science and Technology of Mexico (CONACYT).

NOTES

1. Time-geography is a current of thought started during the 1970s by Torsten Hägerstrand that discussed and highlighted the importance of time in human activities and particularly in mobility. With its introduction, it was defined that time acted as a constant affecting human movement acting also as a constraint (e.g., the impossibility of being in two places at the same time, or traveling continuously without rest or food). This current also introduced the 'space-time path', which illustrates in an x, y, z graphic how a person navigates through the environment in space and time.
2. Geographic information systems can be thought of as systems (and software) capable of storing, retrieving, displaying, mapping, and analyzing any form of information that is geographically referenced.
3. Starting mainly from the Neolithic and in some cases reaching the Bronze Age , different societies in diverse parts of Europe (and the world) built funerary monuments and other constructions of probable symbolic and social significance with large stones, now known as 'megalithic monuments'. In some instances they can be associated with large earthworks and/or settlements. Examples range from Stonehenge (UK) and Carnac (France), to Valencina de la Concepción (Spain).
4. Good relationships between groups may play a fundamental part in the establishment of, for instance, rights of way, or the use of a particular patch of land, and they may define how territoriality is understood and managed between them.

REFERENCES

Alfaro Giner, C. 2001. Vías pecuarias y romanización en la Península Ibérica. In: J. Gomez-Pantoja (ed.), *Los Rebaños de Gerión: Pastores y Trashumancia en Iberia Antigua y Medieval,* 215–232. Madrid: Casa de Velázquez.

Almagro-Gorbea, M. 2005. La literatura tartésica: fuentes históricas e iconográficas. *Gerión* 23(1): 39–80.

Braudel, F. 1966. *La Méditerranée et le Monde Méditerranéen à l'Epoque de Philippe II: Vol. 1. La Part du Milieu.* Paris: Librairie Armand Colin.

Cámara Serrano, J. A. 2001. *El Ritual Funerario en la Prehistoria Reciente en el Sur de la Península Ibérica.* British Archaeological Reports International Series: 903. Oxford: Archaeopress.

Cara Barrionuevo, L. and J. M. Rodríguez López. 1987. Cara Barrionuevo, Lorenzo, and Juana María Rodríguez López. Trashumancia ganadera y megalitos: El caso del Valle Medio-Bajo del río Andarax (Almería). *Crónica del XVIII Congreso Arqueológico Nacional,* 235–248. Seminario de Arqueología.

Carmona Ruiz, M. A. 1993. Vías pecuarias de la Sierra Norte de Sevilla: la presencia de los ganados sorianos en el Concejo de Fregenal durante la baja Edad Media. In: *Camineria Hispánica: Actas del I Congreso de Caminería Hispánica, Pastrana,* 199–210.

Chapman, R. 1979.Transhumance and megalithic tombs in Iberia. *Antiquity* 53: 150–152.

Chapman, R. 1995. Ten years after. Megaliths, mortuary practices and the territorial model. In L. Anderson Beck (ed.) *Regional Approaches to Mortuary Analysis,* Plenum Press. New York and London: 29–51.

Costa Caramé, M. 2011. *La Metalurgia y Sus Repercusiones Economicas, Sociales e Ideológicas en las Comunidades del III y II Milenio cal ANE en el Suroeste de la Peninsula Ibérica.* British Archaeological Reports International Series: 2106. Oxford: Archaeopress.

Criado Boado, F. and J. Vaquero Lastres. 1993. Monumentos, nudos en el pañuelo: megalitos, nudos en el espacio: análisis del emplazamiento de los monumentos tumulares gallegos. *Espacio, Tiempo y Forma, Serie I, Prehistoria y Arqueología* 6: 205–248.

Cruz Berrocal, M. 2005. *Paisaje y Arte Rupestre: Patrones de Localización de la Pintura Levantina.* British Archaeological Reports International Series: 1409. Oxford: Archaeopress.

Davidson, I. 1980. Transhumance, Spain and ethnoarchaeology. *Antiquity* 54: 144–147.

Duque Espino, D. 2005. Resultados antracológicos de los yacimientos de la Coudelaria de Alter do Chão y su integración en las secuencias paleoecológicas y paleoambientales de la Prehistoria Reciente del Suroeste peninsular. *Revista Portuguesa de Arqueologia* 8(1): 21–41.

Estévez Escalera, J. 1983. La fauna del corte 3: aproximación a la fauna del yacimiento de Setefilla. La mesa de Setefilla. Lora del Río (Sevilla). Campana de 1979. In: *Excavaciones Arqueológicas en España 122,* 158–168. Madrid: Ministerio de Cultura.

Fairén Jiménez, S. 2004. ¿Se hace camino al andar? Influencia de las variables medioambientales y culturales en el cálculo de caminos óptimos mediante SIG. *Trabajos de Prehistoria* 61(2): 25–40.

Fairén Jiménez, S., M. Cruz Berrocal, E. López Romero González De La Aleja and S. Walid Sbeinati. 2006. Las vías pecuarias como elementos arqueológicos. In: Ignacio Grau Mira (ed.), *La Aplicación de los SIG en la Arqueología del Paísaje,* 55–68. Alicante: Universidad de Alicante.

Frachetti, M. D. 2006. Digital archaeology and the scalar structure of pastoral landscapes: modeling mobile societies of prehistoric Central Asia. In: T. Evans and P. Daly (eds.), *Digital Archaeology,* 128–147. Routledge: London.

Frachetti, M. D. 2008. Variability and dynamic landscapes of mobile pastoralism in ethnography and prehistory. In: H. Barnard and W. Willeke (eds.), *The Archaeology of Mobility: Old World and New World Nomadism,* 366–396. Cotsen Institute of Archaeology, University of California.

Galán Domingo, E. and A. M. Martín Bravo. 1991–1992. Megalitismo y zonas de paso en la cuenca extremeña del Tajo. *Zephyrus* 44–45: 193–205.

Galán Domingo, E. and M. Ruíz-Gálvez. 2001. Rutas ganaderas, transterminancia y caminos antiguos: el caso del Occidente Peninsular entre el Calcolítico y la Edad del Hierro. In: J. Gómez-Pantoja (ed.), *Los Rebaños de Gerión: Pastores y Trashumancia en Iberia Antigua y Medieval,* 279–311. Madrid: Colección de la Casa de Velázquez.

García Sanjuán, L. 1999. *Los Orígenes de la Estratificación Social: Patrones de Desigualdad en la Edad de Bronce del Suroeste de la Península Ibérica (Sierra Morena Occidental c. 1700–1100 a.n.e/2100–1300 A.N.E.).* British Archaeological Reports, International Series 823. Oxford: Archaeopress.

García Sanjuán, L. 2011. The warrior stelae of the Iberian South-west: Symbols of power in ancestral landscapes. In T. Moore, and Armada Pita, L. (eds.) *Atlantic Europe in the First Millennium BC: Crossing the Divide*, 534–557. Oxford: Oxford University Press.

García Sanjúan, L. and B. Ruíz González, eds. 2009. *Las Grandes Piedras de la Prehistoria: Sitios y Paisajes Megalíticos de Andalucía.* Antequera: Junta de Andalucía.

García Sanjuán, L. and D. W. Wheatley, eds. 2010. *The Palacio III Funerary Complex (Almadén de la Plata, Sevilla, Spain): A New Approach to the Megalithic Phenomenon in Southern Iberia.* Southampton: University of Southampton.

González Álvarez, D. 2007. Aproximación etnoarqueológica a los Vaqueiros d'Alzada: un grupo ganadero trashumante de la montaña asturiana. *Arqueoweb* 8(2).

Hurtado Pérez, V., ed. 2011. *El Asentamiento de el Trastejón (Huelva): Investigaciones en el Marco de los Procesos Sociales y Culturales de la Edad del Bronce en el Suroeste de la Península Ibérica.* Sevilla: Junta de Andalucía.

Hurtado Pérez, V. and L. García Sanjuán. 1996. Prospecciones de superficie en la sierra de Huelva. Campaña de 1992. In: *Anuario Arqueológico de Andalucía 1992,* 237–243. Sevilla: Junta de Andalucía.

Linares Catela, J. A. and C. Odriozola Lloret. 2011. Cuentas de collar de variscita y otras piedras verdes en tumbas megalíticas del Suroeste de la Península Ibérica: cuestiones acerca de su producción, circulación y presencia en contextos funerarios. In: L. García Sanjuán, C. Scarre and D. Wheatley (eds.), *Exploring Time and Matter in Prehistoric Monuments: Absolute Chronology and Rare Rocks in European Megaliths. Proceedings of the 2nd EMSG Meeting (Seville, November 2008),* 335–369. Menga Monographs no. 1. Seville: Junta de Andalucía.

Llergo López, Y. and J. L. Ubera Jiménez. 2008. Cambios en el entorno vegetal de Carmona (Sevilla) desde el Calcolítico hasta el Medievo. *Carel* 6: 2369–2393.

López Sáez, J. A., P. López García and M. Martín Sánchez. 2001. Análisis palinológico del yacimiento arqueológico de Pocito Chico (El puerto de Santa María): el paisaje prehistórico durante el Holoceno Reciente en las marismas de Cádiz. *Cuaternario y Geomorfología* 15(1–2): 45–59.

Martín Galindo, J. L. 1966. La dehesa extremeña como tipo de explotación agraria. *Revista de Estudios Geográficos* 103: 157–226.

Mlekuž, D. in press. Approaching past movement: time to move a little further? In P. Verhagen and S. Polla (eds.), *Computational Approaches to Movement in Archaeology: Theory, Practice and Interpretation of Factors and Effects of Long Term Landscape Formation and Transformation.* TOPOI Berlin: Freie Universitaet Berlin.

Moreno Rey, A. 1998. La estribacion occidental de Sierra Morena. Introducción al medio físico y humano." In L. García Sanjuán (ed.) *La Traviesa. Ritual funerario y jerarquización social en una comunidad de la edad del bronce de Sierra Morena Occidental,* 17–34. Sevilla: Universidad de Sevilla.

Murrieta-Flores, P. A. 2007. Mobility, transhumance and prehistoric landscape: a GIS approach to the archaeological landscape of Almadén de la Plata in Andalucía, Spain. M.Sc. thesis, University of Southampton.

Murrieta-Flores, P. 2010. Travelling in a prehistoric landscape: exploring the influences that shaped human movement. In: J. Frischer, J. Webb Crawford and D. Koller (eds.), *Making History Interactive: Computer Applications and Quantitative Methods in Archaeology (CAA). Proceedings of the 37th International Conference, Williamsburg, Virginia, United States of America, March 22–26, 2009*, 258–276. British Archaeological Reports. Oxford: Archaeopress.

Murrieta-Flores, P. 2011. Travelling through past landscapes: analysing the dynamics of movement during Late Prehistory in Southern Iberia with spatial technologies. Ph.D. thesis, University of Southampton.

Murrieta-Flores, P. 2012. Understanding human movement through spatial technologies: the role of natural areas of transit in the Late Prehistory of South-western Iberia. *Trabajos de Prehistoria, CSIC* 69(1): 114–133.

Murrieta-Flores, P., D. Wheatley and L. García Sanjuán. 2011. Antes de los mapas: navegación y orientación terrestre en la Prehistoria Reciente Ibérica. *PH: Boletín del Instituto Andaluz del Patrimonio Histórico* 19(77): 85–88.

Pellicer Catalán, M. and V. Hurtado Pérez. 1980. *El poblado metalúrgico de Chinflon (Zalamea la Real, Huelva)*. Sevilla: Universidad de Sevilla.

Pérez Macías, J. A. 1983. Introducción al Bronce final en el noroeste de la provincia de Huelva. *Habis* 14: 207–238.

Pérez Macías, J. A. 1999. Nuevos testimonios de ocupación prehistórica en la Sierra de Huelva. *XII Jornadas del Patrimonio de la Comarca de la Sierra. Aracena (Huelva)*, 211–226. Aracena: Diputación Provincial.

Pérez Macías, J. A. 2010. Un asentamiento del III Milenio A.C. en La Lapa (Encinasola, Huelva). *XXII Jornadas de la Comarca de la Sierra. Higuera de la Sierra (Huelva)*, 269–285. Higuera de la Sierra: Diputación Provincial de Huelva.

Piñón Varela, F. 1987. El Cabezo de los Vientos, La Zarcita (Santa Bárbara de Casa): un poblado Calcolítico fortificado en el NE de la provincia de Huelva. Campana de 1985. *Anuario Arqueológico de Andalucía/1985*, 272–278. Sevilla: Junta de Andalucía.

Rodriguez Vidal, J., Alvarez Garcia, G., Alcaraz Pelegrina, J.M., Martinez Aguirre, A., Caceres Puro, L.M., Cantano Martin, M. and Caro Gomez, J.A. 2001. Episodios cuaternarios de sedimentación quimica en la cueva de Los Covachos (Almaden de la Plata, Sevilla). *Geogaceta* (21): 107–10.

Rodríguez Vidal, J., G. Alvarez García, L. M. Cáceres, A. Martínez Aguirre, J. M. Alcaraz, J. A. Lòpez Sàez, J. A. Riquelme, J. M. Recio, M. A. Núñez Granados, J. M. Rodrigo, J. Molina and A. B. Moreno. 2003. Registro sedimentario del cuaternario reciente en la cueva de La Sima, Sierra Norte de Sevilla. *Boletín Sedeck. Sociedad Española de Espeleología y Ciencias del Karst* 4: 80–87.

Ruíz-Gávez Priego, M. and E. Galán Domingo. 1991. Las estelas del suroeste como hito de vías ganaderas y rutas comerciales. *Trabajos de Prehistoria* 48: 257–273.

San Miguel Ayanz, A. 1994. *La Dehesa Española: Origen, Tipología, Caracteristicas y Gestión*. Madrid: Fundación Conde del Valle de Salazar.

Saxe, A. 1970. *Social dimensions of mortuary practices*. PhD Thesis, University of Michigan.

Stevenson, A. C. and R. J. Harrison. 1992. Ancient forests in Spain: A model for land-use and dry forest management in South-west Spain from 4000 BC to 1900 AD. *Proceedings of the Prehistoric Society* 58: 227–247.

Torres Ortíz, M. 2002. *Tartessos*. Madrid: Real Academia de la Historia.

Walker, M. J. 1983. Laying a mega-mith: dolmens and drovers in prehistoric Spain. *World Archaeology 1:Transhumance and Pastoralism* 15: 37–50.

Wheatley, D. W., L. García Sanjuán, P. Murrieta Flores J. and Márquez Pérez. 2010. The megalithic phenomenon in Southern Spain: approaching the landscape dimension. *Oxford Journal of Archaeology* 29(4): 387–405.

12 Space and Time in the Architecture of Prehistoric Enclosures

The Iberian Peninsula as a Case Study

José E. Márquez-Romero & Víctor Jiménez-Jáimez

INTRODUCTORY REMARKS

After decades of research, the existence of an *enclosure idea* in prehistoric Europe is widely accepted nowadays (e.g., Andersen 1997; Burgess et al. 1988; Darvill and Thomas 2001; Gibson 2002; Jorge 2003a; Márquez-Romero and Jiménez-Jáimez 2010; Valera and Evangelista 2010; Varndell and Topping 2002). However, a good number of key issues remain under discussion. The following was originally brought up by Evans (1988) and recently revisited by Whittle (2006: 21): were prehistoric enclosures built with a preconceived plan in mind? Were their particular layouts predesigned?

Tim Ingold's definition of architecture in terms of 'building' and 'dwelling' (1995) played a part in this debate. His work allowed a better understanding of how people organize and transform the world around them, serving as an inspiration for archaeological studies of the prehistoric past (e.g., Jorge 2006). Ingold (1995: 66) defended that, from a 'building' perspective, "worlds are made before they are lived in; or in other words, that acts of dwelling are preceded by acts of world-making". Thus, architecture is predesigned, and thinking comes before building. According to Ingold, however, there is another way to conceive architecture, which he calls 'dwelling'. Following Heidegger, he states that "the activities of building—of cultivation and construction—belong to our dwelling in the world, to the way we are. . . . To build is in itself already to dwell" (Ingold 1995: 76). That is, "design does not precede dwelling, and we do not impose form through the implementation of a preconceived plan on a passive world" (Pollard 2006: 44).

However, these remarkable concepts have to be very carefully used in the study of prehistoric enclosures. Otherwise, as Whittle (2006: 21) points out, they may bring about a new generalization about building that relegates past people "to a kind of agency-less lack of purpose". Moreover, the dwelling perspective cannot explain, by itself, the striking similarities between hundreds of European prehistoric enclosures. Thus his conclusion:

> While many monuments demonstrably go through changes of plan, over extended periods of time, and in that sense perhaps never reached

> final form, I do not see that it need be universally true that each phase of building was carried out without a sense of design. (Whittle 2006: 21)

The text that follows aims to reflect briefly on this on the basis of the archaeological evidence from a specific geographical context: the Iberian Peninsula. More precisely, the Perdigões ditched site (Reguengos de Monsaraz, Portugal), where a team from the University of Málaga is currently conducting fieldwork, has emerged as a perfect testing ground for examining the problem.

It is obvious that understanding space (place, architecture, landscape) and time (events, temporality, chronology) in prehistoric enclosures is crucial to clarifying the issue. Thus, in the first part of this chapter, and after exploring the idea of enclosure, we shall review the various architectural traditions in the creation of said monumental spaces. This will allow us to identify recurrent features in Iberian and European enclosures, as well as characterize what will be referred to as 'family likeness' as a sign of some kind of geographically widespread principles. In the second part of the chapter, we will address the question of time. Relying on the works of Braudel (1986 [1958]) and our own experience at Perdigões, we will try to recognize traces of individual events and long-term processes in prehistoric enclosures. Ultimately, we aspire to moderately balance *dwelling* and *design*, *habitus* and *agency*, and *longue durée* and *courte durée* in the building of Iberian prehistoric enclosures.

THE IDEA OF 'ENCLOSURE'

In the context of human life and social interactions, if a particular *space* receives some sort of meaning or significance, then it becomes a *place* (Tuan 1979). Western European prehistoric enclosures were places that acquired their meaning by virtue of two widespread architectural procedures: cosmogenesis and monumentalization.

To limit something is to create a cosmos. Through the act of cosmogenesis the undefined space becomes ordered and clarified, turning into a place-where-things-happen. In a defined space, there is a simple but revealing dialectic between *inside* and *outside* (Bachelard 1994 [1957]: 250–270). Demarcated spaces help bring human beings together, create identities, and build a sense of community: entering an enclosure and living significant experiences in there must have favored the development of strong interpersonal ties. But, like a coin with two sides, limits also bring about social exclusion. The fact that enclosures were delimited places tells us that a certain change of status—a passage—was required for access to the inside, most likely because the outside was seen as chaotic and asocial by nature.

In Bourdieusian terms, an enclosure is a *social space,* a system of positions in which social agents try to keep or (re)negotiate the social distances between them (Bourdieu 1987: 147–166). For that reason, groups and individuals do not tend to distribute randomly or carelessly across their interior. Quite the contrary, as ethnographic accounts show us, an enclosure is a *microcosmos:* the arrangement of people and things often mimetically reproduces the social order in its entirety. The way the *Mapuche,* indigenous inhabitants of southern Chile and southwestern Argentina, organize their ceremonial centers, with their *rucas* (houses) laid out in a fashion that strictly follows the distribution of groups and territories in the wider landscape, is a good example of this (Andersen 1997: 306).

In addition, the manifest disproportion between the enclosed area of many Late Prehistoric European enclosures and the amount of space that, in principle, would be required for most activities in prehistoric societies suggests that one of the purposes of the construction of prehistoric enclosures was the monumentalization of an open space. Definitely, for most sites, their size and the magnitude of their architectural elements seem to point in that direction. It could be argued, then, that the considerable effort invested could be sanctioning the particular meaning and significance of an enclosure. As a public symbol, social life inevitably structured itself around the monument. Visibility must have contributed to this, as well: monumentality attached importance and preeminence to the place by turning it into a focus of visual attention, a reference in the landscape. All in all, we could say that, just like a tumulus or mound stressed the relevance of a closed space, and a monolith served as a marker for a specific point in the landscape, a prehistoric enclosure monumentalized an extensive, meaningful area.

Finally, we ought to point out that cosmologic and astronomic principles often played an essential role in the layout of these enclosures, adding new dimensions of meaning to each place, and suggesting that its construction and use must have taken place in compliance with astrological prescriptions (Márquez-Romero and Jiménez-Jáimez 2010: 492). This appears to be a constant feature of European (e.g., Bertemes and Northe 2007; Pollard and Ruggles 2001) and Iberian enclosures, with several demonstrated cases in the Southwest (Valera 2008).

All things considered, it should be obvious that Neolithic Europe saw the rise of new conventions or rules that determined how an open space should be monumentalized, whether it be with ditches (ditched/causewayed enclosures), earth (banks), big stones (stone circles), timber posts (timber circles/palisade enclosures), walls (walled enclosures), or a combination thereof (e.g., henges). The act of enclosing was probably an architectural language, shared and intelligible for many European communities, from the Early Neolithic to the Bronze Age (Scarre 2001: 24). We shall focus now on Iberian enclosures, exploring their architectural principles (Figure 12.1).

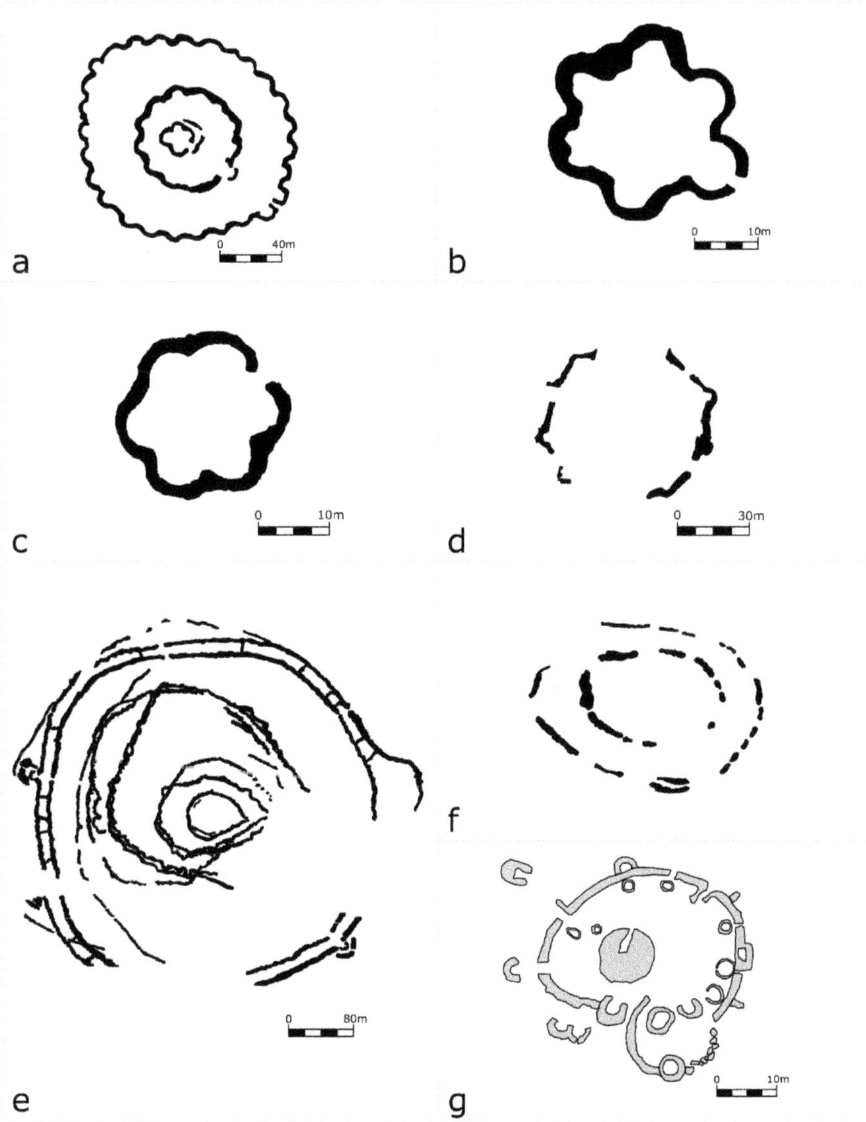

Figure 12.1 Different Neolithic/Copper Age Iberian ditched (a: Xancra; b: Outeiro Alto; c: Santa Vitória; d: Venta del Rapa; e: Perdigões; f: Moscatel) and walled (g: Castelo Velho) enclosure layouts (drawn by the authors from data published on a: Valera and Becker 2011, Figure 3; b: Valera and Becker 2011, Figure 7; c: Valera and Becker 2011, Figure 7; d: Lechuga et al. 2011, Figure 3; e: Márquez-Romero, Valera et al. 2011, Figure 5; f: Delibes and Herrán 2007, p. 149; g: Jorge 2004, Figure 3).

ARCHITECTURAL PRINCIPLES (I): IBERIAN DITCHED ENCLOSURES

In the Iberian Peninsula, most prehistoric enclosures known to date were built in the fourth and third millennia BC. These were ditched enclosures, that is, spaces demarcated by U- or V-shaped ditches dug in the ground. Ditch depth varies, but typically ranges between 1 and 7 m, whereas their width may exceptionally reach 20 m. The layout of such enclosed places is almost invariably circular, although some may have been defined in part by natural features, giving them a more irregular shape. For instance, the inner circle at La Pijotilla made use of a small river to enclose the interior area (Hurtado 2008), Carmona was placed in the border of a plateau (Lazarich et al. 1995), and the Papa Uvas enclosure limits were completed with the coastline (Martín de la Cruz and Lucena 2003). With very few exceptions (e.g., Delibes and Herrán 2007), Iberian ditched enclosures are not as causewayed as their European counterparts. They show, however, well-defined gates or entrances, namely clear gaps in the otherwise continuous ditch, usually no more than three, uncommonly up to five (e.g., Ditch 1 at Perdigões). Some entrances are arranged in a more complex way, following widespread models like that of the 'crab's pincers' (*pince de crabe* in French, *pinza de cangrejo* in Spanish), which can be seen at Fuente de la Mora (Díaz-del-Río 2004), or less frequent designs like that of 'inbrex' type, that has recently been discovered at Perdigões (Márquez-Romero, Valera et al. 2011).

The enclosed surface area is extremely variable, ranging from 0.5 to more than 70 ha, with estimates that surpass 100 ha for the remarkably large sites of Marroquíes Bajos (113 ha; Zafra, Castro and Hornos 2003) and Valencina de la Concepción (around 400 ha; Vargas 2003; Costa et al. 2010), where researchers have not been able to determine the limits and layout of the enclosure(s) as yet. It is not uncommon to see several ditched enclosures constructed in the same place, sometimes arranged in a concentric fashion, other times overlapping one another. In either case, available chronological data point to ditches that share space but not time, since the majority of those ditches date from different centuries or millennia—that is, of course, in the rare case that such features have been successfully dated. This makes us think of a typical ditched site as a sort of a palimpsest rather than a chronological unit. We will come back to this point later.

Normally, the inside area of an Iberian ditched enclosure dated in the fourth or third millennia BC does not show any visible traces of aboveground features: no houses, no buildings, no walls of any kind. Instead, hundreds or even thousands of pits of various shapes and sizes appear scattered all across the enclosed space. Most pits are from 1 to 3 m in diameter and 30 cm to 2 m deep, but there are some (few) examples of larger pits that reach diameters of more than 5 m and a depth of more than 3 m. Banks made of earth are supposed to be part of the enclosure architecture. Although no direct evidence of them has been found yet, results from recent geophysical surveys carried out at Perdigões show an 'empty' strip (i.e., space without pits) that extends parallel to the so-called Ditch 2, which hints at the presence of an inside bank at some point in the past (Márquez-Romero, Valera et al. 2011).

Certain sites, such as Marroquíes Bajos, do indeed possess houses and walls, in addition to ditches and pits, but the alleged simultaneity in the use of both walls and ditches, and really the temporality of the whole site, is currently under intense debate (Zafra, Castro and Hornos 2003).

Another intriguing feature of the Iberian ditched enclosures is an asymmetrical distribution of archaeological deposits. Although the total amount of items (potsherds, prismatic blades and other flint tools, flint knapping waste, broken quern stones, animal and human remains, complete or incomplete, etc.) recorded in every enclosure is in most instances vast, nearly all evidence comes from ditch and pit fillings. This is consistent with the multiple traces of secondary deposition and deliberate filling documented. Currently known data seem to point toward pit and ditch fillings being the outcome of deliberate human actions. Besides, filling assemblages do not always look like something compatible with the modern notion of rubbish, since they usually contain nonrandom associations and dissociations between objects such as human and animal remains (complete or not), potsherds, and so forth.

ARCHITECTURAL PRINCIPLES (II): OTHER IBERIAN ENCLOSURES

To a lesser degree, other architectural techniques were employed to build enclosures in Late Prehistoric Iberia. Palisades, for instance, although rare, are worthy of some attention. Belatedly recognized by European research, the delimitation of a place using wooden fences or walls is now acknowledged as a widespread practice during the Neolithic on the Continent and the British Islands (Gibson 2002). There are only a couple of identified cases in the Iberian Peninsula. One of them is at La Revilla del Campo, located in Spain's *Meseta Central,* where a double palisade circumscribes a small area of only 0.06 ha (Rojo et al. 2008). The other encloses an estimated surface area of 0.15 ha and was discovered at the Portuguese site of Perdigões thanks to a magnetometry survey (Márquez-Romero, Valera et al. 2011).

Stone walled enclosures, on the other hand, constitute a common feature of the Iberian Copper Age (third millennium BC). Very much unlike ditched enclosures, usual topographic locations of walled sites include flat hilltops, cliff edges, or high rocky tors. Variability in size is noticeable: whereas most sites do not surpass 1 ha, there are some monumental exceptions, like Los Millares (6 ha), San Blas (30 ha), or the walled phase at Marroquíes Bajos (see above) (34 ha). They normally comprise more than one wall line, and every wall, built in masonry style, is composed of two faces with gravel filler between them. Sometimes, adjacent to the outer side of the walls are found circular buildings, the so-called 'bastions' and 'towers' we just referred to.

Occasionally, gates or entrances may show morphological resemblances to the aforementioned 'crab's pincers', but making use of stones as raw material and masonry as building technique instead of ditches dug in the ground. The enclosed space is occupied mainly by what appears to be circular huts with stone masonry foundations and floor deposits, whereas walls and roofs could have been built employing perishable materials.

It has lately been pointed out the likely existence of a different kind of enclosure in the Bronze Age of the Iberian Central Plateau, with examples such as Castro de la Plaza, Pico del Águila, or Cuesta de la Horca (Rodríguez and Moral 2007: 779), which may reach up to 17 ha in size. These Bronze Age sites have enclosing walls up to 200 m long, 22 m wide, and estimated heights of 3 to 4 m, composed of irregular stone bricks and unevenly distributed mud. Such enclosures tend to be located on hilltops and cliff edges, so that walls do not define circular closed spaces, like those of earlier enclosures, just by themselves. Rather, they fill the gaps between natural features, like rivers, cliffs, ravines, and such (a model known in France as *éperon barré*). Entrances are few, usually no more than one, and consist of a simple interruption in the wall line. The general scarcity of huts or any other domestic structures inside suggest brief and sporadic stays rather than permanent occupation.

FAMILY LIKENESS: RELATED ENCLOSURES IN A BROADER GEOGRAPHICAL CONTEXT

The scenario outlined above does not differ much from what can be seen in the rest of Western Europe. Most enclosures have features that are individually distinctive, but with a clear 'family likeness' that can be inferred from recurrent patterns in enclosure layout and archaeological deposit formation, namely pit and ditch filling (Márquez-Romero and Jiménez-Jáimez 2010). Only the predominance, in the last decades of the twentieth century, of an overly locally focused archaeology in Spain and Portugal has delayed the general acceptance of such an idea. This assumption is not new. Years ago, A. Whittle (1977: 329) wrote that "enclosures with ditches and banks were a feature of Neolithic cultures in western Europe from the end of the Linear Pottery culture onwards". In our view, after three decades of research, Whittle's statement turned out to be quite accurate. Today, we could argue that western European Neolithic and Copper Age uniqueness lies, above all, in the profusion of enclosures and, most notably, ditched enclosures.

The oldest enclosures known date back to the Early Neolithic (sixth millennium BC) of the Mediterranean area and central Europe. Most of them were ditched enclosures, but their shape was not always a circle, so we can find elliptical, spiral, and even roughly rectangular layouts; ditches were either continuous or causewayed. Traces of houses, both circular and rectangular, have been identified in certain sites, although their spatial and temporal relationship with the enclosing ditches is usually in need of further analysis (Márquez-Romero and Jiménez-Jáimez 2010: 275–277). In southern Italy and Sicily, enclosures were built as early as the first half of the sixth millennium BC (Skeates 2002; Tiné 1983; Whittle 1996). The tradition began a little later in the Iberian Peninsula, but still in the sixth millennium BC, with examples like Mas d'Is in the east (Bernabeu et al. 2003), La Revilla del Campo, a palisade enclosure, in the *Meseta* (Rojo et al. 2008), or Los Cascajos in the north of the Peninsula (García and Sesma 2001). By that time, enclosures were also being constructed in central Europe, in the context of the *Linearbandkeramik* culture (Andersen 1997; Keeley and Cahen 1989; Meyer 2002; Mordant and Mordant 1988; Whittle 1977; 1988; 1996).

Enclosure building continued in central Europe from the fifth to the fourth millennia BC, in line with the development of Neolithic cultures like Cerny, Rössen, Stichbandkeramik, Lengyel, or Michelsberg (Andersen 1997). But it was in the fourth millennium when the tradition of demarcating spaces by way of (interrupted) ditches dug in the ground reached a considerable part of Europe, from the north to the Mediterranean, and from central Europe to Ireland, in parallel with the emergence and spread of megalithic monuments. In Jutland and southern Scandinavia, enclosures are known as *indelukke* (Andersen 1997; 2002; Thorpe 2001); in central Europe as *unterbrochene Erdwerke* (literally 'interrupted earthworks') (Andersen 1997; Meyer 2002; Starling 1988); and in France as *enceintes fossés* (Burnez and Cassen 1966; Joussaume 1988; Scarre 1998). Finally, in the British Islands (Briard 2000; Gibson 2002; Harding 1998; Oswald et al. 2001; Pryor 1998), recent research has shown that the so-called *causewayed enclosures* appeared in the Early Neolithic (c.3800 BC), and prehistoric peoples kept building them for about three hundred years (Whittle, Healy and Bayliss 2011). Generally speaking, ditched enclosures dating from this period do not present traces of houses or walls; quite the contrary, they usually contain only pits. Their shape is circular in most cases, but there are examples of sites delimited by natural features, such as rivers or cliffs, like the well-known archetype of the *éperon barré* ('cordoned off spur') typical of Neolithic France.

It is important to note that, unlike other European regions, where ditched enclosures all but disappeared c.3000 BC, it seems that, in the Iberian Peninsula, sites surrounded by circular ditches were constructed and used for much longer, reaching the last third of the third millennium BC. In Iberia, Valencina de la Concepción (Costa et al. 2010), Papa Uvas (Martín de la Cruz and Lucena 2003), La Pijotilla (Hurtado 2008), and Perdigões (Lago et al. 1998; Márquez-Romero, Valera et al. 2011; Márquez-Romero, Suárez et al. 2011) in the southwest; Martos (Lizcano 1999) and Marroquíes Bajos (Zafra, Castro and Hornos 2003) in the high Guadalquivir valley; Les Jovades and Arenal de la Costa (Bernabeu 1993) in the east; and Moscatel, Casetón de la Era (Delibes et al. 2010), Gózquez de Arriba, Las Matillas, and Fuente de la Mora (Díaz-Del-Río 2004) in the Meseta are some of the most outstanding ditched enclosures from the fourth to the third millennia BC.

The third millennium BC (Copper Age) saw the advent of new architectural principles in the construction of Iberian enclosures. Restricted geographically to the Iberian Atlantic Façade and southeast, stone-walled enclosures include what appear to be circular houses and other domestic features. But they are also one of the most controversial topics in Iberian archaeology. Traditionally perceived as 'fortified settlements' with 'bastions' and 'defensive towers' (Almagro and Arribas 1959; Molina et al. 2004; Zafra, Castro and Hornos 2003), it has been claimed, in the last decade and a half, that they could just be a special form of enclosure (e.g., Jorge 1994; 2003b; Márquez-Romero and Jiménez-Jáimez 2010; Whittle 1996).Among the seventy known walled enclosures, a few of them stand out: Los Millares (Molina et al. 2004) and the later phases at Marroquíes Bajos (Zafra, Castro and Hornos 2003) in the southeast; Los Vientos (Piñón 1987), Castelo de Santa Justa (Gonçalves 1989), San Blas, and Palacio Quemado (Hurtado 2008) in the Guadiana basin; and Castanheiro do Vento and Castelo Velho in the Duero basin (Jorge 2003a). The geographical distributions of ditched and walled enclosures partially overlap each other, making it safe to assume that they somehow shared the same spaces in the Iberian Peninsula for more than 500 years (Márquez-Romero and Jiménez-Jáimez 2010).

Looking at the big picture, the beginning of the end for Iberian enclosures took place in the last centuries of the third millennium BC, when both walled and ditched enclosures saw a progressive decrease in human activity until complete abandonment. After that, there are some examples that date back to the Bronze Age, in the second millennium BC, like the aforementioned walled enclosures in the Meseta (Rodríguez and Moral 2007: 779), and even to the first millennium BC (e.g., López and Suárez 2010), although by then the idea of enclosure and its expression in the Peninsula is much more difficult to trace.

A tradition that lasted almost 4,000 years and covered most of western Europe was obviously subject to regional variability and transformations over time, so it should not be surprising to identify geographical and chronological peculiarities, the addition of new features, and the disappearance of others. But for a long time, at least from the sixth to the third millennium BC, there were comparable ideas of enclosure, ubiquitous architectural patterns, and analogous formation processes.

TIME AND DESIGN IN PREHISTORIC ENCLOSURES

At this point, it should be reasonable to say that we can make inferences only about Iberian enclosures, particularly those dating back to the fourth and third millennia BC, under the assumption that most of them shared certain features with one another and with their European counterparts: circularity, enclosing ditches, scarcity of houses and other domestic features, pits, complex depositional practices, astrological alignments, and so on. This prompts us to think of the existence of a set of common ideas and actions. On the other hand, variability was very marked within the limits of these basic principles. Neither the 'building' nor the 'dwelling' perspective can explain this fact on its own, so we need to find a way to avoid such a dichotomy—which is, by the way, what Ingold intended in the first place.

With that goal in mind, we shall make use of several well-known analytical categories. First, we will borrow from Braudel's time perspectivism (1986 [1958]; 1991) the notions of *histoire événementielle* and *longue durée* in order to discern the materialization of the 'event' and that of the 'structure' (in the Braudelian sense of the word) from the archaeological record at prehistoric enclosures. Second, we will use two concepts that have gained a lot of traction lately in archaeological studies: 'habitus', introduced by Bourdieu, and 'agency'. Our intention is, in a way, to combine 'building' and 'dwelling' in a compromising solution that might be useful, at least in theory.

The Annales School time perspectivism is a familiar reference for many archaeologists (e.g., Harding 2005; Hodder 1987; Lucas 2008; Whittle, Healy and Bayliss 2011). To understand the scope of our approach, it should be noted that the notions of 'event' and 'structure' are not intrinsic to the passage of time, so, in that sense, "the categories upon which time perspectivism relies are surely unhelpful" (Harding 2005: 94): the absolute chronology and the specific sequence of events at a particular site can only be clarified at the moment by way of a coherent radiometric dating program (Whittle, Healy and Beyliss 2011), which clearly exceeds the objectives of this chapter. Having said that, *histoire événementielle* and *longue durée* can actually be valuable to understand the rhythms and scales of time in the history of every site.

Braudel (1986 [1958]: 66) claimed that time comprises a mass of facts, some exceptional, some obscured by repetition, which constitute the domain of *histoire événementielle*. Prehistoric archaeologists have repeatedly searched for particular events in the archaeological record, but the aggregate nature of said record makes it a difficult task (Lucas 2008: 61). Still, most ditched enclosures offer the possibility of getting close to the short span, given the presence of numerous deliberately formed archaeological contexts—that is, ditch and pit fillings.

We shall follow Lucas (2008: 62–63) in his dissection of the act of deposition and its relationship with past events and behaviors. Thus, regarding pit and ditch fillings, we distinguish between things and their arrangement. Concerning the former, the biography (Gosden and Marshall 1999) of each individual object may be different. Artifacts and ecofacts contained, for instance, in a pit are the surviving remains of a number of events that occurred before their deposition in the pit. Archaeologists may or may not be able to reconstruct the biography of each and every one of them in its entirety. But there is one thing they have in common, and that is their involvement in the actions that directly resulted in the assemblage we see today. Hence, if we focus on the organization of things, that is, the depositional context, instead of just the objects, we will find out that there is at least one event that can be archaeologically identified: the act of deposition.

We know that the material remains of events and actions were in continuous movement in the prehistoric past; sometimes, even archaeological deposits were routinely or exceptionally altered or removed by the same people who created them, for a variety of reasons. According to Lucas (2008: 63), a certain degree of irreversibility is a quality of most surviving archaeological contexts: the fact that we can effectively define an assemblage tells us that the actions that produced it were not neutralized by subsequent activities. In our opinion, pits and ditches are, for the most part, closed contexts arranged relatively quickly and in a specific way by deliberate human actions. This is also the reason why only in rare occasions the artifacts and ecofacts in them would be taken somewhere else. This gives their fillings a higher degree of irreversibility than that of the majority of archaeological deposits, which, in turn, brings us back to our original statement about the possibility of reconstructing individual acts of deposition that occurred at prehistoric enclosures. Ditch 1 at Perdigões, where we could observe not only the human-driven character of the filling process, but particular depositional events as well, some of them necessarily executed from within the ditch itself (Márquez-Romero, Suárez et al. 2011: 172), provides a good example (Figure 12.2).

Events are closely related to variability in these assemblages. Sediments and objects are combined in such a variety of ways that the chances of finding two pit or ditch fillings with the exact same composition are extremely slim. This is probably indicative of the multiple circumstances surrounding the depositional act, like its specific time and location, the life history of the items included, the predepositional events, the actors responsible and their purposes, and so forth. Some of these factors belong to the realm of 'agency', that is, the way human actions are enabled within larger social systems (Dornan 2002: 304).

But if we put too much weight into variability, we may not be able to create concepts and therefore not be able to think about the past. Actually, it could be argued that variability here is somewhat deceptive, as it can make archaeologists overlook the aforementioned common depositional patterns. Bourdieu's concept of 'habitus' (1977 [1972]) as a set of guidelines acquired through the experiences and practices of everyday life and not necessarily conscious, allows for an infinity of possible answers within a limited number of principles (Bourdieu and Wacquant 1992: 109). When we are examining the multiple depositional practices, we have to take into account that, as in other spheres of human activity, habitus does not really dictate 'what to do' or 'what can be done'; rather, it determines 'what cannot be done'. In other words, there are general principles that make the existence of unique, unrepeatable gestures or actions—the domain of human agency—possible, but that also prevent random or arbitrary behaviors. Using Chomsky's 'generative grammar' as an analogy, we could say that such depositional practices are the product of some kind of 'rule-governed creativity' (Chomsky 1970: 22).

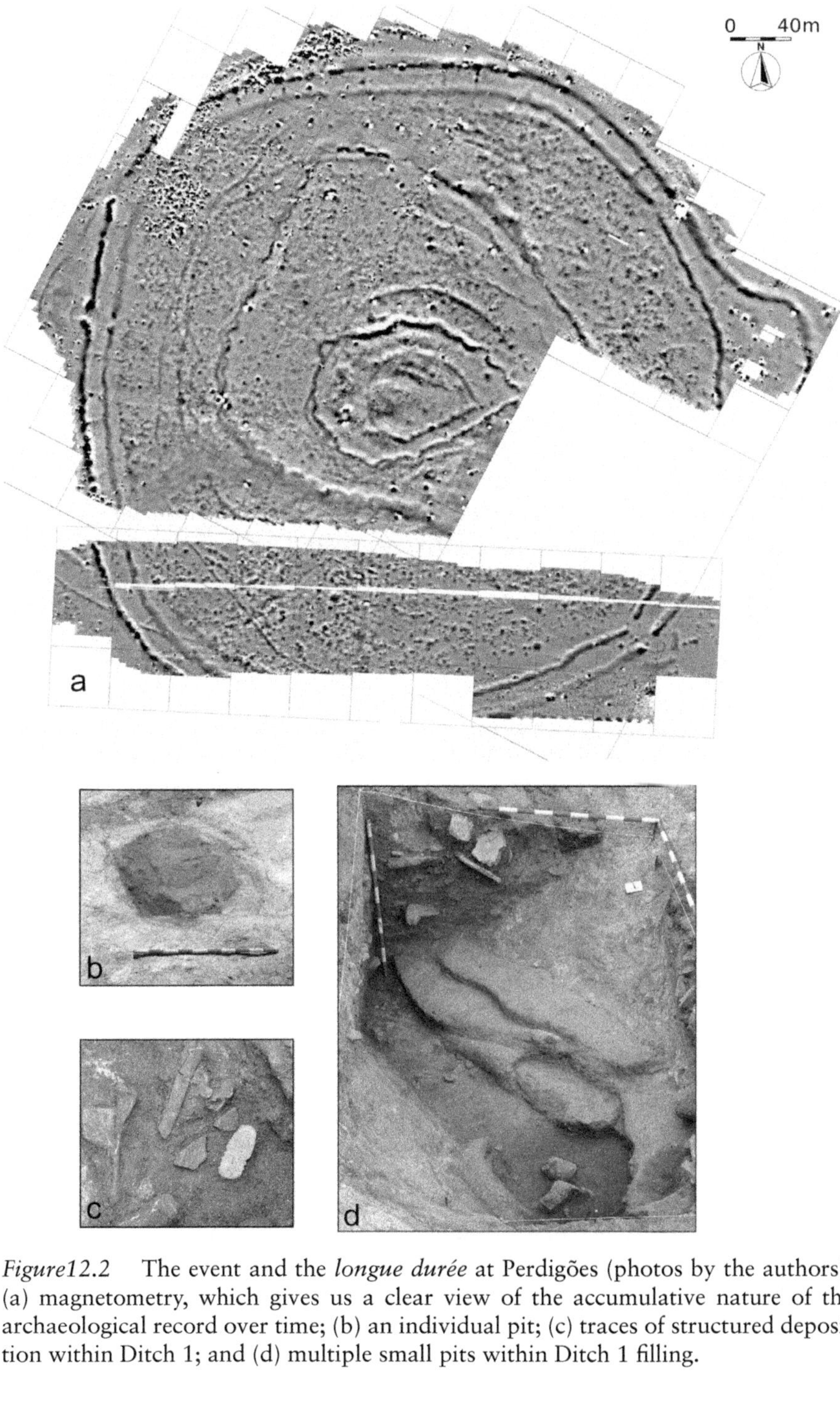

Figure12.2 The event and the *longue durée* at Perdigões (photos by the authors): (a) magnetometry, which gives us a clear view of the accumulative nature of the archaeological record over time; (b) an individual pit; (c) traces of structured deposition within Ditch 1; and (d) multiple small pits within Ditch 1 filling.

The same idea can be applied to enclosure layout, since enclosures are subject to some degree of variability themselves. Every enclosure was constructed in unique circumstances by people with distinctive motivations and under specific conditions. This explains the more or less significant differences in shape, size, and features between them. But whatever those dissimilarities are, there are always limits to what can be done, and those limits are defined by the habitus. It is from this point of view that the recurrence of features such as enclosing ditches, circularity, astrological alignments, pits, deliberate depositions, and so forth—the 'family likeness' we just mentioned—makes sense, even giving us archaeologists the possibility of making cross-cultural, comparative studies like the one carried out by Parkinson and Duffy (2007). The slow pace at which the habitus may change also serves as an explanation of the longevity of the European prehistoric enclosure tradition, while at the same time opening the door to Braudel's notion of *longue durée*.

Getting back to the original issue of preconceived plans in the construction of European prehistoric enclosures, we must question the validity of the 'building' versus 'dwelling' dichotomy, much like what Whittle did, although in the terms we just described. In our view, an archaeological site as a whole is the outcome of many daily or special events and practices, accumulated over time. As such, prehistoric enclosures were in a state of flux: ditches were built and abandoned not much later, entrances were monumentalized and dismantled, more and more pits were dug and backfilled, and so on. In this sense, we agree with Evans's (1988: 89) point that we should think of the "acts of enclosure as an event or activity rather than as a stable or lasting construct". But we also think that some sense of design existed here, as demonstrated by the family likeness found in most of them. The key, for us, is that design decisions were not the sole result of individual or collective agency, but the combined effect of agency and habitus, and therefore, not necessarily a discursively conscious action.

ACKNOWLEDGMENTS

We would like to thank the editors for inviting us to participate in this volume.

REFERENCES

Almagro, M. and A. Arribas. 1959. *El poblado y la necropolis de Los Millares (Santa Fe de Mondújar, Almería)*. Madrid: Biblioteca Praehistorica Hispanica III.

Andersen, N. H. 1997. *Sarup Volumen 1: The Sarup Enclosures*. Moesgaard: Jutland Archaeological Society Publications XXXIII.

Andersen, N. H. 2002. Neolithic enclosures of Scandinavia. In: G. Varndell and P. Topping (eds.), *Enclosures in Neolithic Europe: Essays on Causewayed and Non-causewayed Sites*, 1–10. Oxford: Oxbow Books.

Bachelard, G. 1994 [1957]. *La Poética del Espacio*. Translated by Ernestina de Chapoucin. México: Fondo de Cultura Económica.

Bernabeu, J., ed. 1993. *El III Milenio a.C. en el País Valenciano: Los Poblados de Jovades (Concentaina) y Arenal de la Costa (Ontinyent)*. Valencia: Universidad de Valencia. Bernabeu, J., T. Orozco, A. Díez, M. Gómez and F. J. Molina. 2003. Mas D'Is (Penàguila, Alicante): Aldeas y Recintos Monumentales del Neolítico Inicial en el Valle del Serpis. *Trabajos de Prehistoria* 60(2): 39–59.

Bertemes, F. and A. Northe. 2007. Der kreisgraben von Goseck ein beitrag zum verständnis früher moumentaler kultbauten mitteleuropas, 137–168. In: K. Schmotz (ed.), *Proceedings of the 25th Archaeological Conference of Lower Bavaria*. Rahden: Vorträge des Niederbayerischen Archäologentages.

Bourdieu, P. 1977 [1972]. *Outline of a Theory of Practice*. Translated by Richard Nice. Cambridge: Cambridge University Press.

Bourdieu, P. 1987. *Choses Dites*. Paris: Les editions de minuit.

Bourdieu, P. and L. Wacquant. 1992. *Réponses: Pour une Anthropologie Reflexive*. Paris: Seuil.

Braudel, F. 1986 [1958]. *La Historia y las Ciencias Sociales*. Translated by Josefina Gómez Mendoza. Madrid: Alianza Editorial.

Braudel, F. 1991 [1990]. *Escritos sobre la Historia*. Translated by Mauro Armiño. Madrid: Alianza Editorial.

Briard, J. 2000. *Les Cercles de Pierres Préhistoriques*. Paris: Errance.

Burgess, C., P. Topping, C. Mordant and M. Maddison, eds. 1988. *Enclosures and Defences in the Neolithic of Western Europe*. British Archaeological Reports International Series 43, (ii). Oxford: Archaeopress.

Burnez, C. and H. Cassen. 1966. Les camps neolithiques des Matignons à Juillac-le-Coq (Charente). *Gallia Préhistoire* 9(1): 131–245.

Costa, M. E., M. Díaz-Zorita, L. García Sanjuán and D. Wheatley. 2010. The Copper Age settlement of Valencina de la Concepción (Seville, Spain): demography, metallurgy and spatial organization. *Trabajos de Prehistoria* 67(1): 85–117.

Chomsky, N. 1970. *Current Issues in Linguistic Theory*. The Hague and Paris: Mouton.

Darvill, T. and J. Thomas, eds. 2001. *Neolithic Enclosures in Atlantic Northwest Europe*. Oxford: Oxbow Books.

Delibes, G., M. Crespo, J. Fernández, J. I. Herrán and J. A. Rodríguez. 2010. Un recinto de fosos calcolítico en el valle medio del Duero: el Casetón de la era (Villalba de los Alcores, Valladolid). In: *Actas de las IV Jornadas sobre Patrimonio Arqueológico de la Comunidad de Madrid*, 241–250. Madrid: Dirección General de Patrimonio Histórico.

Delibes, G. and J. J. Herrán. 2007. *La Prehistoria: Biblioteca Básica de Valladolid*. Valladolid: Diputación de Valladolid.

Díaz-Del-Río, P. 2004. Copper Age ditched enclosures in Central Iberia. *Oxford Journal of Archaeology* 23(2): 107–121.

Dornan, J. L. 2002. Agency and archaeology: past, present and future directions. *Journal of Archaeological Method and Theory* 9(4): 303–329.

Evans, C. 1988. Acts of enclosure: a consideration of concentrically-organized causewayed enclosures. In: J. C. Barret and I. A. Kinnes (eds.), *The Archaeology of Context in the Neolithic and Bronze Age: Recent Trends*, 85–96. Sheffield: University of Sheffield, Department of Archaeology.

García, J. and J. Sesma. 2001. Los Cascajos (Los Arcos, Navarra): intervenciones 1996–1999. *Trabajos de Arqueología Navarra* 15: 299–306.

Gibson, A., ed. 2002. *Behind Wooden Walls: Neolithic Palisaded Enclosures in Europe*. British Archaeological Reports International Series, 1013. Oxford: Archaeopress.

Gonçalves, V. S. 1989. *Megalitismo e Metalurgia no Alto Algarve Oriental: Uma Aproximação Integrada.* Lisboa: INIC.

Gosden, C. and Y. Marshall. 1999. The cultural biography of objects. *World Archaeology* 31(2): 169–178.

Harding, J. 1998. An architecture of meaning: the causewayed enclosures and henges of Lowland England. In: M. Edmonds and C. Richards (eds.), *Understanding the Neolithic of North-Western Europe,* 204–230. Glasgow: Cruithne Press.

Harding, J. 2005. Rethinking the great divide: long-term structural history and the temporality of event. *Norwegian Archaeological Review* 38(2): 88–101.

Hodder, I. 1987. The contribution of the long-term. In: I Hodder (ed.), *Archaeology as Long-Term History,* 1–8. Cambridge: Cambridge Universtiy Press.

Hurtado, V. 2008. Los recintos con fosos en la Cuenca Media del Guadiana. *ERA-Arqueologia* 8: 182–197.

Ingold, T. 1995. Building, dwelling, living. In: M. Strathern (ed.), *Shifting Contexts: Transformation in Anthropological Knowledge,* 57–80. London: Routledge.

Jorge, S. O. 1994. Colónias, fortificaçoes, lugares monumentalizados: trajectória das concepções sobre um tema do calcolítico peninsular. *Revista da Faculdade de Letras da Universidade do Porto II Serie* 11: 447–546.

Jorge, S. O., ed. 2003a. *Recintos Murados da Pré-história Recente.* Porto: Faculdade de Arte da Universidade do Porto.

Jorge, S. O. 2003b. Revisiting some earlier papers on the Late Prehistoric walled enclosures of the Iberian Peninsula. *Journal of Iberian Archaeology* 5: 89–135.

Jorge, S. O. 2004. O sítio como mediador de sentido: Castelo Velho de Freixo de Numão: um recinto monumental pré-histórico do Norte de Portugal. In: *Estudos em Homenagem a Luís António de Oliveira Ramos,* 583–611. Porto: Faculdade de Letras da Universidade do Porto.

Jorge, V. O., ed. 2006. *Approaching Prehistoric and Protohistoric Architectures of Europe from a Dwelling Perspective.* Proceedings of the TAG session Sheffield 2005. Porto: ADECAP.

Joussaume, R. 1988. Analyse structurale de la triple enceinte de fosses interrompus a Champ-Durand, Nieul-sur-l'autize, Vendee. In: C. Burgess, P. Topping, C. Mordant and M. Maddison (eds.), *Enclosures and Defences in the Neolithic of Western Europe,* 275–299. British Archaeological Reports International Series 403 (ii). Oxford: Archaeopress.

Keeley, L. H. and D. Cahen. 1989. Early Neolithic forts and villages in NE Belgium: preliminary report. *Journal of Field Archaeology* 16(2): 157–176.

Lago, M., C. Duarte, A. Valera, J. Albergaria, F. Almeida and A. Carvalho. 1998. Povoado dos Perdigões (Reguengos de Mosaraz): dados preliminares dos trabalhos arqueológicos realizados em 1997. *Revista Portuguesa de Arqueologia* 1(1): 45–152.

Lazarich, M., I. Ladrón de Guevara, M. Rodríguez and M. Sánchez. 1995. El yacimiento de El Acebuchal (Carmona, Sevilla): un análisis de las estructuras calcolíticas a través de los escritos inéditos de J. Bonsor e Historiografía. *Spal Revista de Prehistoria y Arqueología* 4: 81–100.

Lechuga, M., M. Soto, F. Pérez, M. J. Díaz, R. Fernández, F.Gómez, J. P. Bellón and J. L. Serrano. 2011. Autovía del Olivar: estudio del poblamiento prehistórico en la Campiña de Jaén. *Memorial Siret. I Congreso de Prehistoria de Andalucía: La Tutela del Patrimonio,* 571–574. Antequera: Junta de Andalucía.

Lizcano, R. 1999. *El Polideportivo de Martos (Jaén): Un Yacimiento Neolítico del IV Milenio a.C. Nuevos Datos para la Reconstrucción del Proceso Histórico del Alto Guadalquivir.* Córdoba: Cajasur.

López, F. and J. Suárez. 2010. La organización y la explotación del territorio del litoral occidental de Málaga entre los siglos VI–V a.C: de las evidencias literarias a los nuevos datos arqueológicos. *Mainake* 32(2): 781–811.

Lucas, G. 2008. Time and the archaeological event. *Cambridge Archaeological Journal* 18: 59–65.

Márquez-Romero, J. E. and V. Jiménez-Jáimez. 2010. *Recintos de Fosos: Genealogía y Significado de una Tradición en la Prehistoria del Suroeste de la Península Ibérica (IV–III Milenios AC)*. Málaga: Servicio de publicaciones de la Universidad de Málaga.

Márquez-Romero, J. E., J. Suárez, V. Jiménez-Jáimez and E. Mata. 2011. Avance a la secuencia estratigráfica del "foso 1" de Perdigões (Reguengos de Monsaraz, Portugal) a partir de las campañas 2009 y 2010. *Menga, Revista de Prehistoria de Andalucía* 2: 157–175.

Márquez-Romero, J. E., A. C. Valera, H. Becker, V. Jiménez-Jáimez and J. Suárez. 2011. El complexo arqueológico dos Perdigões (Reguengos de Monsaraz, Portugal): prospecciones geofísicas. Campañas 2008–09. *Trabajos de Prehistoria* 68(1): 175–186.

Martín de la Cruz, J. C. and A. Lucena. 2003. Visiones y revisiones de Papa Uvas (Aljaraque, Huelva). In: S. O. Jorge (ed.), *Recintos Murados da Pré-história Recente,* 285–306. Porto: Faculdtade de Arte da Universidade do, Porto.

Meyer, M. 2002. Palisaded enclosures in the German Neolithic. In: A. Gibson (ed.), *Behind Wooden Walls: Neolithic Palisaded Enclosures in Europe,* 59–92. Oxford: British Archaeological Reports International Series 1013.

Molina, F., J. A. Cámara, J. Capel, T. Nájera and L. Sáez, L. 2004. Los Millares y la periodización de la Prehistoria Reciente del Sureste. In: *II-III Simposio de Prehistoria Cueva de Nerja, Las Primeras Comunidades Metalúrgicas de la Prehistoria de Andalucía,* 142–158. Nerja: Fundación Cueva de Nerja.

Mordant, C. and D. Mordant. 1988. Les enceintes neolithiques de la Haute-Vallee de la Seine. In: C. Burgess, P. Topping, C. Mordant and M Maddison (eds.), *Enclosures and Defences in the Neolithic of Western Europe,*231-253. British Archaeological Reports International Series 403 (ii). Oxford: Archaeopress.

Oswald, A., Dyer, C. and Barber, M. 2001.*The Creation of Monuments: Neolithic Causewayed Enclosures in the British Isles.* Swindon: English Heritage.

Parkinson, W. A. and P. R. Duffy. 2007. Fortifications and enclosures in European prehistory: a cross-cultural perspective. *Journal of Archaeological Research* 15: 97–141.

Piñón, F. 1987. El Cabezo de Los Vientos, La Zarcita (Santa Bárbara de Casa): un poblado calcolítico fortificado en el N.E. de la provincia de Huelva. Campaña de excavaciones de 1985. *Anuario Arqueológico de Andalucía 1985* 2: 272–278.

Pollard, J. 2006. Solid nature and ephemeral architecture? Understanding the architecture of earlier Neolithic occupation in southern Britain from a dwelling perspective. In: v. o. Jorge (ed.), *Approaching Prehistoric and Protohistoric Architectures of Europe from a Dwelling Perspective,* 41–51. Proceedings of the TAG session Sheffield 2005. Porto: ADECAP.

Pollard, J. and C. Ruggles. 2001. Shifting perceptions: spatial order, cosmology, and patterns of deposition at Stonehenge. *Cambridge Archaeological Journal:* 11(1): 69–90.

Pryor, F. 1998. *Etton: Excavations at a Neolithic Causewayed Enclosure near Maxey Cambridgeshire 1982–7.* London: English Heritage.

Rodríguez, J. A. and S. Moral. 2007. Algunos ejemplos de ingeniería prehistórica poco conocidos: tres poblados amurallados del Bronce medio de la sub-Meseta norte. In: L. Arenilla (ed.), *Actas del V Congreso Nacional de Historia de la Construcción Burgos 7–9 Junio 2007,* 775–780. Madrid: Instituto Juan de Herrera.

Rojo, M. A., M. Kunst, R. Garrido, I. García and G. Morán. 2008. *Paisajes de la Memoria: Asentamientos del Neolítico Antiguo en el Valle de Ambrona (Soria, España).* Valladolid: Universidad de Valladolid.

Scarre, C. 1998. Arenas of action? Enclosures entrances in Neolithic western France c.3500–2500 BC. *Proceedings of the Prehistoric Society* 64: 115–137.

Scarre, C. 2001. Enclosures and related structures in Brittany and western France. In: T. Darvill and J. Thomas (eds.), *Neolithic Enclosures in Atlantic Northwest Europe,* 24–42. Oxford: Oxbow Books.

Skeates, R. 2002. The neolithic enclosures of the Tavoliere, south-east Italy. In: G. Varndell and P. Topping (eds.), *Enclosures in Neolithic Europe: Essays on Causewayed and Non-causewayed Sites,* 51–57. Oxford: Oxbow Books.

Starling, N. J. 1988. The neolithic Höhensiedlungen of Central Germany. In: C. Burgess, P. Topping, C. Mordant and M. Maddison (eds.), *Enclosures and Defences in the Neolithic of Western Europe,* 419–421. British Archaeological Reports International Series 403 (ii). Oxford: Archaeopress.

Thorpe, I. J. 2001. Danish causewayed enclosures: temporary monuments? In: T. Darvill and J. Thomas (eds.), *Neolithic Enclosures in Atlantic Northwest Europe,* 190–203. Oxford: Oxbow Books.

Tiné, S. 1983. *Passo di Corvo e la Civiltà Neolitica del Tavoliere.* Genua: SAGEP.

Tuan, Y.-F. 1979. Space and place: humanistic perspective. In: S. Gale and G. Olsson (eds.), *Philosophy in Geography,* 387–427. Dordrecht: Reidel.

Vargas, J. M. 2003. Elementos para la definición territorial del yacimiento prehistórico de Valencina de la Concepción (Sevilla). *Spal Revista de Prehistoria y Arqueología* 12: 125–144.

Valera, A. C. 2008. Mapeando o cosmos: uma abordagem cognitiva aos recintos da Pré-historia recente. *ERA-Arqueologia* 8: 112–127.

Valera, A. and H. Becker. 2011. Cosmologia e recintos de fossos da pré-história recente: resultados da prospecção geofísica en Xancra (Cuba, Beja). *Apontamentos de Arqueologia e Património* 7: 23–32.

Valera, A. C. and L. Evangelista, eds. 2010. *The Idea of Enclosure in Recent Iberian Prehistory. Proceedings of the XV World Congress (Lisbon, 4–9 September 2006) vol. 36,* British Archaeological Reports International Series 2124. Oxford: Archaeopress.

Varndell, G. and P. Topping, eds. 2002. *Enclosures in Neolithic Europe: Essays on Causewayed and Non-causewayed Sites.* Oxford: Oxbow Books.

Whittle, A. 1977. Earlier Neolithic enclosures in North-West Europe. *Proceedings of the Prehistoric Society* 43; 329–348.

Whittle, A. 1988. Contexts, activities, events: aspect of Neolithic and Copper Age enclosures in central and western Europe. In: C. Burgess, P. Topping, C. Mordant and M. Maddison (eds.), *Enclosures and Defences in the Neolithic of Western Europe,* 1–19. British Archaeological Reports International Series 403 (ii). Oxford: Oxbow Books.

Whittle, A. 1996. *Europe in the Neolithic: The Creation of New Worlds.* Cambridge: Cambridge University Press.

Whittle, A. 2006. The temporality of enclosure. In: V. O. Jorge (ed.), *Approaching Prehistoric and Protohistoric Architectures of Europe from a Dwelling Perspective,*15–24. Proceedings of the TAG session Sheffield 2005. Porto: ADECAP.

Whittle, A., F. Healy and A. Bayliss. 2011. *Gathering Time: Dating the Early Neolithic Enclosures of Southern Britain and Ireland.* Oxford: Oxbow Books.

Zafra, N., M. Castro and F. Hornos. 2003. Sucesión y simultaneidad en un gran asentamieno: la cronología de la macro-aldea de Marroquíes Bajos, Jaén. C. 2500–2000 Cal. ANE. *Trabajos de Prehistoria* 60(2): 79–90.

13 The Four-Dimensional Palace

The Middle Bronze Age Palace of Kabri through Time

Assaf Yasur-Landau and Eric H. Cline

Same as it ever was, same as it ever was,
Same as it ever was, same as it ever was,
Time isn't holding us, time isn't after us . . .

Talking Heads, *Once in a Lifetime*

INTRODUCTION: PALATIAL TIME SLICES

Palaces in the ancient Near East and the Aegean pose numerous challenges for an archaeological treatment. This is primarily because of their size and longevity, which provide startling amounts of synchronic and diachronic data concerning the same structure. It is hard enough to understand activities and behavioral patterns in a single three-dimensional structure. Add to that the numerous changes that long-lived palaces go through over time, and the archaeologist is left with a four-dimensional puzzle, from which most of the pieces are missing, as a result of the limitations of preservation and the extent of the excavated areas.

It is thus utterly understandable that the most common practice for presenting the data for Near Eastern and Aegean palaces is by dissecting them temporally, often into phases or periods. The question of continuity between the phases of strata is often negotiated: is the palace the same palace although it goes through numerous changes, or are these a series of superimposed palaces? This is not a merely semantic question, as the answer reflects thought and notions about continuity in rulership, material culture, and formation processes.

In the case of the largest palace in the Aegean, namely Knossos, broadly excavated and meticulously studied as it is, there are difficulties in understanding questions of architectural continuity. There, the division between the phases is aided by catastrophic events, such as the earthquake ending the Protopalatial period, as well as by the meticulous studies of deposits from the long history of Knossos. These enable a spatial reconstruction of the development of the palace through time, presented in separate time slices (Macdonald 2010; 2012; Tomkins 2012).

Evidently, there is a long history of monumental architecture at the same spot, the Kephala hill, possibly beginning with Early Minoan II period court complexes (Tomkins 2012), though perhaps even earlier. Are these the first phases of the monumental Protopalatial MMIB structure, or its predecessors (see Macdonald 2012)? Furthermore, are the Neopalatial phases at Knossos, each painstakingly attributed with architectural features as well as an artistic program, separate palaces or parts of the same organism (MacDonald 2010)? Evidently, a natural catastrophe separates Protopalatial and Neopalatial phases, and a change of rulership, and at least some cultural change, occurred during the possible 'Mycenaean takeover' of LMII, but is it the same palace?

Naturally, the same problem of the interpretation for a multiperiod series of palatial architectural features has also been encountered by virtually every archaeologist in the Aegean who has been involved with a palatial site (see, for example, La Rosa 2010 and Militello 2012 for Phaistos; Driessen 2010 for Malia; Davis 2010 for Pylos). Similar challenges have also been faced by excavators in the Near East, working to interpret multiperiod palatial structures. Take, for example, the Zimri-Lim palace at Mari, destroyed by Hammurabi of Babylon, which contained elements that originated in the much earlier *Shakanaku* period. These include the Ishtar chapels, which preserved paintings that were more than a century old at the time of the destruction of the palace (Margueron 1995; for a possible similar situation at Qatna, cf. Pfälzner 2007: 36–40).

In contrast to the Aegean and Syro-Mesopotamia, the study of palatial structures in the southern Levant has unfortunately been woefully neglected. In this region, Canaanite palaces have been studied almost exclusively from an architectural point of view (Oren 1992; Wright 1985: 269–281). They have been separated from studies of Canaanite rulership, which either focused on iconography (Beck 2002: 58–93) or utilized the Egyptian Execration Texts or the later Amarna Letters (Goren, Finkelstein and Na'aman 2004).

A similar approach is seen in the *Reallexicon der Assyriologie,* where the social and political aspects of the palaces of the Near East are discussed, but MB and LB palaces from the southern Levant are only briefly surveyed, with only their plans mentioned at best (Miglus 2004: 257–259). In this reality, the series of superimposed monumental structures in Area AA at Megiddo were discussed as separate entities, or as separate palaces each belonging to a distinct stratum (e.g., Oren 1992), and not as a temporal development of the same social construct evolving over time.

In this chapter, we will focus on presenting some data and a few preliminary thoughts concerning the palace at Kabri in a timely manner, or rather, through the lens of time. Simply put, we are interested in the difference between time measured and established by archaeologists and time experienced by the humans who created the archaeological record. We believe that this will eventually help to offer a new approach to time in archaeology and to move beyond traditional approaches.

THE DIFFERENT TEMPORAL FLAVORS OF PALATIAL ACTIVITIES

The usual, somewhat technical, approach to time in the palaces, namely that of separating them into archaeologically recognizable chronological slices, reflects only one aspect of time, that is, large-scale formation processes of the archaeological record. This problem is duly noted by Foxhall (2000: 495), who states: "On the whole, the time scales which archaeologists have traditionally been most comfortable using (e.g. those of artefact typologies, 'periods,' etc.) are longer and less differentiated than the social time scales in which the archaeological record was created in the first place".

In the study of palaces, this common archaeological practice immediately emphasizes the more rare behavioral patterns connected with time, such as foundation, large-scale construction, and large-scale destruction, occurring on time scales of decades and even centuries. In comparison, most activities in the palaces took place on much shorter cycles, that is, of days, months, seasons, and years.

Gosden (1994) suggests understanding time in a twofold way: *measured time* and *experienced* or *human time*. However, it may well be that even the "experienced time" in which the residents and dependents of the Bronze Age palaces acted came in different flavors.

For instance, Damm (2000) and Hurcombe (2000) offered models that may be used for analyzing the degree of specialization according to time and knowledge invested. Damm (2000: 111–114) argued that time has been seen by archaeologists as a linear, objective entity, but it is in fact socially constructed, for the passing of time is dependent on the task performed. Hurcombe (2000) analyzed activities in terms of knowledge skills required and argued that time is different for every activity, ranging from intermittent time, meaning that the process can be interrupted and resumed later, to doubled time, meaning that the process is ongoing and monitored but other tasks can be undertaken—that is, presumably, what we would now call multitasking on the part of the participant.

Construction

Construction events at palaces, for instance, vary greatly in their rhythms and amplitude. No doubt seasonal maintenance activities were needed and included activities such as plastering, cleaning drains, repairing rain damage, waterproofing the roofs, and so on.

The rhythm in which the internal features within the palaces changed may well have been different from one room to another, with changes such as the raising of floors, in the decoration of the room, or in installations and furnishings such as drains, made in each at different times. It may eventually be possible to assess the degree of these changes, which are not connected to an overall change in the plan of the palace in a scale of years, in which case we could approach the concept of time both more holistically and realistically, but this must await further excavation and analysis of the data.

In addition, change in the overall architectural plan of the palace, accompanied by what may be identified as a new building program, may have occurred in intervals of decades or more. In many cases, such change may have been brought about by an alteration of political power, or a need for reconstruction following a natural catastrophe such as an earthquake.

Finally, acts of foundation of new palaces occur very irregularly and, of course, involve a special event necessitating rituals and a feast. Thus, for example, we find a Hittite ritual for the foundation of the Erection of a New Palace, which includes also the deposition of foundation deposits (Goetze 1969: 357–358), and the grand feast held by Assurnasirpal, king of Assyria, upon the inauguration of the palace in Calah (Pritchard 1969: 556–560).

Rituals and Feasts

In terms of rituals and feasts, seasonal festivals and events occurring on an annual basis were part of the agricultural year, creating a sacred calendar that was followed by the community (Bourdieu 1990: 220). Thus, for example, sacred calendars of Late Bronze Age Emar list rituals and various stages of rituals occurring in specific months (Fleming 2000). The mention of the month Plowistos(?) in Tn 316 from Pylos (Palaima 2004: 240–241) may also indicate a series of rituals planned for a specific month.

Other rituals and feasts occur on a multiyear scale, not according to regular intervals, but according to the necessity of the event, sometimes conforming to important events in the life cycle of the palatial elite (for an ethnographic example of the impact of life cycle events on domestic architecture, see Boivin 2000). Examples include installation feasts, such as the installation of the Baal priestess in Emar (Fleming 1992) and the investiture and initiation feast for the wanax mentioned in Un 2 from Pylos (Palaima 2004: 242). Another feast of investiture is seen in the Pylos Ta series, mentioning the inventory of furniture for the investiture of an official by the wanax (Killen 1998; Palaima 1995: 631; Speciale 1999).

Another event that occurs on long, yet irregular, intervals is the death of members of the royal household and their burial in tombs below and by the royal palaces, as indicated by the multiple burials found under the palace of Qatna (Pfälzner 2007: 55–60). The burial feasts that accompanied these events were apparently augmented with *kispum* in some cases; that is, feasts held at regular intervals in memory of the dead ancestors (Fleming 2000: 186–189; Pfälzner 2007: 58).

Maintenance Activities and Production

In addition, maintenance activities, such as cooking (for subsistence purposes) and cleaning (Picazo 1997), were no doubt practiced in the palaces on a daily basis. It is likely that seasonal cleaning was also practiced, together with other seasonal routine maintenance activities such as (re)plastering and so on.

Cooking for feasts and other special events, on the other hand, occurred according to the complex time pattern of these events. And production of textiles, ceramics, and other goods within the palace itself or in nearby palatial workshops could be either seasonal (as in the case of production of wine and olive oil) or year-round (as in the case of spinning and weaving).

THE PALACE AT KABRI IN TIME

We must also note that palaces are often identified in the archaeological record in the Near East, Africa, Asia, and Central and South America, yet the criteria for such identifications are given only infrequently. However, it seems that palaces often share common broad constructional and functional features.

The archaeological manifestation of these functions is commonly thought by scholars to be similar across the ancient Near East. Thus, for example, Margueron (1997: 199) argues that the Near Eastern palace "is an organism that must satisfy specific requirements and whose contents vary little". Every palace should contain royal living quarters that are the private residence of the king and family, official quarters that include courts and a throne room facilitating the public activity of the king, and administrative offices, in charge of the dependent personnel and goods held by the palace (ibid. 199–200). This traditionalist approach correctly recognized the palace as a hyper-complex artifact that answers the sociopolitical needs of the rulership. However, the assumption of similarity between palaces embedded within this view neglects the flexibility and adaptive nature of palatial architecture over time.

We believe that a palace should be viewed as a major adaptive mechanism aimed at facilitating the activities of the government and transmitting ideological messages that, for example, would strengthen the power of the ruling elite of the Kabri polity. As such, it is not enough that the palace architecture and decoration should be tailor-made to fit the needs of the rulers who first built it, but should be updated to answer new political, social, and economic challenges.

We thus view the Middle Bronze Age palace of Kabri (Figure 13.1) as a four-dimensional shadow imprinted on the archaeological record of Tel Kabri by the Canaanite system of rulership that governed Kabri for three centuries without any apparent break or destruction. Based on the results of the 2005 through 2011 seasons of excavation at Tel Kabri and of the accompanying regional survey in the western Galilee, we believe that we can follow the palace's development in the four dimensions of space-time, attempting to gain insights as to the conditions that led to changes in plan and function. As the study of the palace of Kabri is still at its beginning, the observation about events and their frequency in the palace will naturally be biased toward the rarer yet highly visible events of massive constructional change. At the same time, initial observation on more minute changes in

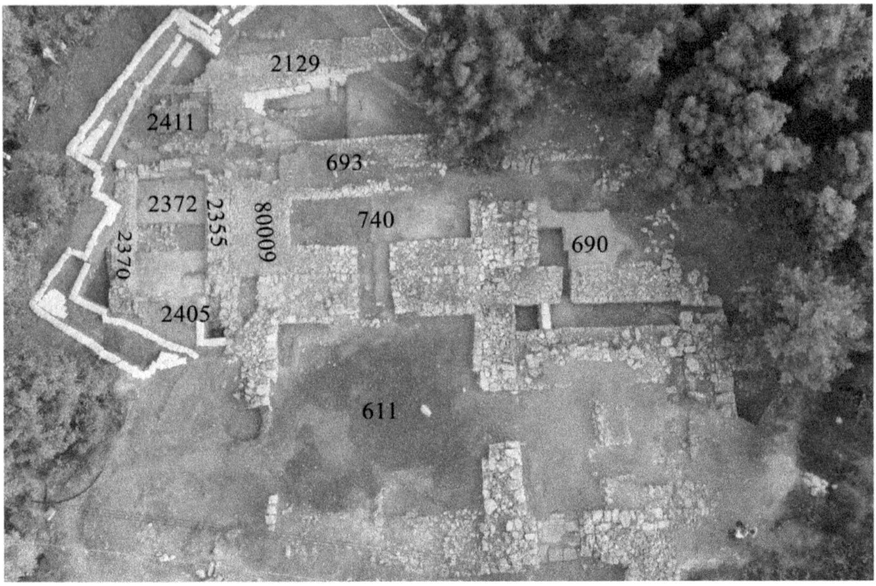

Figure 13.1 Aerial view of Tel Kabri, Israel. (Photograph by SkyView Photography, Ltd).

construction, and evidence of other behavioral patterns within the palace, provides a hope that—with the advance of the study of the palace using household archaeology methods—more details will emerge concerning the daily, common, and yet elusive aspects of the palatial time cycle.

The Kabri Archaeological Project primarily aims at examining the formation of the Canaanite palatial economy, while also investigating the palace using a combination of scientific methods and applying the methodological lessons of household archaeology for the study of domestic assemblages (e.g., Souvatzi 2008; Yasur-Landau, Ebeling and Mazow 2011).[1]

The phasing of the site presented below (e.g., Kabri DW I-VI) represents a first, and preliminary, attempt, as we work through the material. Previous relevant publications from the project, on which the following discussion depends and in which more detailed argumentation can be found, are Cline and Yasur-Landau (2007); Yasur-Landau, Cline and Pierce (2008); Cline, Yasur-Landau and Goshen (2011); Yasur-Landau, Cline and Samet (2011); Yasur-Landau et al. (2012); and Goshen, Yasur-Landau and Cline (2013).

The site was inhabited early in the Middle Bronze Age I (MBI), as is evident from tombs excavated by Kempinski (for full details of Kempinski and Niemeier's excavations, including their original suggestions for the phasing of the site, see the Kabri volume edited by Scheftelowitz and Oren 2002). It seems, however, that the rise of social complexity was gradual, and that for most of the MBI there was no palatial structure at the site. The area upon which the palace was built was occupied in the middle part of the MBI—our phase DW VII—by domestic structures, the remains of which were found under wall 673, the western wall of the palace, as well as east of wall 676/733 in the eastern part of Area DW. Our survey results suggest that at that time, Kabri was but one of a series of unfortified settlements in the western Galilee (Yasur-Landau, Cline and Pierce 2008).

The earliest Middle Bronze Age monumental structure at the site may be attributed to the middle–late MBI, our phase DW VI, perhaps parallel to Aphek phases 2 to 3. The private houses were flattened and covered by a thick constructional fill of earth and rubble, which contained much Early Bronze Age pottery, probably an indication of the origin of the fill. Two massive stone walls creating a long space, named by us 'the corridor', are the major component of this structure. They are preserved to a height of at least 2 meters. The corridor had two narrow openings: to the east, an opening that may have been a window, and to the north, an opening that led to another space, covered with a very thick white plaster floor. Although very little of the structure was excavated by us, it seems that the first palatial, or proto-palatial, structure at Tel Kabri had a fortlike appearance. This brings to mind other fortified structures of the MBI period connected with the rise of complexity in a region that is still preurban, such as the fort at Qiriat Shmona (south) and the MBI structure at Tel Mevorakh (Yasur-Landau, Cline and Pierce 2008).

Not much later, during the transitional MBI–II period, our phase DW V (and roughly equivalent to Kempinski's stratum 4) in the palace area, the proto-palace went through an extensive enlargement and elaboration program (Figure 13.2). This constructional phase was accompanied by thick deliberate fills that filled the corridor as well as the room to its north. The corridor deposit included vast amounts of restorable pottery belonging to the transitional MBI–MBII, their types suggesting some resemblance to a tomb assemblage. Whether it was a ceremonial deposit connected with the renovation project, or the remains of a food offering connected to a still-undiscovered nearby tomb, the filling of the corridor, as a one-time event, and the deposition of its ceramic assemblage, were no doubt important landmarks in the history of the palace.

During the construction, three small rooms were added to the east of the corridor. The deposits of pottery, animal bones, and consecutive beaten earth floors in these rooms are no doubt connected to changes during the life of the DW V palace. The ongoing analysis of the animal bones and pottery will give us more data on the daily activities of palatial food consumption, as they do not reflect any one single event of deposition. It is possible that an early hall, the predecessor of Ceremonial Hall 611, was built to the west of the corridor, as indicated by a floor predating the MBII floors found by Kempinski. Similarly, hall 740 was built to the west of Ceremonial Hall 611.

This was the period during which Tel Kabri was first surrounded by a massive earthen rampart, 25 m thick in some places (Kempinski 2002), an

Figure 13.2 Plan of the northwest part of the palace at Tel Kabri. (Drawn by Dov Porotsky).

act that consolidated the power of the governing rulership on the one hand and improved the security of the entire community on the other. This may well have been a response to the processes of the interpolity rivalry occurring in the western Galilee between Acco and Kabri, which resulted also in the disappearance of small MBI rural sites, and the fortification of secondary settlements such as Achziv and perhaps Avdon (Yasur-Landau, Cline and Pierce 2008).

During the next phase, DW IV, roughly equivalent to Kempinski Stages 3a and 3b, and the Middle Bronze Age II, the palace is further extended to the northeast and south (Figure 13.2). The entire western wing of the palace is repaved with a new set of plaster floors. Our excavations in Area D–North suggest that the palace reached this area.

Furthermore, Kempinski's excavations in Area F (published as the southern part of Area D, Oren 2002), and our adjacent excavations in Area DS-1 and DS-2, indicate that the palace reached this area too during the MBII, covering the remains of transitional MBI–II private houses with a series of monumental structures. The remains in Area DS-2 include a well-built terrace wall with a drain and a flight of steps (Figure 13.3). These led to Area F, which contained a courtyard with a thick plaster floor and platforms paved with drafted stones. A massive brick collapse that covers the entire area

Figure 13.3 Plan of the southern portion of the palace at Tel Kabri. (Drawn by Dov Porotsky).

suggests that there is a massive structure still buried to the east, which was possibly built during this phase.

Sometime during this period, perhaps during the later part, roughly equivalent to Kempinski's Stage 3b (Oren 2002), the palace was decorated with at least two sets of Aegean-style wall paintings, and two sets of Aegean-style floors. These include the miniature wall fresco and the frescoed floor previously discovered by Kempinski and Niemeier in Area DW and the Aegean-style wall painting with a blue background, as well as possibly another painted floor, found in secondary deposition in Area DS during our recent excavations (see full discussion in Cline, Yasur-Landau and Goshen 2011).

The extended period of DW IV may be divided into smaller time scales, as made evident by the repeated repaving of rooms in the western part of the palace. Thus, the fresco floor in Ceremonial Hall 611 (Niemeier and Niemeier 2002) was the last of a series of three floors belonging to this phase, the first white, the second red, and the third polychrome. At least three plaster floors belonging to this phase were also found in the adjacent room 740, and very likely also in room 697, indicating that construction was practiced in varied time scales in the palace. Unfortunately, the use of superimposed, massive plaster floors has left relatively little material culture evidence in situ from the excavated part of the palace.

Whereas the palace of phase DW V was vast, the palace of phase DW IV was even larger and was also more refined, as indicated by the first use at the site of architectural refinements such as entrances with inner recesses, ashlar paving stones and perhaps orthostats (Yasur-Landau et al. 2012), and, above all, Aegean-style painting. In an earlier article, we argued that the repeated choice of the Kabri rulership of Aegean art, reflected in up to four different wall and floor paintings, rather than Syrian-inspired art, was deliberate and calculated (Cline, Yasur-Landau and Goshen 2011). It was a manifestation of the rulers' will to demonstrate their connections with faraway areas that other polities did not possess, on the one hand, and on the other hand, an aspiration to belong to a more "cosmopolitan" Mediterranean narrative. It seems, therefore, that whereas the brutish palace of phase DW V was an expression of the newly acquired dominance of Kabri in the Acco valley, the very large and more refined palace of DW IV may have presented more of an ideological message, that is, an expression of the will of the rulers of Kabri to take part in interregional trade and perhaps politics.

A major renovation project separated phase DW IV from DW III, or Kempinski's Stages 3b and 3c (Oren 2002). It is possible that the one-hall monumental structure in Area DS-1 was built during this period. The function of this structure is still undetermined, yet the material used for the mending of the floor and for the mud bricks included pieces of Aegean-wall paintings belonging to the previous phase.

In Area DW, the walls of room 740 were thickened considerably and the Aegean-style miniature fresco was removed and used as packing material in a newly created threshold. The new floor of room 740 covered a pit into which thousands of pottery sherds had been thrown, belonging to the DW IV phase palace. A hall to the east of Ceremonial Hall 611 was divided into two smaller rooms. Finally, it is possible that Ceremonial Hall 611 itself had orthostats that were removed during this phase.

It is currently unclear what all of these changes meant or why they might have been undertaken. Reasons may have ranged from a change in the ideology of the rulers to a cessation in participation in international or interregional relations to a simple change in rulers, with the new rulers having different taste in interior decoration than the previous ones.

To the north and west of the external wall of the palace, two large features now appear. The first is wall 2129, a 1.8 meter thick wall, built north of wall 673. It is constructed as a series of segments, each stepping out slightly farther to the south from the previous. It may be either the foundations of a curtain wall or, more likely, a causeway.

The second new feature was found during the 2011 season, when the so-called Orthostat Building was uncovered (see Yasur-Landau et al. 2012). This is located immediately to the west of the ceremonial area of the palace at Kabri and is connected to the causeway just mentioned. This is a small building, with an outer size of approximately 6 m by 14 m, which has two rooms positioned one behind the other. The inner faces of the stone wall foundations in both rooms are lined with beach rock orthostats about 25 cm thick, 60 cm high, and between 40 cm and 1.5m long. These orthostat slabs were covered with white plaster, and into their tops square dowel holes were chiseled. The back room of the structure was paved with large drafted and plastered stones and contained at least three restorable storage jars. These jars, as well as the bone assemblage that indicates the consumption of high-quality cuts of meat in the structure and vicinity, suggest that the structure may have been used for recurring events of feasting in connection with, but not within, the palace. It is hoped that the excavation of the courtyard to the west of the Orthostat Building will yield more refuse deposits that may provide further evidence on the nature and timing of these feasting events, but at the moment nothing more can be said.

The reasons for this renovation program are still not understood. The end of Aegean art in the palace of Kabri may be due to a sudden lack of resources or a cessation of connections to Aegean artisans. However, the expansion of the palace on the one hand, and the Cypriot pottery found in its latest destruction deposits on the other, suggest that there was no lack of resources nor a severance of international ties at this time. Thus, we probably need to look elsewhere for an explanation of the change in the decorative program of the palace. It may be as simple as a new political-cultural agenda on the part of the Kabri rulership (Cline, Yasur-Landau and Goshen 2011), or it may be far more complex. We are hopeful that our continued excavations at the site will provide the answer in the near future.

The latest postrenovation palace was destroyed during the late MBII, before the beginning of the Late Bronze Age I period. This dating is indicated by the presence of White Painted Cypriot Middle Cypriot wares and a Chocolate on White Amphora in combination with the absence of final MBII wares such as early Bichrome Wheelmade, canonical Chocolate on White bowls, Cypriot Proto White Painted, and perhaps even early Cypriot White Slip I. In terms of absolute chronology, a destruction date within the first half of the sixteenth century BCE seems a plausible option (see full discussion in Cline, Yasur-Landau and Goshen 2011).

The last event in the palace may well have been a feast in Ceremonial Hall 611, as evident by the large and luxurious Chocolate-on-White amphora found smashed at the corner of the hall. An analysis by Andrew Koh (see Yasur-Landau et al. 2012) of a new fragment of the same amphora found by our team in 2009 indicates that red wine was consumed shortly before the destruction of the palace.

The destruction of the DW III palace deposited evidence pertaining to daily and seasonal activities of spinning and weaving. At least sixty-three loom weights and six spindle whorls, one made from ivory, were found in the western, more official wing of the palace. They may have fallen down from the residential quarters perhaps located on the second floor of the building. Their relatively close clustering may indicate that they belong to one or two looms and were a manifestation of household production, rather than a palatial industry (Goshen, Yasur-Landau and Cline 2013). This activity conforms to Hurcombe's (2000) 'intermittent time', cited briefly above, in that textile production can be interrupted and resumed later. Since we have yet to discover the kitchen of the Kabri palace, as well as the storage area, very little can be said at the moment about the flavors of time connected with maintenance activities in the palace.

CONCLUSIONS

With the continuing of the excavations at the site, our understanding of the palace of Kabri is an ongoing process. In phase DW VI, the proto-palace is a fortlike structure. It grows in size and resembles more closely a proper palace in phase DW V, during the transitional MBI–MBII, at the same time that the Kabri rampart is built. The further growth and elaboration of the phase DW IV palace reflects the dominant status of Kabri in the Acco valley, as well as its overseas connections. Finally, the last phase of the palace, DW III, seems to show less overseas contact in terms of painted decoration, yet its growth continues until its demise during late MBII.

The tentative development scheme of the palace presented here shows development over three centuries—from the late MBI to the late MBII, in large time slices separated by major renovation events. However, the repaving of floors points to the ability to differentiate even thinner time slices, with fewer years between them. Furthermore, even though this is a palace rather than a domestic dwelling, the identification of deposits connected with more daily or seasonal activities in the palace, whether it be textile production or recurring feasting, strongly suggests that further analysis of the remains using lessons learned in the field of household archaeology may give a fuller experience of the various flavors of palatial time.

We believe that we will eventually be able to discuss the data derived from the palace at Kabri more clearly in terms of scales, rhythms, and cycles of time, and in the future contribute further to a new approach to time in archaeology, in part by moving beyond traditional approaches. It will be especially promising if we can further differentiate between the "measured time" and "experienced time" for activities, as suggested by Gosden (1994) and discussed briefly above, and perhaps also begin to substantially utilize models such as those suggested by Damm (2000) and Hurcombe (2000).

In the meantime, however, further excavation and analysis of the palace at Kabri in a timely manner is required.

NOTE

1. The excavations and survey are part of the Kabri Archaeological Project, an interdisciplinary research project funded primarily by the Israel Science Foundation (ISF grant no. 848/10) and the Institute for Aegean Prehistory (INSTAP), as well as by Haifa and George Washington Universities. Scientific collaborators in the project include Andrew Koh from Brandeis University (residue analysis); David Ben Shlomo from The Hebrew University of Jerusalem (petrography); Guy Bar-Oz and Nimrod Marom from the University of Haifa (zooarchaeology); Felix Höflmayer from the German Archaeological Institute in Berlin (radiocarbon); Ruth Shahack-Gross from the Weizmann Institute (geomorphology); and Sariel Shalev from the University of Haifa (XRF analysis). We would like to thank all of our scientific collaborators, as well as our Senior Staff, including Nurith Goshen, Alexandra Ratzlaff, and Inbal Samet, for all of their hard work and efforts on behalf of the Kabri Archaeological Project.

REFERENCES

Beck, P. 2002. *Imagery and Representation: Studies in the Art and Iconography of Ancient Palestine: Collected Articles.* Tel Aviv: Institute of Archaeology, Tel Aviv University.

Boivin, N. 2000. Life rhythms and floor sequences: excavating time in rural Rajasthan and Neolithic Catalhoyuk. *World Archaeology* 31(3): 367–388.

Bourdieu, P. 1990. *The Logic of Practice.* Stanford: Stanford University Press.

Cline, E. H. and A. Yasur-Landau. 2007. Poetry in motion: Canaanite rulerships and Minoan narrative poetry at Tel Kabri. In: S. P. Morris and R. Laffineur (eds.), *EPOS: Reconstructing Greek Epic and Aegean Bronze Age Archaeology. Proceedings of the 11th International Aegean Conference EPOS Reconsidering Greek Epic and Aegean Bronze Age Archaeology, Los Angeles, UCLA, The J. Paul Getty Villa, 20–23 April 2006,* 157–166. Aegaeum 28. Liège: Université de Liège.

Cline, E. H., A. Yasur-Landau and N. Goshen. 2011. New fragments of Aegean-style painted plaster from Tel Kabri, Israel. *American Journal of Archaeology* 115(2): 245–261.

Damm, C. B. 2000. Time, gender and production: a critical evaluation of archaeological time concepts. In: M. Donald and L. Hurcombe (eds.), *Gender and Material Culture in Archaeological Perspective,* 110–122. London: Palgrave Macmillan.

Davis, J. L. 2010. Pylos. In: E. H. Cline (ed.), *The Oxford Handbook of the Bronze Age Aegean,* 680–689. New York: Oxford University Press.

Driessen, J. 2010. Malia. In: E. H. Cline (ed.), *The Oxford Handbook of the Bronze Age Aegean,* 556–570. New York: Oxford University Press.

Fleming, D. E. 1992. *The Installation of Baal's High Priestess at Emar: A Window on Ancient Syrian Religion.* Atlanta: Scholars Press.

Fleming, D. E. 2000. *Time in Emar: The Cultic Calendar and the Rituals from the Diviner's House.* Winona Lake, IN: Eisenbrauns.

Foxhall, L. 2000. The running sands of time: archaeology and the short-term. *World Archaeology* 31(3): 484–498.
Goetze, A. 1969. Ritual for the erection of a new palace. In: J. B. Pritchard (ed.), *Ancient Near Eastern Texts,* 357–358. Third Edition. Princeton: Princeton University Press.
Goren, Y., I. Finkelstein and N. Na'aman. 2004. *Inscribed in Clay: Provenance Study of the Amarna Tablets and Other Ancient Near Eastern Texts.* Tel Aviv: Institute of Archaeology, Tel Aviv University.
Gosden, C. 1994. *Social Being and Time.* Oxford: Blackwell.
Goshen, N., A. Yasur-Landau and E H. Cline. 2013. Textile production in palatial and non-palatial contexts: the case of Tel Kabri. In: M.L. Nosch, H. Koefoed, and E. Andersson (eds.), *Textile Production and Consumption in the Ancient Near East: archaeology, epigraphy, iconography*, 45–53. Oxford: Oxbow.
Hurcombe, L. 2000. Time, skill and craft specialization as gender relations. In: M. Donald and L. Hurcombe (eds.), *Gender and Material Culture in Archaeological Perspective,* 88–109. London: Palgrave Macmillan.
Kempinski, A. 2002. History of the site and its environs: I. The Late Neolithic to the end of the Persian Period. In: N. Scheftelowitz and R. Oren (eds.), *Tel Kabri: The 1986–1993 Excavation Seasons,* 449–452. Tel Aviv: Institute of Archaeology, Tel Aviv University.
Killen, J. T. 1998. The Pylos Ta tablets revisited. *Bulletin de Correspondance Hellénique* 122: 421–422.
La Rosa, V. 2010. Phaistos. In E. H. Cline (ed.), *The Oxford Handbook of the Bronze Age Aegean,* 582–598. New York: Oxford University Press.
Macdonald, C. 2010. Knossos. In E. H. Cline (ed.), *The Oxford Handbook of the Bronze Age Aegean,* 529–542. New York: Oxford University Press.
Macdonald, C. 2012. Palatial Knossos: the early years. In: I. Schoep, P. Tomkins and J. Driessen (eds.), *Back to the Beginning: Reassessing Social and Political Complexity on Crete during the Early and Middle Bronze Age,* 81–113. Oxford: Oxbow Books.
Margueron, J. C. 1995. Mari: A Portrait in Art of a Mesopotamian City-State. In: J.M. Sasson (ed.), *Civilizations of the Ancient Near East*, II, 885–899. New York: Hendrickson Publishers.
Margueron, J. C. 1997. Palace. In: E.M. Meyers (ed.), *The Oxford Encyclopedia of Archaeology in the Near East,* 197–200. *New York and Oxford: Oxford University Press.*
Miglus, P. 2004. Palast. B. Archäologisch. *Rivista dell'Istituto Nazionale di Archeologia e Storia dell'Arte* 10(3–4): 233–276.
Militello, P. 2012. Emerging authority: a functional analysis of the MM II settlement of Phaistos. In: I. Schoep, P. Tomkins and J. Driessen (eds.), *Back to the Beginning: Reassessing Social and Political Complexity on Crete during the Early and Middle Bronze Age,* 236–272. Oxford: Oxbow Books.
Niemeier, B. and W.-D. Niemeier. 2002. The frescoes in the Middle Bronze Age palace. In: N. Scheftelowitz and R. Oren (eds.), *Tel Kabri: The 1986–1993 Excavation Seasons,* 254–298. Tel Aviv: Institute of Archaeology, Tel Aviv University.
Oren, E. D. 1992. Palaces and patrician houses in the Middle and Late Bronze Ages. In: A. Kempinski and R. Reich (eds.), *The Architecture of Ancient Israel,* 105–120. Jerusalem: Israel Exploration Society.
Oren, R. 2002. Area D. In: N. Scheftelowitz and R. Oren (eds.), *Tel Kabri: The 1986–1993 Excavation Seasons,* 55–72. Tel Aviv: Institute of Archaeology, Tel Aviv University.
Palaima, T. G. 1995. The last days of the Pylos polity. In: R. Laffineur and W.-D. Niemeier (eds.), *Politeia: Society and State in the Aegean Bronze Age,* 623–632. Liège: Université de Liège.

Palaima, T. G. 2004. Sacrificial feasting in the Linear B documents. In: J. C. Wright (ed.), *The Mycenaean Feast. Hesperia* 73(2): 217–246.

Pfälzner, P. 2007. Archaeological investigations in the royal palace of Qatna. In: D. M. Bonacossi (ed.), *Urban and Natural Landscapes of an Ancient Syrian Capital: Settlement and Environment at Tell Mishrifeh/Qatna and in Central-Western Syria,* 29–64. Udine: Forum.

Picazo, M. 1997. Hearth and home: the timing of maintenance activities. In J. Moore and E. Scott (eds.), *Invisible People and Processes: Writing Gender and Childhood into European Archaeology,* 59–67. London: Leicester University Press.

Pritchard, J. B. 1969. The banquet of Ashurnasirpal II. In J. B. Pritchard (ed.), *Ancient Near Eastern Texts,* 556–560. Third Edition. Princeton: Princeton University Press.

Scheftelowitz, N. and R. Oren, eds. 2002. *Tel Kabri: The 1986–1993 Excavation Seasons.* Tel Aviv: Institute of Archaeology, Tel Aviv University.

Souvatzi, S. 2008. *A Social Archeology of Households in Neolithic Greece.* Cambridge: Cambridge University Press.

Speciale, M. S. 1999. La tavoletta PY Ta 716 e i sacrifici di animali. In: V. La Rosa, D. Palermo and L. Vagnetti (eds.), *Epi Ponton Plazomenoi: Simposio Italiano di Studi Egei Dedicato a Luigi Bernabò Brea e Giovanni Pugliese Carratelli. Roma, 18–20 febbraio 1998,* 291–297. Roma: Scuola archeologica italiana di Atene.

Tomkins, P. 2012. Behind the horizon: reconsidering the genesis and function of the 'First Palace' at Knossos (Final Neolithic IV–Middle Minoan IB). In: I. Schoep, P. Tomkins and J. Driessen (eds.), *Back to the Beginning: Reassessing Social and Political Complexity on Crete during the Early and Middle Bronze Age,* 32–80. Oxford: Oxbow Books.

Wright, G. R. H. 1985. *Ancient Building in South Syria and Palestine.* Leiden: E. J. Brill.

Yasur-Landau, A., E. H. Cline, N. Goshen, N. Marom and I. Samet. 2012. An MB II Orthostat Building at Tel Kabri, Israel. *Bulletin of the American Schools of Oriental Research,* 367: 1–29.

Yasur-Landau, A., E. H. Cline and G. A. Pierce. 2008. Middle Bronze Age settlement patterns in the western Galilee, Israel. *Journal of Field Archaeology* 33(1): 59–83.

Yasur-Landau, A., E. H. Cline and I. Samet. 2011. Our cups overfloweth: "Kabri goblets" and Canaanite feasts in the Middle Bronze Age Levant. In W. Gauß, M. Lindblom, R. A. K. Smith and J. C. Wright (eds.), *Our Cups Are Full: Pottery and Society in the Aegean Bronze Age: Papers Presented to Jeremy B. Rutter on the Occasion of His 65th Birthday,* 382–392. Oxford: Oxbow Books.

Yasur-Landau, A., J. R. Ebeling and L. B Mazow, eds. 2011. *Household Archaeology in Ancient Israel and Beyond.* Culture and History of the Ancient Near East, vol. 50. Leiden: Brill.

14 Semiotic Approaches for the Study of the Urban Environment of the Late Bronze Age Settlement of Akrotiri on Thera

Konstantinos Athanasiou

INTRODUCTION

The aim of this chapter is to study and analyze the urban spaces of the prehistoric settlement of Akrotiri on the Aegean island of Thera during the Late Bronze Age in order to come a step closer to understanding its social and political structure.

Before proceeding with the analysis, it would be useful to clarify some key concepts that are used throughout this chapter. The first concept is that of 'space'. According to Pellegrino (2006: 89), "'space' is the field where the phenomena are being articulated. Through this articulation, space takes its form and its meaning".

In this study, two kinds of phenomena are being explored: social and political. The articulation of space with social phenomena will produce the *social space,* whereas the articulation of space with political phenomena will produce the *political space*. Regarding the former, it is difficult for someone to conceive space without its social content and conversely, a society to be conceived without the spatial element. According to Lefebvre (1974), space and time are inextricably interrelated. The relationship between space and society must be understood as an interactive process in which people produce and transform spaces, while at the same time they are influenced by these spaces in various ways. The political space can be defined as the space that is produced and formed through competitive activities. The political space is directly connected to a dominant group of people. This group of people create built environments that serve to produce and reproduce their sociopolitical positions and in doing so they project and represent their values symbolically in space. Through this representation of hegemonic values, space contains and keeps the social relationships in a state of cohesion and coexistence (Crang 2009; Lefebvre 1974; Pellegrino 2006).

The understanding of the basic principles on which a settlement is articulated provides researchers with valuable spatial, architectural, political, anthropological, and sociological connotations. As an architect, I am approaching the prehistoric built space with theoretical and methodological tools that are mainly used for the study of the modern built environment. This approach presents a different perspective from the archaeological one and possibly offers interesting observations to the researcher of the prehistoric urban environment. In my quest for the appropriate research tools, a variety of social semiotic techniques are being used in order to investigate the role of urban space and architecture in relation with the sociopolitical environment during the Late Bronze Age Akrotiri on Thera.

The main reason for the selection of the social semiotic vehicle for the study of the urban environment of Akrotiri lies in the structural ability of this semiotic method to correlate form with substance. This means that the research object can be analyzed at multiple levels. Therefore, the remains of a settlement automatically obtain meanings that extend beyond the materiality of architecture. The 'lifeless ruins' become the carrier of a series of cultural, sociopolitical, and anthropological connotations that can be studied and, possibly, interpreted. In addition, the application of a social semiotic theory, and especially the application of the Systemic Functional Theory—the semiotic toolset that will be presented in this chapter—contributes to the main objective of this volume: the investigation of prehistoric space and time not as two different dimensions of the same research object, but as interrelated aspects that are inextricably tied. When these aspects are examined together, a whole new series of connotations is revealed that otherwise would remain hidden.

THE SEMIOTIC TOOLSET: SYSTEMIC FUNCTIONAL THEORY

The semiotic pillar of this research and the toolset that is applied at the prehistoric settlement in question for the deduction of spatial, social, political, and anthropological connotations is Systemic Functional Theory (SFT). SFT constitutes the methodological framework of this study, and a series of semiotic tools are being unfolded under its rules and requirements. All these tools comprise different semiotic approaches that are oriented toward the basic premises of SFT. SFT is based on the seminal work of the linguistic and social semiotician Michael Halliday. His theory was met with great success and renown in the scientific community. As a consequence, its basic principles were diffused in many different arts, sciences, and crafts, such as music, journalism, and image analysis (Stenglin 2004).

SFT was first introduced and applied to three-dimensional spaces by Maree Kristen Stenglin in her work, entitled *Packaging Curiosities: Towards a Grammar of Three-Dimensional Space* (2004). Halliday's theory is inspired by the semiotic theory of Hjelmslev. Its main premise is that a semiotic methodology has to be open to other methodologies with or without semiotic background (Halliday 1986; also, Stenglin 2004). Stenglin (2004; 2009) exploits this by incorporating in her methodology both semiotic and nonsemiotic theories for analyzing the built environment. The combination of SFT with structural and post-structural tools (e.g., Hillier's Space Syntax and Michael Foucault's concept of Heterotopias) has also been applied to the study of the urban network of Akrotiri (Athanasiou 2011).

The specific semiotic methodologies that have been incorporated in the present study of the urban space of Akrotiri will be further discussed later. Specifically, the semiotic methodology of Kristen Stenglin, which is based mainly on Halliday's SFT, will be presented in more detail. The semiotic approaches of Gunther Kress and Theo van Leeuwen (Kress and van Leeuwen 1996) will work supplementarily to create a solid social multisemiotic methodological framework, which can provide the researcher with a powerful interpretation tool for the study of the urban environment.

SFT provides a grammar for space by using three communicative functions: (a) the ideational function of space; (b) the interpersonal function of space; and (c) the textual function of space. Each one of these functions consists of several semiotic tools oriented to the analysis and interpretation of the built environment. Here, a selection of these tools will be presented that focuses on the interrelation of the concepts of built space and time. More information about the application of SFT can be found in Stenglin (2009), as well as in Athanasiou (2011).

THE IDEATIONAL FUNCTION OF SPACE

The ideational function deals with ways of construing representations of human experience and has two aspects: one is concerned with the *field,* and mainly with the identification and classification of spaces, whereas the other is concerned with two types of structures—serial and orbital.

Field

According to Stenglin (2009: 38), "field can be defined as a set of activity sequences oriented to some global institutional purpose". When the notion of field is applied to the built environment, it means that this environment can be analyzed as a range of activities and processes. The notion of 'field' is very close to the creation of a functional categorization that studies space diachronically. In addition to the diachronic identification of the activities that take place in space, 'field' is concerned with identifying the objects involved in these activities.

Serial and Orbital Structures

Serial and orbital structures are important for the patterning of ideational meanings of space. They can be used to organize and describe not only single spaces and buildings, but also spatial patterns that can be part of an urban environment. As far as serial structures are concerned, it should be noted that meaning is accumulated sequentially as the visitor moves. Serial structures tend to be dynamic and linear where the element of time is always present. The more spaces a person crosses through a period of time, the more stimuli this person receives from the built environment. The other tool for analyzing ideational meanings in three-dimensional space is orbital structure. Stenglin (2009: 41) notes that this "semiotic tool construes a space or a series of interconnected spaces as having a core. With orbital structure, a space, or a series of interconnected spaces, are organized around a nucleus and satellite configuration".

THE INTERPERSONAL FUNCTION OF SPACE

Stenglin argues that the term 'interpersonal' indicates that the meanings constructed through the organization of space extend beyond the materiality of architecture. The spaces people occupy in their everyday lives provide the setting for all their activities and affect their ways of behavior and interaction with other people. They are also designed to evoke feelings. Some spaces make a person feel comfortable, protected, and secure whereas other spaces evoke a range of negative feelings. For example, the organization of some spaces could make someone feel dominated; others can make someone feel stifled, whereas others can make a person feel vulnerable and exposed.

Interpersonally, the grammar of three-dimensional space is concerned with Binding and Bonding. Binding regards the relationship between space and emotion, whereas Bonding explores the patterns of interaction between the occupants of a space (Stenglin 2009). Binding organizes the interpersonal relationships constructed between a space and its users into two main categories: security and insecurity. In particular, whether we feel secure or insecure inside a space appears to be strongly dependent on how firmly that space closes in on or opens up around us. Bonding is concerned with the disposition of visitors in relation to spaces, and its basic function is to align people into groups with shared dispositions. There are at least three tools that materialize bonding in the third dimension: bonding icons, hybridization, and classification and framing, developed by the sociologist Bernstein (Stenglin 2004; 2009).

The Interpersonal Function: Binding Choices for Insecurity

There are two Binding choices for constructing a relationship of insecurity between a space and its users. The first set of choices establishes a smothering relationship by making the person feel too restricted in a space. The second set of choices constructs a relationship of insecurity by making users feel vulnerable and exposed.

The Interpersonal Function: Binding Choices for Security

There are two sets of choices for constructing a relationship of security between a building and its users. The first set of choices establishes a relationship of security by designing spaces in which users feel Bound. In the *bound* spaces, the user feels comfortable, safe, and protected. This is accomplished by clearly delineating boundaries around the bound space. The second set of choices establishes a relationship of security by making the user feel Unbound or free. For reasons of space only the unbound space aspect will be presented here. Although the exploration of the urban spaces of Akrotiri in terms of bound space can offer interesting observations, the application of the unbound space tool has the merit that it incorporates the aspect of time, which is crucial to the understanding of the relation between the built environment and the way in which the people at Akrotiri would experience space around them while moving within it.

Stenglin (ibid.) states that by loosening the enclosures that restrict a space, an unbound enclosure constructs a free relationship with its occupant. The feeling of freedom can be the result of various and different factors. It is dependent on sociopolitical, cultural, psychological, and other parameters. However, regardless of the above parameters, if freedom is tightly related with someone's ability to act, then the unbound spaces can be correlated with the number of a person's choices. The more choices someone has, the more intense the feeling of freedom is. Characteristic spaces with these qualities in the urban fabric of a town are crossroads and squares.

The Interpersonal Function: Bonding—Bonding Icons

Two types of icons are important: icons with a rallying function and icons with a privileging function. The first type includes flags, logos, colors, and memorabilia, whereas the latter type includes references to other people, places, and ideals.

The Interpersonal Function: Bonding—Hybridization

Bonding can be realized by the hybridization of space. A space can be characterized as hybrid on two occasions. The first is when hybridization simply means that one space is designed to serve many functions. The second is when a particular space through some events or changes that happened inside that space contributes to the 'recontextualization' of the values of a group of people. So, for example, a square where a revolution took place may carry new and different connotations about the square to a group of people from that time onward. Thus, it becomes a carrier of these new values and symbolizations and leads to the unification of the people into a new communality.

The Interpersonal Function: Bonding—Classification and Framing

Given that bonding is associated with aligning and realigning people into groups with shared dispositions, it is important to develop theoretical tools for analyzing how people interact with one another within space. To gain such insights, Bernstein's work on classification and framing helps us explain patterns of social interaction.

Bernstein (1975: 37) claims that classification is about power. Strong classification means "things must be kept apart", whereas weak classification means they should be "put together". According to Stenglin (2009), things can refer to objects, people, or spaces. In a strongly classified space, for instance, there are many rooms with discrete and separate functions, and the objects found in each room are not interchangeable—they are strongly linked to the function of space. Conversely, a weakly classified space tends to have a single space with many functions, and the objects are highly interchangeable.

Framing, as applied to space by Bernstein, has two interrelated dimensions. First, it refers to the strength of the physical boundaries around a space, and second, it refers to the impact of such physical boundaries on the social *inter*-action between participants within a space. For example, prisons have strongly classified spaces with relatively weak framing in order to monitor prisoners better.

THE TEXTUAL FUNCTION OF SPACE

The textual function of space is related to the organization of information into a meaningful whole. As Stenglin (2009: 55) states, "built environments, like written and spoken texts, unfold in time. They also unfold in space". The tools used are theme–rheme, information value, framing (by Kress and van Leeuwen 1996), salience, and path–venue. Below I address three of these tools that contribute most to the inference of sociopolitical connotations: theme–rheme; information value; and framing (by Kress and van Leeuwen).

Theme–Rheme

Theme–rheme can explore the organization of space, particularly the flow of information in a space or in a series of interconnected spaces. Halliday (1985: 36) refers to theme as "the element which serves as the point of departure of the message". To the spatial description of the built environment, 'theme' is the point, the space or the pattern of spaces that aims to orient people in a space or in a series of spaces, whereas the rest of the built environment is defined as 'rheme'. This procedure is called 'logogenesis'. An identification of the 'logogenesis' of the free spaces of Akrotiri is attempted below, that is, an identification of the basic spatial configuration that is repeated throughout the settlement and functions as the theme.

Information Value

Information value as developed by Kress and van Leeuwen (1996) explores semiotic patterns for better description and analysis of images. By transferring and applying this tool to the built environment, we try to find polar or bipolar patterns that allow for better organization of the spatial information

(Kress and van Leeuwen 1996; Stenglin 2004). In a spatial polar composition, a focal point links the spatial elements and joins them together. Characteristic polar patterns are the statue in the middle of a square, or a square and the surrounding buildings. The second category that information value examines is the bipolar patterns. These patterns have as a main characteristic the division of an entity into two contrasting or opposing choices (Stenglin 2009). So, for example, a bipolar pattern can be related with what is positioned as 'given' and what is 'new' in space. Information value is a very powerful semiotic tool and in this chapter will be used to demonstrate the way in which a bipolar pattern can offer invaluable insight into the urban structure of Akrotiri.

Framing

Framing, as originally conceptualized by Kress and van Leeuwen (1996), refers to the strength of the dis/connection of elements in a visual composition. The presence or absence of framing elements—objects, spaces, and others—divide or unite the space, signifying unity or division. For example, a prison exterior wall consists of a frame that divides space not only in a material and functional way, but also semantically. If this wall is built with a particular stone, it will visually connect the prison with its convict inhabitants and, in doing so, will point to their distinctive social status.

THE SITE, THE TIME, AND THE DESTRUCTION

Akrotiri is situated on the island of Thera, in the Cyclades, and is one of the most important settlements of the Aegean Late Bronze Age. It was destroyed by the tremendous eruption of the volcano of Thera approximately 3,500 years ago and was buried under many meters of volcanic ash, in a similar way as the town of Pompeii in Italy was buried under the ashes of Mount Vesuvius. However, this tragic destruction was the main reason for its excellent preservation. As the archaeologists gradually excavated Akrotiri, they revealed not only a very clear part of the urban fabric of the town at its time of destruction, but also a society of affluence, as reflected by its impressive architecture and the rich findings.

At the time of its destruction, Akrotiri was already a settlement with rich history that counted several centuries of continuous life, as shown in (Table 14.1).

Neolithic and Early Bronze Age (Early Cycladic) findings and Middle Bronze Age (Middle Cycladic) structures have left their imprints in the urban network of the settlement (Doumas 1995; 2008; Palyvou 2005; Sotirakopoulou 2008). Approximately fifty years (Palyvou 2005) before the fatal eruption of Thera's volcano, a very strong earthquake damaged the town, severely destroying a great number of its buildings. (This is known as *seismic*

Table 14.1 Chronological table of major phases and events on the Cyclades, Thera, and Crete.

BCE	Cyclades	Akrotiri	Crete
3000	Late Neolithic	First traces of occupation	Early Minoan
2700	Early Cycladic	Cemetery below the LC Cenotaph Square	
2000	Early Cycladic III	Sacrificial fires below the LC Cenotaph Square	Middle Minoan
1900	Middle Cycladic		
1550	Late Cycladic I	Seismic destruction	Late Minoan I
1500	Late Cycladic II	Volcano eruption—destruction of Akrotiri	Late Minoan II
1450	Late Cycladic III		Late Minoan III

destruction.) However, the town was rebuilt upon the semi-destroyed one. The fallen debris was dumped in the open spaces, resulting in the rising of the street level and the drastic rearrangement of the ground floors of the buildings. All this material was also used for the formation of many terraces throughout the settlement. The new town was impressive, and its spectacular remains reveal an advanced architectural thinking by the builders, as was presented in the seminal work of Palyvou (2005) on Akrotiri, entitled *Akrotiri Thera: An Architecture of Affluence 3,500 Years Old.* As is shown below, the same advanced thinking seems to be manifest in the arrangement of the town's buildings and open spaces.

Akrotiri, a coastal settlement, could be envisaged as a dense cluster of cubelike buildings, many of which had two or three stories, with flat roofs, mounted on the slope of a hill looking at the sea. The plan of the settlement, at least the excavated part of it, seems to follow a loose form of grid. The street net and the free public spaces follow more or less the direction of north–south and east–west axes. The buildings follow the same orientation, with the exception of the so-called West House. Some buildings may have been freestanding, but the majority of them were arranged in blocks (Palyvou 1986; 2005).

The open public spaces, including the street pattern, cover a large portion of the town in comparison with other settlements on Cyclades and Crete. The streets, according to Palyvou, can be divided in four main categories: the *main streets, the secondary streets, the narrow alleys,* and the *blind alleys.* Akrotiri has three main streets, two of which run through the settlement from north to south and one from east to west. The north–south axis consists of the 'Telchines' and 'Daktylon' main street extending to over 120 m in length, and is paved with large flat stones. 'Koureton' street runs through Akrotiri from east to west. It intersects with Telchines street south of Sector

Gamma at Xeste 3 square (Figure 14.1). The secondary streets are all those that seem to function mostly for their surrounding areas. All of these intersect with the main ones; they are narrower (1–1.20 m) and shorter. Narrow alleys are actually gaps between buildings that are too narrow (0.60–1 m) to be used for traffic. Their primary use was to accommodate the basic needs of particular buildings, for example to provide them with adequate air and light, especially their rooms that do not overlook streets or squares. In addition, the narrow alleys were used for the advanced drainage system of the settlement. Blind alleys, according to Palyvou, are an indication of the infill development of the settlement. They were meant to provide light and air to the buildings.

What might impress a visitor at Akrotiri is the large number of squares, compared with the size of the settlement. A square is defined by Cowan (2005) as a landscaped or a paved space enclosed wholly or partly by buildings. The squares at Akrotiri vary in size, shape, and ground plan. It is not clear whether they were laid out as part of a preconceived town plan or were the result of the construction of the surrounding buildings. In any case, their importance was apparently great, and therefore their consideration is vital for the interpretation of the organization of the urban fabric of Akrotiri. Another element that can be found everywhere in the settlement is the 'dentation' of the building facades, that is, a series of recesses or projections on the exterior walls of the buildings, which, according to Palyvou, is related to the form of all the open spaces of the settlement. As will be further discussed below, dentation has played a significant role on many levels of the urban life of the town (Palyvou 1986; 2005).

The most important square of the town was probably 'Cenotaph Square' (Palyvou 2005). It is located in the middle of the excavated settlement between Telchinon and Daktylon streets and must have been a central public space. Cenotaph Square seems to be a place that had continuous life for many centuries. According to Doumas (1995), during the Early Cycladic period a cemetery was located there. Moreover, near the 'cenotaph', Marinatos revealed a rectangular construction: the "sacrificial fire" that dates back to the end of the Early Cycladic period (Marinatos 1999a). During the Middle Cycladic period, the Akrotiri people erected a stone pile—or tomb, according to Doumas (1995—at the top of which they built the 'cenotaph'. The 'cenotaph' is a rectangular construction 1 by 1.50 m made of slabs that contained seven Early Cycladic marble figurines. The square acquired its final form after the *seismic destruction,* having as its main characteristic the 'cenotaph', which was found embedded in it. Doumas (1995; 2008) argues for the importance of the square as functioning as an 'avaton' for the Akrotiri people during the Late Cycladic period.

Another very important square was that of the 'Double Horns'. It is located southeast of Cenotaph Square and owes its name to the double horns of consecration that were found in it. As Palyvou notes (2005: 38), "this area of the town has the most regular town plan arrangement, with

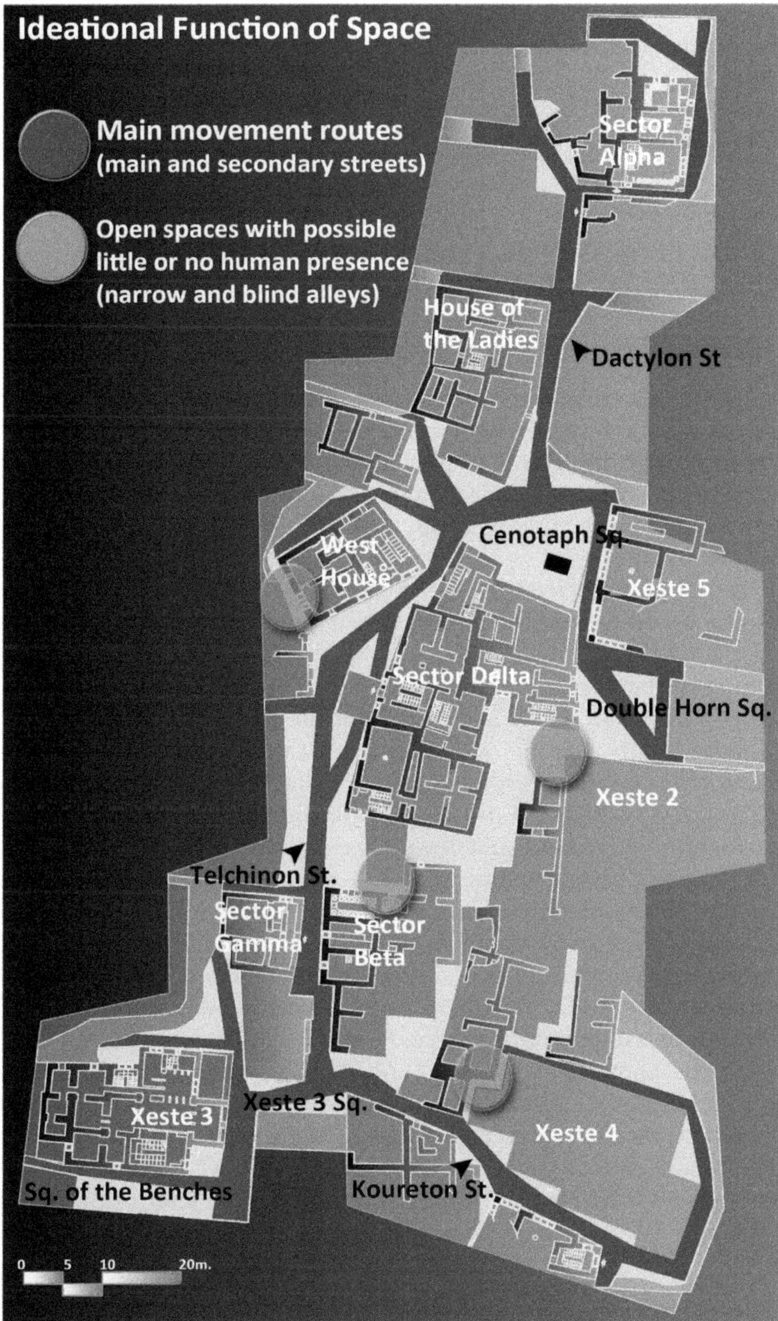

Figure 14.1 General plan of the settlement and the ideational function of space: Field tool.

streets crossing at right angles and neatly indented facades bordering a perfectly rectangular public square". The squares of Xeste 3 and of the Benches both border Xeste 3, a very important building of an apparently public or religious character. They are located in the southwestern part of the excavation and possibly near the prehistoric waterfront and port.

SYSTEMIC FUNCTIONAL THEORY: PRESENTATION AND APPLICATION TO THE URBAN ENVIRONMENT OF AKROTIRI

The Ideational Function of Space: Field

As explained earlier, field is related to the identification and classification of the spaces. Concerning the unbuilt areas of Akrotiri, there are spaces that were used for the movement of people (main and auxiliary streets), others used for the gathering of people (squares), and yet others used for basic functions and needs of the buildings such as ventilation, lighting, and the sewage system (Figure 14.1).

Of special interest are the objects that have been found in the unbuilt areas. The Early Cycladic figurines in Cenotaph Square denote that this may also have served ritual or religious needs. That square is a place where the linking of the past, as exemplified by the cemetery, with the present, as attested by the cenotaph, is apparent, and the ritual or religious character of the place is very possible. The presence of the 'Sacrificial Fires' and the Early Cycladic cemetery further points to a diachronic and 'holy' character of the square. Although at a primary, denotating level, some spaces are quite easy to identify in terms of function—for example, main streets are used for the circulation of people within Akrotiri—other spaces are quite ambiguous and have to be examined within a general context. Again, the Square of the Benches provides a good example. Whereas on a first level it might denote the gathering of people, this may not be its only function. Xeste 3, the building north of this square, must have had a public or religious character. The remains of a possible sacrifice in the House of Stone Benches (the building that gave the name to the square) enhance the religious character of the square. Hundreds of animal horns have been found in a room along with a golden ibex carefully placed in a case. The benches outside the building, and maybe the entire square, may have served the second—connotating—purpose, which is tied to the religious and public character of the buildings (Alexopoulos 2008; Boulotis 2005; Doumas 1995; 2008).

The Ideational Function of Space: Serial and Orbital Structures

The unbuilt spaces of Akrotiri are organized in a serial way, since the access from one space to another is linear. When a person was moving through the open spaces of the town, the meaning of its surrounding spaces was accumulated sequentially. At a second level, however, we notice that the organization of each open wide space tends to be orbital. The existence of many squares and the widenings of streets, especially at the nodes where two or more routes are crossed, results in the creation of large or small nuclei. Most of the entrances of the buildings are placed in these nuclei. So, not only do these spaces compose the meeting point of the town routes; they also constitute the first open space that a person encounters when leaving a building.

The Interpersonal Function of Space: Binding Choices for Insecurity

At Akrotiri, there are open spaces that can make a person feel too restricted. All the blind spaces, which are very narrow in proportion to the great height of the buildings that surround them, may possibly establish a smothering relationship with the people. Similarly, but to a lesser extent, all the narrow alleys that are scattered across the settlement could be included (the narrow alley between the West House and the Anchor House, the small alley between Xeste 4 and Building I, etc.—green circles in Figure 14.1). A main characteristic of these spaces is that their geometry did not promote human activity in them. They are supplementary spaces, and their purpose was to provide light and air to the buildings that surround them.

The Interpersonal Function of Space: Binding Choices for Security: The Unbound Tool

The element of unbound spaces with the parallel feeling of freedom of the individual is very strong and impressive at Akrotiri. The open spaces of the town are formed along a series of narrow parts and widenings. These are in the form of squares or widenings of the streets and are created at the intersection of two or more paths. All these open spaces offer a plethora of choices to the user, which, in turn, can create a strong feeling of freedom. Specifically, there are at least thirteen such spaces in Akrotiri, as is shown in Figure 14.2. This number could increase, if we add the crossroads that lead to small alleys and blind alleys and the open wide spaces that border the site (like the square of Ship Procession north of West House). Eleven of these thirteen unbound spaces are on the three main streets of the settlement. The frequency of these spaces along these main streets is impressive. Along the north–south axis (Telchinon and Daktylon streets) one unbound space occurs every 15 m, whereas along the east–west axis (Koureton street) the frequency increases, with one unbound space occurring every 10 m. This means that a person who crosses Koureton street has every 10 m the freedom of choice to change his or her direction. Other unbound spaces worth mentioning are Cenotaph and Xeste 3 squares. These two spaces were the most important nodes in the settlement since a person starting from these squares had the freedom to choose between five different routes to go.

Figure 14.2 Interpersonal function of space: Binding tool, Security in Unbound spaces.

The Interpersonal Function of Space: Bonding—Bonding Icons

In the open spaces of Akrotiri, both types of symbols can be found. As rallying symbols one can consider all religious or ritual symbols that are spread throughout the settlement. The cenotaph in the homonymous square, Xeste 3, The House of the Stone Benches, where a pile of sacrificial horns has been found, and the square of the Double Horns, where horns of consecration have been recovered, are some of the places where rallying icons can be found. All Xeste buildings can be considered as privileging symbols, as they influence the nearby free spaces, symbolizing the values and ideals of their occupants or their social status. Another source of possible privileging icons might be the 'religious institutions' in Akrotiri. There are evidences of two different religious practices in the town. The first one is based on Doumas's interpretation of the cenotaph and may have its roots in a prehistoric Cycladic religion. This religious practice is spatially evident in Cenotaph Square, where several Cycladic figurines and the 'Sacrificial Fires' dating to the Early Cycladic period have been found. The second is of Minoan origin. Throughout the settlement, numerous religious Minoan elements are evident: horns of consecration, wall paintings, a lustral basin, and ritual or religious objects are some of these elements that show a tight religious relation between Akrotiri and Crete. If we assume that the 'religious institutions' in Akrotiri also had political power, then every visible religious element would be considered as a privileging icon.

The Interpersonal Function of Space: Bonding—Hybridization

At Akrotiri there are open spaces that can be described as hybrids according to both of the aforementioned dimensions of hybridization. Regarding multiple functions, Cenotaph Square and the Square of the Benches are two places in which at least two functions can be recognized: one that is concerned with the gathering of the people and another relating to a ritual or religious character. Two other possible hybrid spaces are found to the east of Sector Alpha (east of the room A1) and to the west of Sector Delta (west of the room D16). The rooms in these buildings share a common feature: a very big window that establishes a direct relationship between the streets and the rooms. Moreover, Doumas (1983) notes that these windows may have supported trading activities between the occupants of the buildings and the people of the settlement. If this hypothesis is correct, the streets that were next to the windows served for commercial activities, in addition to their main function as a passage for pedestrians.

Regarding the second dimension of hybridization, spaces serving to recontextualize the value of one field to another are very difficult to be recognized at Akrotiri because of the lack of sufficient archaeological evidence. However, there is a possible such place: Cenotaph Square. It is likely that the cenotaph symbolized, among other values, the power of the ancestors and the protection that they could bestow on the town, given not only that the town itself survived seismic destruction and was rebuilt after it, but that the surrounding space was revamped in order to emphasize the importance of the town. The cenotaph area was probably charged with ancestral memories. Possibly, the cenotaph with the figurines, artifacts comprising part of the history of the town, connected the town occupants with the past, strengthened their local identity, and could be assumed to have been acting as a source of inspiration and courage in times of hardship. Assuming that the worshipping of the ancestors was part of the religious life of the community, then the space obtains more values, such as sanctity.

The Interpersonal Function of Space: Bonding—Classification and Framing

In the case of open spaces at Akrotiri, it can be observed that most of them are weakly classified, since the great number of squares and openings weakens the strength of the spaces. However, there are areas in which the classification is strong. The blind alleys served only for ventilation and lighting, whereas the narrow streets were mainly used for the quick passing of the people and the separation of the buildings. Similarly, the framing of the borders of the open spaces of Akrotiri is also weak, even in the narrow streets and blind alleys, since the buildings have openings that are oriented toward these.

The Textual Function of Space: Theme–Rheme

As has already been mentioned, Palyvou (1986) claims that the very characteristic dentation of the buildings has played an important part in the design of the form of Akrotiri's open spaces. This results in a variety of shapes of open space. Analyzing the network of free spaces at Akrotiri, one can find the basic spatial elements that constitute the theme. These elements are a crooked line and two rectangular united convex spaces that have the aforementioned crooked line (Figure 14.3) as their sides. This spatial pattern is found almost everywhere in the settlement and can be considered as the point of departure for the geometric description of the settlement. Thus, if this spatial configuration is the theme, then the rheme comprises all the free spaces of Akrotiri.

The Textual Function of Space: Information Value

The bipolar pattern that will be examined here is that of 'high-low'. Usually 'low' symbolizes the real, the actual and the human, whereas 'high' symbolizes the ideal and the divine. In Minoan religion, there exist connotations

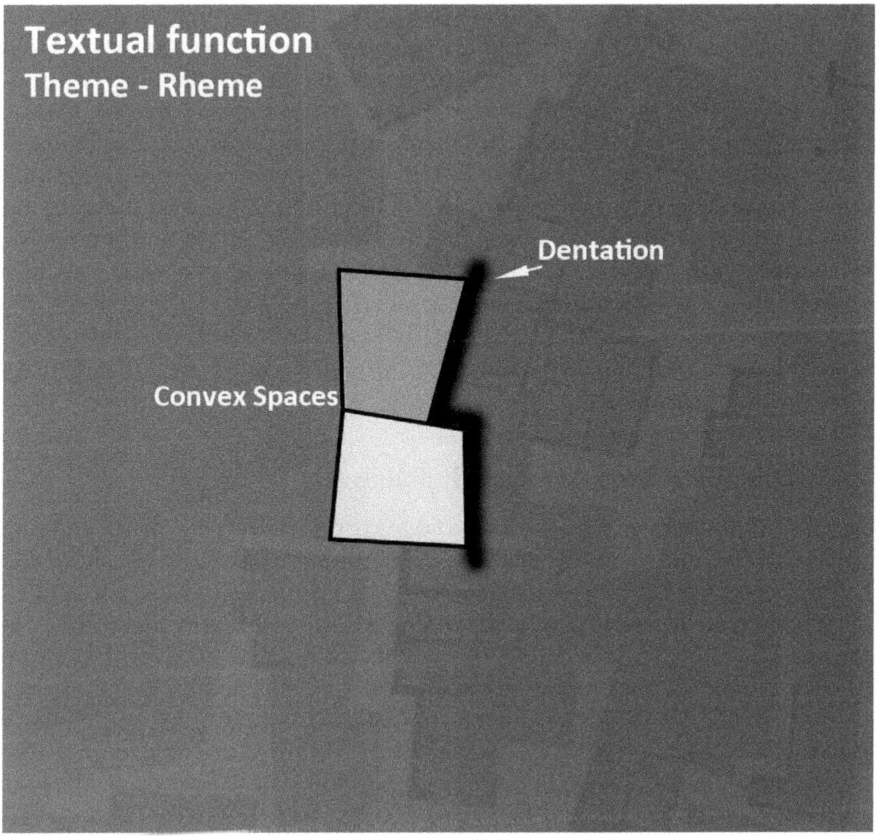

Figure 14.3 Textual function of space: Theme–Rheme tool.

that 'high' and 'low' were related to the above concepts. The peak sanctuaries and the cave sanctuaries, sacred places that are located respectively on the peaks of the mountains and in the caves of Crete, constitute the best proof of the above statement (Gesell 1985).

Similar connections between these two concepts can be identified in Akrotiri in both the wall paintings and the urban layout of the town. The wall painting of Potnia Thiron ('Mistress of Animals') from Xeste 3 is the best example: the goddess is seated atop a platform whereas the worshiper is placed on a lower level. Regarding the urban environment, a 'high' morphological element that implies sanctity has been found at Akrotiri. Horns of consecration have been found at the Square of Double Horns as well as by the entrance of the House of Ladies. These horns were placed on the edge of the flat roofs of the buildings (Alexopoulos 2008; Boulotis 2005; Gesell 1985; Palyvou 2005). If the horns of consecration represent the first element of this bipolar relation (the 'high'), then we can find the second element (the 'low') in Cenotaph Square and more precisely in the cenotaph with the Early Cycladic figurines. However, the Early Cycladic figurines are not related to the Minoan religion. All this, then, represents an interreligious relationship between the two elements (horns of consecration and Early Cycladic figurines). Possibly at Akrotiri the ideal, the 'high', and the new are connected with the Minoan religion, whereas the earthly and the traditional are related to connotations emitted by the cenotaph. Moreover, comparison between the two squares can provide valuable observations and interesting points for consideration. Cenotaph Square has an open and loose structure without clear borders, whereas the Square of Double Horns is probably the best organized area of the settlement. It is possible, then, that this urban organization is related to the influence of Minoan civilization in the ideology of the people of Akrotiri.

The Textual Function of Space: Framing

Such framing patterns can be found at Akrotiri. They operate as unifying spatial, social, and symbolic elements. As framing, patterns of this kind can be considered all the intracommunal building regulations that were applied during the construction of the buildings. All the small and blind alleys of the town constitute the outcome of this kind of regulation. For the creation of such spaces, intracommunal cooperation was required by the occupants of two or more buildings. A very characteristic example of this kind of regulation is the blind alley located between the north and the south building of Sector Beta. In this, at least three windows were opened so that the respective rooms could be better aired and lit. A second communal regulation was associated with the entrances of the buildings, almost all of which were formed in a similar way. There was a window next to the door that provided adequate lighting to the staircase and was always located in the entrance space lobby. The rule was the creation of a space at the entrance that provided the suitable conditions for the movement of people and adequate lighting of the building.

However, the most important intracommunal regulation seems to be the dentation of the buildings. As has already been mentioned, the dentations are found everywhere in the town and are responsible for the shape of the open spaces (Figure 14.4). Palyvou (1986) notes that dentation also served symbolic needs. If dentations are responsible for the form of the majority of the open spaces of Akrotiri, then they are indirectly also responsible for the social relations that are formed in these spaces. In the widenings of the streets and the squares, small or big nuclei were created, within which the social life of the town was taking place. Summing up, the people of Akrotiri were using the dentation to meet a series of needs: practical, such as the ventilation and lighting of the rooms; symbolic, such as to emphasize the facade of the buildings when these facades were overlooking important public spaces; aesthetic, such as to create shadows and to display the volumes

Figure 14.4 Textual Function: Framing tool (by Kress and van Leeuwen), Dentation of the buildings.

of the internal configuration of the buildings; and finally, social, namely to create spaces that promoted social interaction.

CONCLUSIONS

SFT places the human being at the core of research, as he or she roams the space and receives sometimes positive and sometimes negative signals sent by the built environment. However, SFT is not a method without limits, and therefore it should be treated with caution, especially when applied to prehistoric settlements. Its only application outside architecture to date has been in a historic setting, in a society where the way the subject feels under certain circumstances is more or less known. The attempt to understand how the Theran people may have felt is definitely risky and might lead to ambiguous results. However, part of the merit of this method is that it provides the researcher with semiotic tools that might help toward this direction.

Although SFT as applied in Akrotiri is clearly a spatial study, the temporal element is present, subtly most of the time, in many of its tools. In both the ideational and interpersonal function of space are explored the possible feelings a person would have while moving in the streets of the town. However, the concept of movement can be considered as a practice of time in space. With the aid of the *serial structures,* the meanings and the information that a person would receive accumulated additively and were sequentially observed, whereas the tool of the *unbound spaces* in Akrotiri revealed the great number of choices that a person would have while moving in the streets and squares of the town. In addition, and at a different level, the tools of *field, hybridization,* and *bonding icons* explore the strength of urban elements during a period of time. Every urban feature is a possible carrier of meaning that can be studied semiotically. The most important urban feature at Akrotiri seems to be Cenotaph Square, a prominent place with a long history charged with memories of the past.

The study of Akrotiri with the semiotic tools that SFT offers revealed aspects of everyday town life, and therefore aspects of the society that might otherwise be difficult to detect. First, the open spaces of the settlement were analyzed and recomposed, highlighting a landscape that can be described as 'spatial stimulant', characterized by homogeneity and simultaneously by the diversity of the structured and the unstructured volumes—mainly due to the ever-present dentation of the buildings. This impression is complemented by the sense of freedom, as indicated by the large number of choices that a person would have during his or her movement across the settlement. Second, the exploration of various symbols in the urban environment revealed the important role of religion and of community rituals. The settlement was full of such symbols that were acting as small or large poles. These poles were designed either to rally the residents around the object of notation or to highlight a social group, an idea, a religion, or a function.

The 'spatial stimulant' feeling that a person would probably have experienced in Akrotiri was intensified by the symbolic dimension of the town, thus creating a unique place. Responsible for this particular spatial effect might have been the mixing of the Cycladic with the Minoan element. The coherence, the unity, and the homogeneity of society can be supported by the spatial pluralism and the diversity of the urban forms.

Regarding the political spaces of Akrotiri, these should be sought in areas where there is a presence of a hegemonic social class or a dominant group of people. The values that these people could promote are projected and represented symbolically in space. Through the representations of hegemonic values, space would keep social relationships in a state of cohesion and coexistence. A space with such characteristics possibly was the Square of Double Horns. This square is perhaps the best organized area in Arkotiri, and the horns of consecration that were found there may indicate the dominant group of people or social class: a clergy who was influenced by the Minoan religion or a hegemonic group of people who are related to Crete or a combination of these two. Spaces that include privileging icons can be considered as related to the political life of the town, including Xeste buildings or spaces with religious activity. Assuming that Minoan clergy had a prominent position in the social life of Minoan cities, the latter spaces would play an important part in the town's life.

The spatial study of Akrotiri also revealed an aspect that may be directly related to the political situation of the town at the time just before the devastating volcanic eruption. A series of intracommunal building regulations and 'rules of good neighborhood' between people have been identified. These rules and regulations defined more or less the form of the urban fabric. Cooperation on a building level or even on a settlement level may be not only a sign of social cohesion, but also a reflection of a political situation that had at its core the universal cooperation of all the society's members. The lack so far of archaeological evidence for the existence of a building that may have functioned as the seat of a ruler or of any space associated with a centralized political power leaves open the possibility of an egalitarian political structure. It is probable that the power was diffused to all or to various social actors and that the governing of the town was dictated as much by hierarchical as by heterarchical structures. The heterarchical organization of Minoan society is not a new idea, and there have been many arguments supporting it (Driessen 2010a; 2010b; 2010c; in press; Hamilakis 2002, Schoep 2002).[1]

In conclusion, in this chapter an attempt has been made toward investigating specific aspects of the urban fabric of the town of Akrotiri. Utilizing several semiotic tools within the framework of SFT, prehistoric space and time were explored. This exploration provided us with a series of spatial, social, anthropological, and political inferences. I hope that I have contributed toward a better understanding of the urban space of Akrotiri and how it influenced the town's society and its people.

ACKNOWLEDGMENTS

This chapter is based on my doctoral thesis, *Comparative Study of the Urban Fabric of Akrotiri, Thera, with the Aid of Modern Technologies* (Athanasiou 2011, in Greek). Neither the thesis nor this chapter would have been feasible without the great help and guidance of my Ph.D. supervisor, Clairy Palyvou.

NOTE

1 The shaping of the built environment through intracommunal building regulations and 'rules of good neighborhood' is very likely to comprise a horizontal-heterarchical structure. The lack of centralized authority supports the idea of the existence of heterarchical structures that are spatially imprinted. Moreover, in my opinion, there isn't a spatially hierarchical structure that indicates a seat of central power in Akrotiri.

REFERENCES

Alexopoulos, G. 2008. To iero kai to miaro: symboliki kai litourgiki epiveveosi mias kinotitas. "Kerata kathosioseos" kai agogi limaton sto Akrotiri Thiras. In: C. Doumas (ed.), *Akrotiri Thiras: Trianta Xronia Ereunas (1067 – 1997). Epistimoniki sinantisi 19–20 Dekemvriou 1997*, 387–422. Athens: En Athinais Archeologiki Eteria.

Athanasiou, K. 2011. Sigritiki Meleti tou Poleodomikou Istou tis Proistorikis Polis tou Akrotiriou Thiras, me ti Simvoli Sighronon Tehnologion. Ph.d. dissertation, Aristotle University of Thessaloniki.

Barber, R. 1994. *Oi Kyklades tin Epohi tou Chalkou*. Athens: Emporiki Trapeza tis Elladas.

Bernstein, B. 1975. *Class, Codes and Control*, Vol. 3. 2nd ed. London: Routledge.

Boulotis, C. 2005. Ptihes thriskeftikis ekfrasis sto Akrotiri. *Als, Periodical Publication of the Society for the Promotion of Studies on Prehistoric Thera* 3: 20–75.

Cowan, R. 2005. *The Dictionary of Urbanism*. Norfolk: Streetwise Press.

Crang, M. 2009. Spaces in theory, spaces in history and spatial historiographies. In: B. Kümin (ed.), *Political Space in Pre-industrial Europe*, 249–265. Farnham: Ashgate.

Doumas, C. 1983. *Thera, Pompeii of the Ancient Aegean: Excavations at Akrotiri 1967–79*. London: Thames and Hudson.

Doumas, C. 1987. I Xeste 3 kai oi kyanokefaloi stin texni tis Theras. In: L. Kastrinaki, G. Orfanidou and N. Yannadakis (ed.), *Eilapini: Tomos Timitikos Gia ton Kathigiti Nikalao Platona*, 151–159. Herakleion: Dimos Herakleiou.

Doumas, C. 1992. *The Wall Paintings of Thera*. Athens: Thera Foundation.

Doumas, C. 1995. Anaskafi Akrotiriou Thiras. *Praktika Arhaiologikis Eterias 1992*, 147: 176–188.

Doumas, C. 2008. Chambers of mystery. In: N. Brodie, J. Doole, G. Gavalas and C. Renfrew (eds.), *Horizon: A Colloquium on the Prehistory of the Cyclades*, 165–175. Cambridge, UK: McDonald Institute for Archaeological Research.

Driessen, J. 2010a. Beyond collective: the Minoan palace in action. Uncorrected draft for the volume on the January 2010 Sheffield workshop in honor of K. Branigan, edited by M. Relaki and Y. Papadatos. http://uclouvain.academia.edu/JanDriessen/Papers/441482/Beyond_the_collective._The_Minoan_Palace_in_Action

Driessen, J. 2010b. Chercher la femme: identifying Minoan gender relations in the built environment. Draft for the volume edited by U. Maschek Gunkel and D. Panagiotopoulos, *Minoan Realities: Proceedings of a Workshop held in Heidelberg, November 2009*. http://uclouvain.academia.edu/JanDriessen/Papers/441481/Chercher_la_femme._Minoan_gendered_architecture

Driessen, J. 2010c. Spirit of place: Minoan houses as major actors. In: D. J. Pullen (ed.), *Political Economies of the Aegean Bronze Age: Papers from the Longford Conference, Florida State University, Tallahassee, February 22–24, 2007, 35–65.* Oxford and Oakville: Oxbow Books.

Driessen, J. in press. A matrilocal house society in Pre- and Protopalatial Crete? In: I. Schoep, P. Tomkins and J. Driessen (eds.), *Back to the Beginning: Reassessing Social and Political Complexity on Crete during Early and Middle Bronze Age.* Oxford: Oxbow.

Gesell, G. C. 1985. Town, palace and house cult in Minoan Crete. *SIMA 57*. Göteborg: Paul Åströms Förlag.

Halliday, M. A. K. 1986. Interview—M. A. K. Halliday, May 1986, by G. Kress, R. Hasan and J. R. Martin. http://sfs.scnu.edu.cn/halliday/show.aspx?id = 67&cid = 101 (read on July 1, 2012).

Hamilakis, Y. 2002. Too many chiefs? Factional competition in Neopalatial Crete. In: J. Driessen, I. Schoep and R. Laffineur (eds.), *Aegaeum* 23, *Monuments of Minos, Rethinking the Minoan Palaces,* 179–200. Liége: Université de Liége et UT-PASP.

Kress, G. and van Leeuwen, T. 1996. *Reading Images: The Grammar of Visual Design.* 2nd ed. London and New York: Routledge.

Lefebvre, H. 1974. *The Production of Space.* Oxford, UK and Cambridge, MA: Blackwell.

Marinatos, N. 1984. *Art and Religion in Thera: Reconstructing a Bronze Age Society.* Athens: I. Mathioulakis & Co.

Marinatos, S. 1999a. *Excavations at Thera I–III.* 2nd ed. Athens: The Archaeological Society in Athens.

Marinatos, S. 1999b. *Excavations at Thera V–VI.* 2nd ed. Athens: The Archaeological Society in Athens.

Palyvou, C. 1986. Notes on the town plan of Late Cycladic Akrotiri, Thera. *The Annual of British School of Archeology* 81: 179–195.

Palyvou, C. 2005. *Akrotiri Thera: An Architecture of Affluence 3,500 Years Old.* Philadelphia: INSTAP Academic Press.

Pellegrino, P. 2006. *To Noima tou Chorou.* Athens: Typothito.

Schoep, I. 2002. The State of the Minoan Palaces or the Minoan Palace-State? In: J. Driessen, I. Schoep and R. Laffineur (eds.), *Aegaeum* 23, *Monuments of Minos, Rethinking the Minoan Palaces,* 15-33. Liége: Université de Liége et UT-PASP.

Sotirakopoulou, P. 2008. Akrotiri, Thera: the Late Neolithic and Early Bronze Age phases in the light of recent excavations at the site. In: N. Brodie, J. Doole, G. Gavalas, and C. Renfrew (eds.), *Horizon: A Colloquium on the Prehistory of the Cyclades,* 121–135. Cambridge, UK: McDonald Institute for Archaeological Research.

Stenglin, M. K. 2004. Packaging curiosities: towards a grammar of three-dimensional space. Ph.D. dissertation, University of Sydney.

Stenglin, M. K. 2009. Space odyssey: towards a social semiotic model of three-dimensional space. *Visual Communication* 8(1): 35–64.

15 Discussion: Space–Time and the Mediterranean in Contexts

Historical, Interdisciplinary, and Interpretive

Stephanie Koerner

My broad aim with this chapter is to use several connotations of 'space and time' in the humanities as windows into challenges and prospects for utilizing the remarkable archaeological evidence of ancient Mediterranean cultures to create "integrative conceptual and analytic frameworks" (Hadji and Souvatzi, in this volume, henceforth HS), for studying conjunctures in long-term and large-scale processes from perspectives offered by local contexts. Emphasis falls on interdisciplinary, historical, and interpretive connotations of space and time, which contribute to the variety among the chapters in the volume and relate to prospects for integrating contextual and comparative perspectives. I will conclude with some suggestions about areas of overlap between recent innovations in classical studies and Mediterranean prehistoric archaeology.

CONTEXTUALIZING CONNOTATIONS OF SPACE AND TIME

According to Athena Hadji and Stella Souvatzi, many obstacles face using "new information coming from prehistoric sites across the Mediterranean" to improve approaches to space and time in archaeology, including

- notions that see "space" as a self-evident "backdrop for social interaction" and "time" in terms of "chronological sequences" in a "unilinear scheme";
- obstacles that the precedence given to "the long term and large spatial scale" pose for appreciating that "real life" occurs on diverse "space and time scales, including short-term small scale social agency"; and
- gaps between areas of specialization and opposed theoretical traditions.

These concerns touch on a wide range of issues. For many archaeologists working on Mediterranean prehistory since the 1960s, it has been difficult to imagine a time without polemical debates over sharply opposed approaches. Of course, things were not previously static. However, the range of issues debated, and the depth of change in the field's internal and external contexts, compare with circumstances that led to the parting of the ways of Enlightenment and Romantic perspectives on the notion of a so-called science of man; the histories of art and science; and the question of whether modernity should be interpreted as a triumph or a tragedy. In *Before Civilization* (1973), Colin Renfrew noted that innovations in dating methods were transforming archaeology's relationships to history, classical studies, anthropology, and developments in science and technology (cf. Clarke 1973; Dyson 1981; Renfrew 1980; Snoddgrass 1985); calling for critical perspectives on traditional assumptions; and creating an "alarming void" between "a mass of well-dated artifacts, monuments and cultures" and a lack of "connecting interpretations of how these things came about, and of how culture change took place" (Renfrew 1973: 109). He said that this "new situation" forced archaeologists to "go beyond" hitherto predominant paradigms' assumptions and explore "cultures and peoples in their own right, seeing 'events' of European prehistory as the result of local processes" (ibid.).

In the 1990s, Ian Morris (1994: 3) argued that "the sense of 'crisis' in Classical Studies and archaeology" should not be interpreted as "a conflict between theoretical and traditional archaeology" or "between old and young", but as "part of the general collapse of intellectuals' attempts to define what the west is and should be". Revisiting Renfrew (1973) and Morris (1994) is useful for several reasons. First, it enables us to consider sharply diverging orientations over the last decades from perspectives offered by wider 'crises' over relationships between the history of the humanities and those of what Bruce Trigger (1984) called "nationalist, colonialist, and imperialist" ideologies. The second half of the twentieth century saw a series of 'crisis over representation' in tandem with the Cold War, postcolonial conflict, and world-wide critique of the exploitation of the so-called Third World. Emphasis fell on problems with representations of cultural diversity, which eclipsed or helped to legitimate social inequalities and injustice based on problematic notions of race, class, and gender. Attention was drawn to the ways in which hitherto 'taken for granted' schemes (or meta-narratives) for classifying, chronologically ordering, and explaining cultural diversity "denied the coevalness" of the west and so-called people without history (Fabian 1983; White 1987; Wolf 1982). Images of the Mediterranean had long played key roles in the history of so-called antiquarian practices, and they were incorporated into cultural evolutionist and diffusionist models of supposed space and time relationships between regions of the 'dawning of civilization' and those described as marginal or more 'primitive' (Bintliff 1984; Pinsky and Wylie 1989; Trigger 1989).

The history of such images and interpretations raises the question of whether the Mediterranean is "a space, a place, or a concept" (HS). It figured already in antiquity among the key motifs incorporated into cosmological portrayals of the scope of space and time itself (Funkenstein 1986; Gruen

2011). During the nineteenth and early twentieth centuries, generalizations about Mediterranean prehistory and Classical Greek and Roman cultures played major roles in assumptions that prehistoric archaeology's main task was to move answers to such questions as the following further back in time: "What was so distinctive about Europe that encouraged the development of modern capitalism there and nowhere else, and what were the origins of that distinctive developmental sequence" (Rowlands 1981: 147).

In fields focused on Mediterranean antiquity, much concern has been expressed about these assumptions' impacts on relationships between archaeology and the study of the languages, literatures, and cultures of Greek and Roman antiquity (e.g., Alcock and Osborne 2007; Blake and Knapp 2005; Shanks 1996). According to Hayden White (1987: 55), many problems with traditional notions of 'historical representation' cannot be reduced to such technical issues as that "the term history refers both to an object, and to a representation of that object". Deeper problems are due to notions that divide humanity into two parts: one said be 'historical' because it has writing documents; the other supposedly 'unhistorical' because it does not (cf. Rowlands and Gledhill 1977). These problems are not about whether oral communication might be more transient (e.g., Ong 1982). They have been created by highly problematic assumptions about the 'modes of thought' and 'collective representations' of so-called 'primitive' cultures as one of the features these share is denial (and/or fear) of history, contingency, and even time (e.g., Eliade 1971 [1959]).

Put another way, it is one thing to worry about sufficient evidence (for instance, for incorporating multiple scales of analysis) (e.g., Bailey 2007; HS and others in this volume). It is a quite different issue to treat the absence of written documents as evidence for "people without history" (Wolf 1982); and equate cultural differences with the supposedly temporal distances represented by step-ladder models of cultural evolution and diffusionist arrows (Trigger 1989). Such schemes deny the coevalness of diverse historical trajectories (Fabian 1983; Latour 1993) in ways that compare with Hegel's (1928[1837]: 97–98, cf. 1956 [1824) belief that the reason why the word 'history' refers to both the *res gestae* and the *historia rerum gestarum* is that everything of historical or philosophical significance has been documented in writing.

Revisiting Renfrew and Morris is also useful for relating a number of apparently irresolvable disagreements over methodological and theoretical issues mentioned in this volume to wider divisions between the most influentially opposed responses to 'crises over representation' in the physical sciences during the late 19th and early 20th centuries and in the humanities more recently (as mentioned above). According to David Clarke (1973), discoveries being made with the use of contemporary scientific technologies revealed anomalies in hitherto predominant archaeological methodologies archaeology experienced a crisis of *naiveté* ('loss of innocence') that called for "new paradigms" (cf., Kuhn 1970 [1962]; Renfrew 1980; Binford and Sabloff 1982). Many obstacles are posed for addressing even such seemingly technical difficulties as "regional compartmentalization" (HS) by communication gaps rooted in disagreements over 'processual' and 'post-processual'responses to such arguments in archaeology (e.g., Dyson 1981; Patrik 1999 [1985]; Snoddgrass 1985; Barrett 1999 [1988], 1994, 2000; Preucel 1991; Renfrew and Zubrow 1994; Johnson 1999; Murray 1999; Dobres and Robb 2000; Knapp and Blake 2005; and HS; Meegan; Stravrides and others in this volume). Disagreements over the following questions have played key roles.

- Can or should archaeology be modelled on physical sciences?
- Is cultural diversity and change mental or material, internal or external, isolation or interconnectedness, infrastructure or superstructure top down or bottom up, and so on?
- Should modernity be interpreted as a triumph or tragedy?

Quite a number of decades ago, C.P. Snow (1962 [1959]) argued that problems with relationships between "scientists" and "non-scientists" were turning science and the arts into mutually uncomprehending "two cultures". Despite many problems with Snow's argument, the notion of local versions of 'two cultures' is useful for characterising a number of areas of polemic, which shaped the wider contexts of the above mentioned disagreements (e.g., Galison and Stump 1994; Koerner 2007). Michael Friedman characterised the highly relevant example of the polemic at the heart of the parting of the ways of 'analytic' and 'continental' philosophical traditions in such terms While the latter characterised the analytic tradition as disregarding themes as "the meaning of life, the nature of humanity, the character of a good society . . . in favor of an obsession with specific technical problems;" the analytically inclined objected that their opponents ignored "clarity of method and cooperative cumulative progress in favor of deliberate and almost willful obscurity (Friedman 2000:1). One of the most remarkable aspects of numerous analogous cases – a theme that we return to in this chapter's conclusions - is the extent to which the tenacity of seemingly irreconcilable polemic hinges upon shared presuppositions, for instance about nature and culture, and space and time; and "notions that see science as timeless Truth" and art as either "mere individual statement" (Jones and Galison 1998: 21) or esoteric expression of artistic 'Genius' (Most 1986).

Third, revisiting Renfrew and Morris throws light on how such debates have impacted approaches to space and time. Few have devoted more attention to this theme than Geoffrey Bailey (1983; 1987; 2007). Although Bailey focuses mainly on time, his general insights apply to space and time. Bailey (2007: 198) is particularly concerned about divisions between those who stress the importance of time depth to archaeology's distinctive methodological advantages, and those who focus on "the active role of artifacts and material structures in human action and interaction", as well as associated disagreements over

- what sorts of knowledge archaeologists should produce (testable hypotheses and universalizable explanations or interpretations of the unique qualities of particular events);
- the merits of focusing on long- or short-term processes; and
- whether time should be seen as a "natural process" or "cultural representation" (Bailey 1983: 166–170).

Bailey's concerns compare with those expressed by this volume's editors about obstacles that studies working at "cross purposes" pose for integrating "multiple scales" (HS). Bailey believes that appreciating the complementarity of contrasting views reveals connections between conditions of possibility for archaeological research; and for human experiences extending "beyond individual lives and perceptions" (Bailey 2007: 198). For Bailey (2007: 203) and Laurent Olivier (1999; 2001), envisaging multiple scales as a "palimpsest" facilitates seeing different perspectives as complementary connotations of "time perspectivism". This idea compares with several areas of contrast and overlap among chapters in this volume (e.g., HS; Chapman; Stavrides; Harkness; Meegan; Watkins; Düring; Skeates; Miller Bonney; Marketou; Murrieta-Flores; Márquez-Romero and Jiménez-Jáimez; Athanasiou; Yasur-Landau and Cline). Bailey foregrounds the following connotations:

- Different phenomena are best studied at different time scales because they "operate over different time spans and at different temporal resolutions".
- Whereas "different time spans and resolutions" can impose restrictions, using these together with others can enhance "our perception and understanding of the world".
- "Our observation of time is conditioned by particular cognitive and symbolic representations that are specific to particular cultures, states of brain evolution, forms of social organization, or world views. We could call this the subjective definition of time perspectivism, an exploration of the ways in which different people, both past and present (including archaeologists), have thought about the time dimension and their place within it" (Bailey 2007: 201–202).

These connotations compare with the interdisciplinary, historical, and interpretive foci of this chapter and indicate why a recurrent theme in this and several chapters in the volume is that these connotations imply interrelated prospects.

INTERDISCIPLINARY CHALLENGES AND PROSPECTS

Several concerns expressed in this volume compare with Julian Thomas (*Time and Identity: An Interpretive Archaeology,* 1996) and Gavin Lucas (*The Archaeology of Time,* 2005) (e.g., HS; Meegan; Watkins; and Yasur-Landau and Cline in this volume). Chronologies "relative or absolute" that present "time as a linear phenomenon" and "single vision" (Lucas 2005: 14) result in depicting history as a series of stages and time as a fixed category, external to societies and social relationships (Thomas 1996: 34–35). This results in ignoring the historical nature of societies and reducing space and time to structures imposed from "outside by the historian" (ibid.). These problems relate to interests expressed in this volume in the dynamics of "social agency" and the question: "If agency is important in understanding society in the short run", can or should it be included in research on processes that take place over diverse time spans (Dobres and Robb 2000: 11–12)?

For Thomas (1996: 35), examples of these problems can be found in the most influential works of the Annales School of historiography, Fernand Braudel's two-volume study, *The Mediterranean and the Mediterranean World in theAge of Philip II* (1966 [1949]); and in works that employ its tripartite scheme in ways that obscure "interplay of structure and agency" and the "dialectical movements between environment and economy". Braudel's *Mediterranean* has played major roles in promoting "structural history" and interdisciplinary approaches to multiscalar phenomena (e.g., Bintliff 1991; Clarke 1985; see also Chapman; and Márquez-Romero and Jiménez-Jáimez in this volume). The work's strengths make consideration of its limitations useful for appreciating the relevance of such insights in this volume as Chapman's that "whereas we cannot study infrastructure without superstructure, equally we cannot do the reverse". Braudel's *Mediterranean* explores a variety of ecological, social, and cultural processes in this very large area during the fifteenth through eighteenth centuries in terms of three time scales:

- the short-term time span of events or microhistory—a scale Braudel was hesitant about relating to his objections to 'meta-narratives' about famous rulers and great battles;
- an intermediate rate of change, which Braudel called the scale of "conjunctures" and envisaged in terms of models derived from social and political science and economics; and
- that of the *longue durée,* which Braudel derived from geology, biology, and climatology, and anthropology's notions of "collective representations" or mentalities.

Braudel's work has been the focus of much critique and misunderstanding of its original motivations (Clarke 1985). For instance, Hexter (1972: 533, see also, 1971)said that it did not solve the problem he believed to have been Braudel's main concern: relating historical continuities to rapid change. Actually, Braudel was mainly interested in long-term processes, and he envisaged both nature and culture largely in such terms, with consequences for the dynamics of "structure and agency" (Thomas 1996). Whereas Braudel's perspective on nature came from models of long-term geological, climate, and biological processes, he envisaged culture largely in terms of a conception of the *histoire des mentalités* as long-term and large-scale. His *mentalités* compare with Durkheim's (1938 [1895]) "collective representations" or "custom bound behaviors" that occur automatically and repetitively. For Braudel (as for Durkheim), the persistence of such "collective psychology" (even over centuries) could be studied quantitatively. In Braudel (1966 [1949]: 18), a mentality is a 'structure'—an empirical entity that "can distort the effect of time"; "operate simultaneously as a support and an obstacle"; and become an "envelope" that human experience rarely escapes.

These aspects of Braudel's work relate to a paradox with models that treat human agency as a supposedly timeless category or that define it in terms of the idea of altogether context-independent individuals' capacities for rationality. Such notions paradoxically both ignore human agency's historical nature and treat it, at the same time, as the primary node through which history operates. This paradox is not restricted to either side of the nature–culture debates. Whereas models departing from the nature pole see human possibilities as bounded by nature, those departing from the culture pole see human experience as constrained by culture (and impede understanding how culture could have come into being at all—e.g., Ingold 2000; Ingold and Hallem 2007). For Tim Ingold (2000), a critical problem is that—instead of exploring the materially embedded, embodied, and historically contingent ways in which human being cocreate culture, tradition, and themselves—such models perpetuate cultural theories that attribute human behavior to designs that are passed from one generation to the next as the content of traditions. These problems relate to another paradox with static notions of space and time. This is the clash between widely shared assumptions about modernity; and the tenacity of Aristotelian images of the world and science. On the one hand, few notions are shared more widely than those that see rejecting Aristotelian (384–322 BC) cosmology and appreciating contingency as distinctively modern (e.g., Blumenberg 1983; Lévi-Strauss 1973). On the other hand, on both sides of the "real" versus "constructed" polemic, like "the return of the repressed, the supra- and sublunary spheres of Aristotelian cosmology crop up in a new guise, crystalline nature encircling mutable culture" (Daston 2000: 3; see also Latour 1993; Prigogine 1997; and cf. Aristotle 1984, *Metaphysics,* 1027a: 20–27).

Whereas a detailed exploration of this problem cannot be pursued here, attention needs to be paid to its consequences. These include the tenacity of static images of space and unilinear models of time (e.g., Lucas 2005; Murray 1999; Thomas 1996; also HS; Harkness; Márquez-Romero and Jiménez-Jáimez; Watkins; Athanasiou; and others in this volume) and treatments of change as an anomaly rather than as an intrinsic property of reality (Prigogine 1997). This relates to problems with envisaging societies as prior to temporal contingency (Thomas 1996: 34–35) and with notions that see space, time, society, and human agency (as well as such things as 'the Bronze Age') as fixed categories and explanatory constructs. As several chapters in this volume show, notions that see space, time, societies, and human agency as context-independent categories (functions of the unchanging essence of things) lead to disregarding historical contingency; prioritizing hierarchical classification; and treating numerous dyads (such as perceiving subjects versus extended things; culture versus nature, mental versus material, science versus art, space versus time, etc.) as if they were somehow synonymous (Rorty 1979). Examples include approaches that treat history as a perceptual experience, which exists in the minds of individual subjects (as cognitive 'content' of 'mental states'); and approaches that treat history as an 'extended thing' that occurs in different forms (such as typology, band, tribe, chiefdom, state), or cultures of periods and places (such as the categories Neolithic Britain, Bronze Age Denmark, medieval France, Renaissance Italy, Modern Europe, Classical Antiquity, prehistoric Mediterranean, etc.).

These difficulties bring into sharp relief the relevance for fresh approaches to space and time of replacing supposedly context-independent notions of interdisciplinarity with approaches that stress the value of interdisciplinary contextual and comparative perspectives. There is huge diversity among the approaches and case studies represented in this volume. This is surely in part due to wonderful materials coming from diverse parts of the "fragmented yet connecting geography" of the Mediterranean and its becoming a "locus for the exchange of cultural traits" between three continents (HS; Stavrides; Watkins; Miller Bonney; Düring; and Marketou in this volume). But there are also remarkable overlaps. Likewise here materials play important roles—for instance, their being products of "networks of interaction on many different scales of space and time" (e.g., HS; Harkness; Meegan; Marketou; Murrieta-Flores; Márquez-Romero and Jiménez-Jáimez; Athanasiou; and Yasur-Landau and Cline), including interaction of physical and organic processes (e.g., Chapman; Skeates; and others in this volume). Yet some of the most interesting contrasts and overlaps (such as those noted at the end of the editors' introduction) might be understood as relating to authors' choices in combining different connotations of space and time "perspectivism" (Bailey 2007). The novelty of these patterns is difficult to overstate for several reasons. One is the light it throws on contributions that perspectivist (Bailey 2007; or 'relational') conceptions of space and time can make to linking "superstructures to infrastructures" and "top-down to bottom-up" approaches (e.g., Chapman in this volume) from perspectives offered by local contexts. Second, these patterns reveal the importance to new "integrative conceptual and analytic frameworks" (HS) of replacing notions that see contextualization and comparison as mutually exclusive options. Several chapters illustrate the relevance for developing alternatives of bringing different connotations of space-time together in ways that envisage interdisciplinarity (and interpretation) as recontextualization (cf. Bintliff and Gaffney 1986). Such a notion of interdisciplinarity may be extremely useful for developing jointly contextual and comparative approaches to studying conjunctures in long-term and large-scale processes from perspectives offered by local contexts (e.g., Miller Bonney; Düring; and Skeates in this volume).

One reason why Braudel's "layer cake" model eclipses the dynamics of "agency and structure" (Thomas 1996: 18) and "infrastructure and superstructure" (Chapman, in the present volume) is that it treats the disciplinary sources of its time scale models as distinct and largely timeless aspects of reality. By contrast, the alternatives outlined above create new prospects for combining insights that "real life" takes place on diverse "space and time scales, including short-term small-scale social agency" (HS) with comparative perspectives on possibly analogous situations or instantiations (e.g., Baxandall 1972; 1985; Boas 1887; Geertz 1973; 1983; Most 1994).

The expression 'social agency' (HS) brings fresh light to relationships between issues of "infrastructure and superstructures", "bottom-up to top-down" (e.g., Chapman in this volume), and that of intentionality (cf. Renfrew and Zubrow 1994). The idea that intentionality (together with thought and action, free will, morality, ethics, and many other features of human agency and societies) is an interdisciplinary issue involving conceptions of space and time is of considerable antiquity. Aristotle, for example, developed a variety of perspectives on temporality, including his distinction between the 'time of the world' (and especially, that of the eternal essence of timeless categories of things, on which the requirements of a 'science' or 'episteme' hinged) and the 'time of the soul' (as a phenomenological concept). Aristotle envisaged human experience of time as composed of past, present, and future in ways that cannot be identified in quantitative terms (Ricoeur 1988). As with human experiences of time, they are phenomena, conceptual categories, objectified in images, practices, and forms of material culture that impact "personal experience", "the fortunes of collectivities" (Thomas 1996: 33), and memory (all themes that reoccur throughout this volume). This means that linking these connotations of time (or time perspectives) is an interdisciplinary task (which meant, for Aristotle, Kant, Husserl, and many others, a philosophical one).

The concept of "social agency" has major advantages for addressing problems with images of human agency, which are based on the dichotomy between the supposedly timeless, placeless, disembodied capacity of an individual subject for rationality versus the rest of the world (e.g., Knorr-Cetina and Cicourel 1981; Pompa 1990; Schatzki, Knorr-Cetina and Von Savigny 2001). For Hannah Arendt (1989 [1958]; 1977 [1961]) such images have resulted in one of contemporary culture's most problematic features: a "privatized ethics" that sees an atomistic individual subject as the source of all meaning and value; reduces social life to contractual structures; and situates ethics on the margins of the arts and science. 'Social agency' is useful for creating alternatives without running risks of speculating about individual 'mental states' and 'collective representations'; and of abandoning intentionality altogether. The idea enables us to focus on the roles played in the dynamics of infra- and superstructures (or internal and external contexts) of the embeddedness of agency in material, embodied, and social circumstances. But we are not forced to abandon questions about how humans can have discrepant experiences, come to share understandings, and interact (Arendt 1989 [1958]). All of these questions relate to the issues at stake with such problems as the dichotomy of people supposedly 'with and without history' and others mentioned here at the onset (Fabian 1983; White 1987; Wolf 1982) and unilinear chronologies (HS).

Archaeologists have developed conceptions of social agency that show the relevance for addressing these issues of linking contextual and comparative perspectives around relational (interdisciplinary) approaches to space and time as a palimpsest (e.g., Barrett 1994; Gosden 1994; Thomas 2004; and Skeates; Murrieta-Flores; Yasur-Landau and Cline, among others in this volume). Revisiting Edmund Husserl's (1990 [1917]; 1970 [1936]) phenomenological approach to ethics in the dynamics of human life-worlds helps illustrate this point. Husserl's approach built on insights derived from Immanuel Kant's work on human agency, ethics, and the phenomenological (contextual, historical) nature of space and time (e.g., Kant 1992 [1770]; 1998 [1781], 1997 [1788], 1955 [1790], 1970 [1795]). However, for Husserl (1990 [1917]) (as for many others since), the metaphysical (*a priori*) presuppositions that underwrote Kant's approach to these themes are quite problematic. Husserl sought to replace notions that see human beings as atomistic, interchangeable nodes through which the dynamics of nature and culture operate, with a conception of a 'life-world' that calls for interdisciplinary orientations. Life-worlds are prisms (or palimpsests) of diverse fields of experience. These have physical, organic, embodied, social, and ethical dimensions that humans experience as sources of meanings, values, and their own identities. As with Bailey's (2007) and Olivier's (1999) notion of a "palimpsest", these fields cannot exist without one another. However, social–ethical–intentional fields are prior to the others in terms of contributions to what Thomas (1996) might call their "historical natures".

This relates to this volume's editors' emphasis on the contextual nature of social agency, and questions about including agency in studies of longer and larger scale historical processes of change as well as continuities generated by material culture, communication systems, and memory (e.g., Dobres and Robb 2000; and Düring; Skeates; and Marketou in this volume). In Husserl, local events and human expressions can and have played crucial roles on these scales. Thus, if local expressions of social agency create social–ethical–intentional fields, they can impact processes on the very scales on which the historical natures of meanings, values, and societies operate.

HISTORICAL CHALLENGES AND PROSPECTS

These themes relate to questions about the deeply historical nature of the roles that the Mediterranean has had in the long-term history of symbolic interpretations of space and time, which form the roots of modern cosmologies. Recent decades have seen much critical attention focused on problems with the roles played in the humanities by such key timbers as the dichotomies of nature–culture, statics–dynamics, mind–body, individuals–artifacts, moderns–premoderns, science–art, and so on. The concept of "historical epistemology" (Gosden 1999; Nickels 1989; Rheinberger 1997, 2010; Toulmin 1990) has become widely used to describe jointly contextual and comparative approaches to such categories. According to the biologist, philosopher, and historian of science Hans Jorg Rheinberger (2010), the task of "historical epistemology" is to explore hitherto taken-for-granted categories in relation to their roles in historically contingent 'cultural systems'. It examines the specific conditions *under which* and the cultural means *with which* things and experiences are made into the objects, which enable people to understand, communicate about, and intervene in the world (Rheinberger 2010). Christopher Gosden (1999) argued for such approaches to relationships between anthropology and archaeology in order to replace notions that see these fields (and their categories) as timeless universals by approaches that stress their historically contingent natures as 'cultural systems' (cf. Geertz 1973;1983).

Anthropology and archaeology have made crucial contributions to historical epistemology, especially through studies of the roles symbolic interpretations of space and time have played in the construction of the so-called Other (e.g., Fabian 1983; Spriggs 1977; Wolf 1982). The question Hadji and Souvatzi raise about the Mediterranean ("Is it a space, a place, or a concept?") foregrounds the usefulness of historical epistemology for exploring its roles as a cosmological image of space and time (e.g., Davis and Kennedy 2009; Lovejoy and Boas 1965; Morris 1994; Schnapp 1993). The term "image" (or "picture", such as a map) bears stressing. White (1978: 110) notes that we rarely understand what we are being told of different times and places without images. The importance of maps to studying the historical contingency of cosmological representations of space and time has long been recognized (e.g., Harley and Woodward 1987). Writing on case of 'spaces', John Pickles notes:

> Mapping technologies and practices have been crucial to the emergence of modern 'views of the world', Enlightenment sensibilities and contemporary modernities. And, indeed, this is perhaps the crucial issue: maps provide the very conditions of possibility for the worlds we inhabit and the subjects we become. (Pickles 2004: 4)

Pickles's argument raises questions about symbolic roles that images of the Mediterranean have played together with other key timbers of what

Toulmin calls the "modern cosmopolis" (cosmos + polis) in the histories of the various dichotomies mentioned above that are being debated today in archaeology and the humanities. Few scholars' work is likely to be more useful than Toulmin's for addressing such questions. Toulmin studied mathematics, physics, and philosophy at Cambridge, where he became interested in the work of Ludwig Wittgenstein and in historical approaches to conceptual problems. Starting with his *Introduction to the Philosophy of Science* (1953), Toulmin made vital contributions to studies of the historical contingency of such seemingly self-evident categories as those of space and time; the mental and the material; nature and culture; and others shared by the most influentially opposed responses to 'crises over representation' (Toulmin 1961; 1990; 2001; Toulmin and Goodfield 1961; 1962). For Toulmin, the locally situated nature of disciplinary ideas and ideals meant that in order "to understand particular historical developments, there is no alternative but to pursue detailed investigations" (Rheinberger 2010: 60). Here is space only to mention the elements of Toulmin's book *Cosmopolis: The Hidden Agenda of Modernity* (1990) that are most relevant for our immediate purposes, namely, its contextual and comparative approaches to cosmological images of space and time; the importance of the 'myth of the clean slate' to modern cosmology; and the history of situations that have eclipsed alternative possibilities.

The topics that concern Toulmin's *Cosmopolis* (1990) are neither unique nor part of a unilinear trajectory. They are instantiations (possibly analogous situations) that demand linking contextual and comparative approaches. Toulmin used examples from diverse fields (archaeology, anthropology, classical studies, art history, theology, etc.) throughout his works, and in *Cosmopolis,* to suggest that perhaps from the very beginnings of large-scale societies there have been images of linkages between *cosmos* and *polis,* in which cultural interpretations of space and time have played central roles (1990: 67). Emphasis falls, however, on discontinuities in the long-term history of key timbers of modern cosmology. Thus, he stressed that it may not have been until "Alexander the Great broadened the Greek horizon beyond its former preoccupation with single cities" that we find philosophers attempting to "fuse the natural and social orders into a single unit—the structure of Nature reinforces a rational Social Order" (Toulmin 1990: 68; e.g., Plato [424-347 BCE], *Republic*, 1930, 1935) Moreover, it may have been only under circumstances of crises that

- claims were made about the state of emergency of contemporary affairs (and even the human condition, in general) making it necessary to demolish everything that went before and quest for altogether context-independent certainty from a clean slate;
- utopic and dystopic interpretations of space, time, and the structures, forces, and composition of nature and society diverged around such dichotomies as the sublunary versus celestial spheres; realms of *is* versus

ought; and the oral versus the written (e.g., Eusebius of Caesaria's [c.260–339] *Historia ecclesiastica;* Augustine of Hippo's [c.354–430] *De civitate dei* [*The City of God*]) (Funkenstein 1986); and
- the Mediterranean came to figure among the key motifs in pictorial depictions of these dichotomies.

Throughout the Middle Ages, the Mediterranean was at the center of the world history maps (*mappi mundi*) that portrayed how the most important people, places, and things were believed to be distributed across sacred and profane space and time. In early Renaissance times, critiques of these images and interest in ancient Mediterranean material culture contributed to the 'revolution in historical reasoning' and other roots of the 'Scientific Revolution' (Collingwood 1949; Funkenstein 1986). The story did not end there. Interpretations of Classical antiquity played central roles in the parting of the ways of Enlightenment and Romantic interpretations of modernity as either a triumph or a tragedy, and their impacts on the histories of archaeology and the humanities in general. Toulmin's approach to these developments is novel and relevant. It shifts attention away from generalized notions of 'crises over representations' toward contextual and comparative perspectives on the circumstances of supposed settlements. Although these vary in many details, the situations that have figured most centrally in the history of the 'two cultures' of the sciences and the humanities are comparable in at least one crucial respect. This is the decisive role played by claims about the necessity of breaking away from everything before and the 'quest of certainty' (to borrow Dewey's [1929] expression) from 'scratch' (or a *tubula rasa,* Locke 1975 [1689]).

Ilya Prigogine drew attention to the usefulness of Toulmin's insights for challenging static notions of space and linear time schemes. A number of anthropologists have stressed the relevance of Prigogine's work on "chaos theory" and nonlinear dynamics to such tasks (Allen 1989; Lucas 2005). It bears mention that Prigogine (1997) experienced objections to his work throughout his career, which evidenced the clashes we noted earlier between assumptions about the importance to science and modernity of awareness of contingency, and the extent to which notions that see science as concerned with timeless truth (and/or reversible or changeless change) hinge on Aristotelian cosmology (cf. Daston 2000: 3). For Prigogine, such problems related to widespread disregard of

- the fact that "real life" occurs in local contexts (cf. HS);
- the importance of human experiences of contingency to language, culture, and history; and
- the relevance of the indeterminacy of nature for the roles of science in society (Nicholis and Prigogine 1989).

For Prigogine, Toulmin's work reveals remarkable parallels between the historical circumstances surrounding Descartes's quest of certainty and

Einstein's. In these cases and many others (e.g., Shapin and Schaffer 1985), the two core elements of the 'modern cosmopolis' (the myth of the clean slate and quest of certainty) came to figure among authoritative solutions to conflict and uncertainty (Prigogine 1997: 185). The notion of 'quest of certainty' has never been dreamed up in a social vacuum. It reoccurred as a response to 'crises over representation' (in both the epistemic and the ethical sense of the term), for instance, in the midst of the Thirty Years' War (1618–1648), the twentieth century's World Wars, and Cold War and postcolonial conflicts. Moreover, research over the last decade on such situations suggests that the history of such supposed settlements of 'crises over representation' may also be the history of the categories that perpetuate the 'two cultures' polemic, as well as of interpretations of cultural differences as supposed distances in time envisaged as 'progress' (Fabian 1983; Latour 1993; Latour and Weibel 2002; 2005; White 1987; Wolf 1982).

Toulmin's aims were never restricted to critique. The expression 'hidden agenda' in his book's title refers to innovations in the arts and humanities that have figured as importantly among the roots of modernity as the so-called Scientific Revolutions. They form likewise the roots of such novel innovations as Bailey's (1983; 2007) model of connotations of "time perspectivism", Ingold's (2000) concept of a 'taskscape', Thomas's concept of 'material culture', Geertz's concept of 'thick description', and others represented in this volume.

INTERPRETIVE CHALLENGES AND PROSPECTS

> Going beyond dualism opens up an entirely different landscape, one in which states and substances are replaced by processes and relations; the main question is not any more how to objectify closed systems, but how to account for the diversity of the processes of objectification. (Descola and Pálssen 1996:12)

The idea that a relational conception of space and time has interdisciplinary, historical, and interpretive connotations emerged in Immanuel Kant's later works (the *Critique of Pure Reason* 1998 [1781], the *Critique of Practical Reason* 1997 [1788], the *Critique of the Faculty of Judgement* 1955 [1790], and associated essays. Starting in the 1770s, Kant explored the intuitional, cognitive, and phenomenological nature of space and time; and arguments that (as is the case with categories, in general) space and time are not reducible to things in themselves (Hatfield 2006). They share properties with things but are human interpretations of our experiences of them as phenomena (Kant 1992 [1770]: 403). Numerous questions relating to these ideas have been debated ever since: Are space and time 'real' or cognitive 'ideals'? Are they substances, or properties of substances and/or forces? Do they depend on relationships between objects, or are they independent?

What is the relationship between space and time ? How do they relate to the human mind? And finally, how might answers to these questions intersect with one another?

In Kant, these questions have direct bearing on the following ones: (1) What can I know? (2) What should I do? (3) What dare I hope? (4) What is it to be human? (Kant *Schriften* 1902, vol. 9:25, my translation). Throughout the works noted above and related publications, Kant stressed the importance of the phenomenological nature of space and time for linking answers to these questions to one another. For Kant, one of the remarkable things about humans is that they do not experience the world as disorganized sensory input but find it intelligible. We experience the world in ways that are not identical with either the things themselves (which exist independent of our experiences) or with the various ways in which we interpret them (in concepts, practices, symbolic representations, etc.). Our experiences of the world in and through both are crucial for our capacities as humans capable of consciousness, intentional behavior, and the cultural practices that constitute societies historically. There would be no basis for *sensus communis* (common sense), no conditions for the possibility of ethics and moral interaction. Expressed in Kant's terms, we would lack the freedom and creative capacities needed to treat one another not simply as means to personal ends, but as fellow human beings.

For Kant, such orientations were relevant for addressing some of the most pressing ethical challenges facing his time. He believed that, together with new forms of media and communication, they could become the basis of critique on the part of publics around the world of the senseless "plunder" of remote corners of the earth by the supposedly "civilized" (Kant 1970 [1795]; Werbner 2008). Unfortunately, many of the advantages of Kant's insights became eclipsed by efforts to replace the 'metaphysical' foundations of these aspects of his work by what became the most influentially opposed paradigms of the 'linguistic' turn in the humanities, and the parting of the ways of the analytic and continental philosophical traditions mentioned here earlier.

According to Thomas (1996: 18), learning that the categories we have taken for granted are not eternal principles but products of contingent circumstances raises the question of "what is to become of material culture?" Such dichotomies as those of "culture/nature and mind/body" "reduce material culture to an essence" and give us the impression that we are forced to choose between "the realm of ideas or biological processes" (Thomas 1996: 19). For Thomas alternatives need to be grounded in appreciating the historical natures of cultures, societies, and human agencies, and the interpretive prospects this creates. Referring to Hans Georg Gadamer's (1975: 264) work, Thomas explains:

> Archaeology investigates the past through the medium of material things. . . . Yet it is increasingly clear that we do not simply reconstruct

> the way that things were. Instead we establish a relationship between the past and the present. . . . Considered in this way, the perceived distance between the past and the present is not so much a barrier to understanding as a productive space. (Thomas 2004: 1)

Thomas's perspective compares with Bailey's (2007) arguments for replacing notions that we are forced to choose between different connotations of "time perspectivism". Such arguments go against the grain of notions that relativism must be grounded in beliefs that different cultures, languages, and societies are mutually closed to one another. Geertz called these notions into question as follows:

> The truth of the doctrine of cultural relativism (or historical relativism—it is the same thing) is that we can never apprehend another people's or another period's imagination neatly, as though it were our own. The falsity of it is that we can therefore never genuinely apprehend it at all. We can apprehend it well enough, at least as well as we apprehend anything else not properly ours; but we do so not by looking behind the interfering glosses that connect us to it but through them. (Geertz 1983: 44)

The last decades have seen growing interest in the bearing such orientations have on pressing contemporary problems. According to the anthropologist Anne-Belinda Preis (1996: 286), because "in the years to come, some of the most crucial intellectual, moral and ideological battles about human rights issues are likely to turn on their cross-cultural intelligibility and justifiability, a radically new and far more dynamic approach to culture is needed" (see also Council of Europe 2005). For Hadji and Souvatzi, the ecological diversity and cultural plurality of the ancient Mediterranean is highly relevant for addressing such challenges. Erich Gruen's study, *Rethinking the Other in Antiquity* (2011), supports this argument. Gruen (2011: 3) notes that, according to some scholars, Mediterranean antiquity can be summarized by the expression "Worlds at War" and is responsible for "the 2,500-Year Struggle between East and West" (Pagden 2008). For Gruen, such views eclipse the diversity of ways in which ancient cultures envisaged their histories. Even (indeed, perhaps, precisely) the cultures that were most preoccupied with identity "had much more nuances and complex opinions of other people" (Gruen 2011: 3). Whereas it may be easy to "gather individual derogatory remarks (often out of context), it is a mistake to interpret them as "a blanket of xenophobia and ethnocentrism, let alone racism" (ibid.). There is abundant evidence that "ancient societies, while certainly acknowledging differences among people (indeed occasionally emphasizing them) could visualize themselves as part of a broader cultural heritage, could discover or invent links with other societies and could couch their own historical memories in terms of a borrowed or appropriate past" (Gruen 2011: 4).

SOME CONCLUDING SUGGESTIONS

In *Modern Critical Theory and Classical Literature* (Sullivan and De Jong 1994), J. P. Sullivan notes that the "discipline of Classics, as academically defined, comprehends the study of ancient Mediterranean culture", with attention to "Greece and Rome, and their interactions with contiguous areas". He stresses the impacts on the field's ideas and goals of insights drawn from critical theory and, especially, anthropologically oriented approaches to culture. Gruen is not an isolated case. He figures among a growing number of specialists in classical studies who are exploring themes that compare with the research foci in Mediterranean prehistoric archaeology that Hadji and Souvatzi see as related directly to issues of space and time, namely "identity, memory, mobility, connectivity, meaning, experience, architecture and representation, materiality, culture contact and colonialism, innovation and tradition, and transition and transformation" (HS).

One of the responses to "crises over representation" has been the tendency to avoid connections, which had been obscured by the assumptions that motivated cultural evolutionary and diffusionist models of relationships between developments in the ancient Near East, the Mediterranean, and Europe, which we touched on earlier. This tendency has had powerful impacts on relationships between research on connections between developments in these regions, which Chapman and others mention in this volume. It has had analogous impacts on relationships between classical studies and Mediterranean prehistoric archaeology. It would be a major mistake to underestimate the importance of avoiding risks of anachronism, and its consequences for a number of the problems we examined from perspectives offered by revisiting Renfrew (1973) and Morris (1994) at the onset. However, such risks can be avoided by distinguishing connotations of space-time perspectivism (Bailey 2007), integrating contextual and comparative approaches, and using other methodological tools represented in this volume. What bears mention is that there may be areas of overlap between innovations in classical studies and Mediterranean prehistoric archaeology, which relate directly to problems posed by the 'two cultures' polemic for approaches to space and time in archaeology. For instance, much attention has focused on problems with the notions of science as "timeless Truth" that both sides of the polemic share (Jones and Galison 1998: 21). On the other hand, we are only beginning to examine equally problematic notions of 'art' as mere individual statement (ibid.) or as esoteric expression of artistic 'Genius'. This makes it useful to conclude with reference to an example of the sorts of studies that can provide useful points of departure for examining the history of such notions of 'art', as well as for using innovations in classical studies in Mediterranean prehistoric archaeology in ways that facilitate contextual and comparative approaches to ancient Mediterranean cosmological interpretations of space and time.

Few scholars have made more valuable contributions to such prospects than Glenn Most. I will conclude this chapter with reference to his study of one of the most frequently cited passages in an "Ode" that the fifth-century BCE poet Pindar was commissioned by his patron to write to celebrate the victory of an athlete his patron supported. Most's (1986) study is remarkable in numerous respects. It challenges the most influential interpretations and translations of the passage and the roles these misinterpretations and mistranslations have played in the history of the generalizations about 'art' on which much of the 'two cultures' polemic has hinged. Most focuses particular attention on such problems with these mistranslations as

- attributing notions to Pindar that are not otherwise evidenced until in Plato's work in the fourth century BCE;
- portraying Pindar as isolated from both his immediate and wider cultural contexts, and attributing notions to him that none of his contemporaries shared; and
- depicting Pindar as having made claims about his work that that would not have made sense for a poet seeking the admiration and associated commission payments of patrons (notable examples include mistranslations, which portray Pindar as claiming that his poems are too esoteric and sophisticated for everyday people to understand without the help of expert critics).

For Most, a satisfactory translation demands examining connections between the expressions Pindar used and the ways in which these were used in his wider fifth-century cultural context. Most's findings concern expressions in Pindar that figure centrally among the most widely shared connotations of the term 'translation' in everyday usage and in works by important poets and playwrights. The range of connotations of 'translation' in Pindar's Ode and contemporary sources shows how important it was for Pindar to relate his Ode to patrons' and audiences' values and cosmological orientations, including those having to do with their cultural perspectives on space and time. Fifth-century BCE Greek expressions that referred to 'translation' had a wide variety of connotations. However, close inspection reveals that, once basic meanings are contextualised, the extant evidence can be sorted into a small numbers of groups. Interestingly (relating to considerations in this volume of 'social agency' and possible ancient Mediterranean conceptions of space, time, and agency), the expression refers to "an agent that performs any act of translation of signification from one kind of language in which it is invisible or entirely unintelligible into another kind in which it is visible and intelligible" (Most 1986: 308). The main groups of connotations can be summarized as follows:

- One is a 'translation' of communication of gods and other cosmological forces into human forms of communication.
- A related connotation is a secularized variant that stresses forces of nature rather than either humans or the gods. For example, Hippocratic

texts argued that what some interpreted as 'the sacred disease' (epilepsy) is not supernatural but caused by imbalanced fundamental organic substances, or humors.

- A third connotation refers to transferring meaning from silent thought to its expression, as in interpretations of spoken discourse as 'pronunciation'.
- Another connotation is translation of the language of one group (spoken, text, notation) into that of another, which makes the ideas in the former language accessible to speakers of the latter. This compares with today's conventional conceptions of translations as mediating from one to another language.
- In addition, there are connotations that treat a 'herald' or prediction as a form of translation. This is not surprising, in light of abundant ancient Greek and cross-cultural evidence that a herald or prediction involves transferring meanings from a domain in which these would not be intelligible to one in which they are intelligible (Most 1986: 308–310).

Few issues recur more often in this volume than that of connections between the roles that cultural understandings of space and time have played in ancient societies and continue to play in approaches to their study; and what Bailey (2007: 220) describes as the "genuine paradox" that "we cannot work out what tools we need until we know what sort of phenomena are there in the longer-term record to investigate, and we cannot investigate those different phenomena until we have some tools to do it with. And to solve that paradox we will need to work at both simultaneously". The work of Gruen and Most described above illustrates the importance to their work of such orientations. As with numerous chapters in this volume, it illustrates the usefulness for addressing this paradox of interdisciplinary frameworks for linking contextual and comparative approaches to one another from perspectives offered by local contexts.

ACKNOWLEDGMENTS

It has been a great pleasure and honor to have been invited by Athena Hadji and Stella Souvatzi to be involved in this wonderful project. I am likewise very indebted with this work to Glenn Most and Julian Thomas.

REFERENCES

Alcock, S. E. and R. Osbourne, eds. 2007. *Classical Archaeology.* Oxford: Blackwell.

Allen, P. 1989. Modelling innovation and change. In: S. E. van der Leeuw and R. Torrence (eds.), *What's New? A Closer Look at the Process of Innovation,* 258–280. London: Unwin Hyman.

Arendt, H. 1977 [1961]. *Between Past and Present*. New York: Penguin Books.
Arendt, H. 1989 [1958]. *The Human Condition*. Chicago: University of Chicago Press.
Aristotle (384–322 BC). 1984. *The Complete Works of Aristotle,* ed. J. Barnes. Princeton: Princeton University Press.
Bailey, G. N. 1983. Concepts of time in Quarternary history. *Annual Review of Anthropology* 12: 165–192.
Bailey, G. N. 1987. Breaking the time barrier. *Archaeological Review from Cambridge* 6: 5–20.
Bailey, G. 2007. Time perspectives, palimpsests and the archaeology of time. *Journal of Anthropological Archaeology* 26: 198–223.
Barrett, J. 1994. *Fragments from Antiquity: An Archaeology of Social Life*. Oxford: Basil Blackwell.
Barrett, J. 1999 [1988]. Fields of discourse: reconstituting a social archaeology. In: J. Thomas (ed.), *Interpretive Archaeology,* 23–32. Leicester: Leicester University Press.
Barrett, J. 2000. A thesis on agency. In: M. A. Dobres and J. Robb (eds.), *Agency in Archaeology,* 61–68. London: Routledge.
Baxandall, M. 1972. *Painting and Experience in Fifteenth Century Italy.* Oxford: Clarendon Press.
Baxandall, M. 1985. *Patterns of Intention: On the Historical Explanation of Pictures*. New Haven: Yale University Press.
Binford, L. P. and J. Sabloff. 1982. Paradigms, systematics, and archaeology. *Journal of Anthropological Research* 38(2): 137–153.
Bintliff, J. L., ed. 1984. *European Social Evolution: Archaeological Perspectives*. Bradford: University of Bradford.
Bintliff, J. L., ed. 1991. *The "Annales" School and Archaeology.* Cambridge: Cambridge University Press.
Bintliff, J. L. and Gaffney, C. F., eds. 1986. *Archaeology at the Interface: Studies in Archaeology's Relationships with History, Geography, Biology and Physical Science*. Oxford: British Archaeological Reports.
Blake, E. and A. B. Knapp. 2005. *The Archaeology of Mediterranean Prehistory.* Oxford: Blackwell.
Blumenberg, H. 1983. *The Legitimacy of the Modern Age,* translated by R. M. Wallace. Cambridge, MA: MIT Press.
Boas, F. 1887. The occurence of similar inventions in widely separate areas. *Science* 91: 485–486.
Braudel, F. 1966 [1949]. *The Mediterranean and the Mediterranean World in the Age of Philip II,* translated by S. Reynolds. New York: Harper and Row.
Clarke, D. 1973. Archaeology: the loss of innocence. *Antiquity* 47: 6–18
Clarke, S. 1985. The Annales historians. In: Q. Skinner (ed.), *The Rise of Grand Theory in the Human Sciences,* 177–198. Cambridge: Cambridge University Press.
Collingwood, R. G. 1949. *The Idea of History.* Oxford: Oxford University Press.
Council of Europe. 2005. The Value of Cultural Heritage in Contemporary Society. http://www.coe.int/t/dg4/cultureheritage/heritage/identities/default_en.asp (June 12, 2012).
Daston, L., ed. 2000. *Biographies of Scientific Objects*. Chicago: University Chicago Press.
Davis, G. and K. Kennedy. 2009. *Medieval and Renaissance Art: People and Possession*. London: Victoria and Albert Publishing.
Descola, P. and G. Pálssen, eds. 1996. *Nature and Society: Anthropological Perspectives*. London: Routledge.
Dewey, J. 1929. *The Quest of Certainty: A Study of the Relation of Knowledge and Action*. London: George Allen and Unwin Ltd.

Dobres, M.-A. and J. Robb, eds. 2000. *Agency in Archaeology.* London: Routledge.
Durkheim, E. 1938 [1895]. *The Rules of the Sociological Method,* translated by S. Solovay and J. Mueller. New York: Free Press.
Dyson, S. L. 1981. A classical archaeologist's response to the 'New Archaeology'. *Bulletin of American Schools of Oriental Research* 242: 713–720.
Eliade, M. 1971 [1959]. *Cosmos and History: The Myth of the Eternal Return.* Princeton: Princeton University Press.
Fabian, J. 1983. *Time and the Other: How Anthropology Creates Its Object.* New York: Columbia University Press.
Friedman, M. 2000. *The Parting of the Ways: Carnap, Cassirer and Heidegger.* Peru, IL: Open Court.
Funkenstein, A. 1986. *Theology and the Scientific Imagination from the Middle Ages to the Seventeenth Century.* Princeton: Princeton University Press.
Gadamer, H. G. 1975. *Truth and Method.* London: Sheed and Ward.
Galison, P. and D.J. Stump, eds. 1994. *The Disunity of Science. Boundaries, contexts, and power.* Stanford: Stanford University Press.
Geertz, C. 1973. *The Interpretation of Cultures.* New York: Basic Books.
Geertz, C. 1983. *Local Knowledge: Further Essays in Interpretive Anthropology.* New York: Basic Books.
Gosden, C. 1994. *Social Being and Time.* Oxford: Basil Blackwell.
Gosden, C. 1999. *Anthropology and Archaeology: A Changing Relationship.* London: Routledge.
Gruen, E. S. 2011. *Rethinking the Other in Antiquity.* Princeton: Princeton University Press.
Harley, J. B. and D. Woodward, eds. 1987. *Cartography in Prehistoric, Ancient, and Medieval Europe and the Mediterranean.* Chicago: University of Chicago Press.
Hatfield, G. 2006. Kant on the perception of space and time. In: P. Guyer (ed.), *Kant and Modern Philosophy,* 61–93. Cambridge: Cambridge University Press.Hegel, G. W. F. 1956 [1824]. *The Philosophy of History.* New York: Dover Publications. Hegel, G. W. F. 1928 [1837]. *Vorlesungen über die Philosophie der Geschichte,* edited by H. Glockner. Stuttgart: Fr. Frommanns,
Hexter, J. H. 1971. *Doing History.* Bloomington: Indiana University Press.
Hexter, J. H. 1972. Fernand Braudel and the *Monde Braudellien. Journal of Modern History* 44: 480-539.
Husserl, E. 1990 [1917]. *On the Phenomenology of the Consciousness of Internal Time (1893–1917),* translated by J. B. Brough. Dordrecht: Kluwer.
Husserl, E. 1970 [1936]. *The Crisis of European Science and Transcendental Phenomenology.* Evanston, IL: Northwestern University Press.
Ingold, T. 2000. *The Perception of the Environment: Essays in Livelihood, Dwelling and Skill.* London: Routledge.
Ingold, T. and E. Hallem, eds. 2007. *Creativity and Cultural Improvisation.* Oxford: Berg Publishers.
Johnson, M. 1999. *Archaeological Theory: An introduction.* Oxford: Blackwell Publishers.
Kant, I. 1992 [1770]. On the Form and Principles of the Sensible and the Intelligible World [Inaugural Dissertation]. In: *Theoretical Philosophy: 1755-1770,* Volume 2, translated by D. Walford and R. Meerbote. Cambridge: Cambridge University Press.
Kant, I. 1998 [1781]. *Critique of Pure Reason,* translated by P. Guyer and A.Wood. Cambridge: Cambridge University Press.
Kant, I. 1997 [1788]. *Critique of Practical Reason,* translated by M. Gregor. Cambridge: Cambridge University Press.
Kant, I. 1955 [1790]. *Critique of the Faculty of Judgement,* translated by J. H. Bernard. London: Henry G. Bohn.

Kant, I. 1970 [1795]. Perpetual peace: a philosophical sketch. In: H. Reiss (ed.), *Kant's Political Writings,* 93–130. Cambridge: Cambridge University Press.
Knapp, A. B. and E. Blake, eds. 2005. *The Archaeology of Mediterranean Prehistory.* Oxford: Blackwell.
Knorr-Cetina, K. and A. V. Cicourel, eds. 1981. *Advances in Social Theory and Method: Toward an Integration of Micro- and Macro-sociologies.* London: Routledge and Kegan Paul.
Koerner, S. 2007. Philosophy and Archaeology. In: A. Bentley and H. Mascher (eds.), *Handbook in Humanities Theory*, 351-371. London: Altamira Press.
Kuhn, T. 1970 [1962]. *The Structure of Scientific Revolutions.* 2nd ed. Chicago: Chicago University Press.
Latour, B. 1993. *We Have Never Been Modern.* Cambridge, MA: Harvard University Press.
Latour, B. and Weibel, P. eds. 2002. *Iconoclash: Beyond the Image Wars in Science, Religion and Art.* London: MIT Press.
Latour, B. and P. Weibel. 2005. *Making Things Public: Atmospheres of Democracy.* Cambridge: MIT Press.
Lévi-Strauss, C. 1973. *Tristes Tropiques,* translated by J. and D. Wrightman. New York: Penquin Books.
Locke, J. 1975 [1689]. *An Essay on Human Understanding,* edited by J. L. Axtell. Cambridge: Cambridge University Press.
Lovejoy, A. O. and G. Boas. 1965. *Primitivism and Related Ideas in Antiquity.* New York: Octagon Books.
Lucas, G. 2005. *The Archaeology of Time.* London: Routledge.
Morris, I. 1994. *Classical Greece: Ancient Histories and Modern Archaeologies.* Cambridge: Cambridge University Press.
Most, G. 1986. Pindar, O.2. 83–90. *Classical Quarterly* 36(2): 304–316.
Most, G. 1994. Simonides' Ode to Scopas in Contexts. In: J. P. Sullivan and I. J. F. de Jong (eds.), *Modern Critical Theory and Classical Literature,* 127–152. Leiden: E. J. Brill.
Murray, T., ed. 1999. *Time and Archaeology.* London and New York: Routledge.
Nicholis, G. and I. Prigogine. 1989. *Exploring Complexity: An Introduction.* New York: W. H. Freeman and Company.
Nickels, 1989. Integrating the science studies disciplines. In: S. Fuller, M. de Mey and T. Shinn (eds.), *The Cognitive Turn: Sociological and Psychological Perspectives on Science.* Dortrecht: Kluwer.
Olivier, L. 1999. The Hochdorf 'princely' grave and the question of the nature of archaeological funerary assemblages. In: T. Murray (ed.), *Time and Archaeology,* 109–138. London and New York: Routledge.
Olivier, L. 2001. Temps de l'histoire et temporalités des matériaux archéologiques: à propos de la nature chronologique des vestiges matériels. *Antiquités Nationales* 33: 189–201.
Ong, W. 1982. *Orality and Literacy.* London: Methuan.
Pagden, A. 2008. *Worlds at War: The 2,500-Year Struggle between East and West.* Oxford: Oxford University Press.
Patrik, L. E. 1999 [1985]. Is there an archaeological record? In: J. Thomas (ed.), *Interpretive Archaeology: A Reader,* 118-144. Leicester: Leicester University Press.
Pickles, J. 2004. A *History of Spaces: Cartographic Reason, Mapping and the Geocoded World.* London: Routledge.
Pinsky, V. and M. A. Wylie, eds. 1989. *Critical Traditions in Archaeology.* Cambridge: Cambridge University Press.
Plato [ca. 424-347] 1930. *Republic*, Books 1-5, edited and translated by P. Shorey. Cambridge MA: Loeb Classical Library.

Plato 1935. *Republic*, Books 6-10, edited and translated by P. Shorey. Cambridge MA: Loeb Classical Library.
Pompa, L. 1990. *Human Nature and Historical Knowledge: Hume, Hegel and Vico*. Cambridge: Cambridge University Press.
Preis, A.-B. 1996. Human rights as cultural practice: an anthropological critique. *Human Rights Quarterly* 18(2): 286–315.
Preucel, R. W., ed. 1991. *Processual and Post-processual Archaeologies: Alternative Ways of Knowing the Past*. Carbondale: Center for Archaeological Investigations of the Southern Illinois University.
Prigogine, I. 1997. *The End of Certainty: Time, Chaos and the New Laws of Nature*. London: Free Press.
Renfrew, C. 1973. *Before Civilization: The Radiocarbon Revolution and Prehistoric Europe*. Cambridge: Cambridge University Press.
Renfrew, C. 1980. The great tradition and the great divide: archaeology as anthropology. *American Journal of Archaeology* 84(3): 287–298.
Renfrew, C. and E. B. W. Zubrow, eds. 1994. *The Ancient Mind: Elements of Cognitive Archaeology*. Cambridge: Cambridge University Press.
Rheinberger, H.-J. 1997. *Toward a History of Epistemic Things: Synthesizing Proteins in the Test Tube*. Stanford: Stanford University Press.
Rheinberger, H.-J. 2010. *On Historicizing Epistemology*. Stanford: Stanford University Press.
Ricoeur, P. 1988. *Time and Narrative*, Vol. 3. Chicago: University of Chicago Press.
Rorty, R. 1979. *Philosophy and the Mirror of Nature*. Oxford: Basil Blackwell.
Rowlands, M. 1984. Conceptualizing the European Bronze and Iron Ages. In: J. Bintliff (ed.), *European Social Evolution: Archaeological Perspectives*, 147–149. Bradford: University of Bradford Press.
Rowlands, M. and J. Gledhill. 1977. The relation between archaeology and anthropology. In: M. Spriggs (ed.), *Archaeology and Anthropology: Areas of Mutual Interest*, 143–159. Oxford: British Archaeological Reports.
Schatzki, T., K. Knorr-Cetina and E. Von Savigny, eds. 2001. *The Practice Turn in Contemporary Theory*. London: Routledge.
Schnapp, A. 1993. *The Discovery of the Past*, translated by I. Kinnes and G. Varndell. London: British Museum Press.
Shanks, M. 1996. Style and the design of a perfume jar from an archaic Greek city state. In: R. W. Preucel and I. Hodder (eds.), *Contemporary Archaeology in Theory*, 364–393. Oxford: Blackwell Publishers.
Shapin, S. and S. Schaffer. 1985. *Leviathan and the Vacuum Pump: Hobbes, Boyle and the Experimental Life*. Princeton: Princeton University Press.
Snoddgrass, A. M. 1985. The new archaeology and the classical archaeologist. *American Journal of Archaeology* 89(1): 31–37.
Snow, C. P. 1962 [1959]. *The Two Cultures and the Scientific Revolution*. Cambridge: Cambridge University Press.
Spriggs, M., ed. 1977. *Archaeology and Anthropology: Areas of Mutual Interest*. Oxford: BAR Supplementary Series 19.
Sullivan, J. P. and I. J. F. De Jong, eds. 1994. *Modern Critical Theory and Classical Literature*. Leiden: E. J. Brill.
Thomas, J. 1996. *Time and Identity: An Interpretive Archaeology*. London: Routledge.
Thomas, J. 2004. *Archaeology and Modernity*. London: Routledge.
Toulmin, S. 1953. *An Introduction to the Philosophy of Science*. London: Huchinson's University Library.
Toulmin, S. 1961. *The Fabric of the Heavens: The Development of Astronomy and Dynamics*. Chicago: University of Chicago Press.

Toulmin, S. 1990. *Cosmopolis: The Hidden Agenda of Modernity.* Chicago: University of Chicago Press.
Toulmin, S. 2001. *Return to Reason.* Chicago: University of Chicago Press.
Toulmin, S. and J. Goodfield. 1961. *The Discovery of Time.* Chicago: University of Chicago Press.
Toulmin, S. and J. Goodfield. 1962. *Architecture of Matter.* Chicago: Chicago University Press.
Trigger, B. 1984. Alternative archaeologies: nationalist, colonialist, imperialist. *Man*(NS) 19: 355–370.
Trigger, B. 1989. *A History of Archaeological Thought.* Cambridge: Cambridge University Press.
Werbner, P., ed. 2008. *Anthropology and the New Cosmopolitanism.* Oxford: Berg.
White, H. 1978. *Tropics of Discourse: Essays in Cultural Criticism.* Baltimore: Johns Hopkins University Press.
White, H. 1987. *The Content of Form.* Baltimore: Johns Hopkins University Press.
Wolf, E. 1982. *Europe and the People without History.* Berkeley: University of California Press.

Contributors

Konstantinos Athanasiou graduated from the Polytechnic School of Aristotle University of Thessaloniki as an architect. He also completed his Ph.D. dissertation entitled *Comparative Study of the Urban Fabric of Akrotiri, Thera with the Aid of Modern Technologies*. He has also worked as an architect in various archeological sites and research projects and has presented his work at many conferences around the world.

Robert Chapman is Emeritus Professor of Archaeology at the University of Reading. His research interests include the development of inequalities in human societies, evidential reasoning in archaeology, and historical materialist approaches to the understanding of past and present societies. His fieldwork has been conducted on the Balearic Islands and in southeast Spain, where he enjoys long-term collaboration with colleagues from the Universitat Autònoma de Barcelona. His publications include *Archaeologies of Complexity* (2003).

Eric H. Cline has published numerous books and articles on the topic of biblical archaeology and the history of ancient Israel. He is currently Professor and Chair of the Department of Classical and Near Eastern Languages and Civilizations at The George Washington University, in Washington, DC. He is the codirector of the Tel Kabri Archaeological Project.

Bleda S. Düring is Assistant Professor in Near Eastern Archaeology at Leiden University. His research focuses on Neolithic Turkey and on early imperialism in Syro-Mesopotamia. He has recently published *The Prehistory of Asia Minor* (Cambridge University Press, 2011). Bleda is director of the Cide Archaeological Project, a survey in the Turkish Black Sea region. He was recently awarded a European Research Council Starting Grant for a project entitled Consolidating Empire: Reconstructing Hegemonic Practices of the Middle Assyrian Empire.

Athena Hadji teaches Prehistoric Archaeology for the Program *Plato's Academy*, launched by the National University of Athens and also an e-learning program on Prehistoric Aegean Art. On occasion, she curates art exhibitions. She holds a Ph.D. in Mediterranean Archaeology from the University of California Berkeley. She has taught Greek Mythology, Archaeology, Museum Studies, History of Art, and Anthropology of Space and Art in Berkeley, Greece and Cyprus. She has contributed to journals, edited volumes, and international conferences with subjects ranging from Early Cycladic art to cultural heritage. Her current research projects are Early Cycladic sculpture and contemporary Athenian graffiti.

Rachel Harkness is a Post-Doctoral researcher in the Department of Anthropology at the University of Aberdeen, Scotland on the ERC-funded project *Knowing from the Inside: Anthropology, Art, Architecture and Design*. Before this she held the Mellon Sawyer Postdoctoral Fellowship at the Institute for Advanced Studies in the Humanities at the University of Edinburgh, running the Sawyer Seminar Series *Embodied Values: Bringing the Senses Back to the Environment* and worked as a researcher at the Universities of Strathclyde and Glasgow on subjects related to the Chemical and Offshore Industries, the Circular Economy and the Environment. Her PhD in Social Anthropology considered the theory and practice of eco-building in the United States and Scotland.

Víctor Jiménez-Jáimez currently holds a two-year post-doctoral Marie Curie Intra-European Fellowship at the University of Southampton. For the last eight years he has worked on prehistoric European ditched enclosures and cultural formation processes of the archaeological record, as a result of which the book *Recintos de Fosos: Genealogía y Significado de una Tradición en la Prehistoria del Suroeste de la Península Ibérica (IV–III Milenios AC)* was published in 2010. He is also a member of the research team at Perdigões (Portugal).

Stephanie Koerner's research and teaching at the University of Manchester have complementary foci: Pre-Columbian and Latin American arts and cultures; contextual and comparative approaches to art and science's histories; and the roles of the humanities in the dynamics of pedagogical institutions and wider human affairs. In tandem with publishing in these areas, she designs courses in these areas for the university's Archaeology and Art History and Visual Studies programs, and is working on a series of edited volumes to accompany them.

Toula Marketou is Curator of Antiquities in the Greek Ministry of Culture, Twenty-Second Ephorate of Antiquities in Rhodes. Her publications are based on numerous excavations undertaken in the wider area of Ialysos on Rhodes and Serayia on Kos and shed light on many aspects of the Bronze Age period in the Dodecanese and the east Aegean.

José Enrique Márquez-Romero is *Profesor Titular* (Senior Lecturer, Associate Professor) at the University of Málaga (Spain). His research interests include prehistoric enclosures, pit sites, and megalithic monuments in the Iberian Peninsula. He has published multiple books and articles on these matters. He is currently conducting fieldwork at the ditched site of Perdigões (Portugal). He has also worked on cultural heritage knowledge dissemination, of which *The Dolmens of Antequera: Official Guide to the Archaeological Complex* is a good example.

Eimear Meegan is a doctoral researcher in University College Dublin's School of Archaeology. Her current research, kindly supported by the UCD John Hume Institute for Global Irish Studies, explores the role of choice as a driver of social change in prehistoric island societies, both within the Mediterranean and elsewhere in Europe. She is particularly interested in the development of archaeological theory and the materiality of production in the prehistoric world.

Emily Miller Bonney received her Ph.D. in Ancient Art and Archaeology from the Institute of Fine Arts, New York University. Her research focuses on Bronze Age Crete and the application of methods from phenomenology, landscape archaeology, and other theoretical approaches. She has published on the Early Bronze Age tholoi of south central Crete.

Patricia Murrieta-Flores holds a European Research Council post-doctoral position at the University of Lancaster and is member of the Archaeological Computing Research Group at the University of Southampton. She specializes in spatial technologies, GIS, high-performance computing, and other applications of technology to archaeological and historical research. She is co-director, field supervisor and scientific adviser to several archaeological projects in Spain and Mexico.

Robin Skeates is a reader in the Department of Archaeology at Durham University, UK. He is also general editor of the *European Journal of Archaeology.* He is a specialist in the prehistory of Italy and the Central Mediterranean region, with particular interests in material, visual, and sensual culture, and well as in museum and heritage studies. His latest book is *An Archaeology of the Senses: Prehistoric Malta* (Oxford University Press).

Stella Souvatzi received her Ph.D. in Archaeology from the University of Cambridge. She is Adjunct Professor at the School of Humanities, Hellenic Open University and currently holds a British Institute at Ankara—Koç

University Research Center for Anatolian Civilizations Senior Residential Fellowship at Istanbul. She has taught in Britain, Greece, and Cyprus and has conducted extensive archaeological fieldwork all over Greece. Her research interests include the Neolithic archaeology of Greece, eastern Mediterranean, and southeastern Europe, and the theory of archaeology and anthropology. Her publications include *A Social Archaeology of Households in Neolithic Greece: An Anthropological Approach* (Cambridge University Press, 2008).

Stavros Stavrides is assistant professor at the School of Architecture, National Technical University of Athens, Greece, where he teaches graduate courses on housing design, as well as a postgraduate course on the meaning of metropolitan experience. He has published numerous articles as well as five books on spatial theory, including *The Symbolic Relation to Space* (1990), *Advertising and the Meaning of Space* (1996), *The Texture of Things* (with E. Cotsou, 1996), *From the City-Screen to the City-Stage* (2002, National Book Award), *Suspended Spaces of Alterity* (2010), and *Towards the City of Thresholds* (in English, 2010). His research is currently focused on forms of emancipating spatial practices.

Trevor Watkins taught Near Eastern prehistory at the University of Edinburgh for many years. He has directed excavation and survey work on Cypriot Neolithic sites, has worked in northeast Syria, and in salvage excavations in north Iraq. Latterly, he worked at the aceramic Neolithic and Epi-palaeolithic site of Pınarbaşı near Çatalhöyük in central Turkey. Since 1990, he has been engaged in interdisciplinary research on the nature of the Epi-palaeolithic and early Neolithic communities of southwest Asia, their architecture, and the extensive networks of exchange and symbolism that they shared.

Assaf Yasur-Landau has published numerous articles and volumes on the archaeology of ancient Israel. He is currently Senior Lecturer in the Department of Maritime Civilizations and Senior Researcher at the Leon Recanati Institute for Maritime Studies at Haifa University. He is the codirector of the Tel Kabri Archaeological Project.

Index

For Product Safety Concerns and Information please contact our EU
representative GPSR@taylorandfrancis.com
Taylor & Francis Verlag GmbH, Kaufingerstraße 24, 80331 München, Germany

www.ingramcontent.com/pod-product-compliance
Ingram Content Group UK Ltd.
Pitfield, Milton Keynes, MK11 3LW, UK
UKHW022116190926
475702UK00013B/1471
* 9 7 8 1 1 3 8 3 7 7 8 4 4 *